Probabilistic and Fuzzy Approaches for Estimating the Life Cycle Costs of Buildings

Editor

Edyta Plebankiewicz

MDPI • Basel • Beijing • Wuhan • Barcelona • Belgrade • Manchester • Tokyo • Cluj • Tianjin

Editor
Edyta Plebankiewicz
Cracow University of Technology
Poland

Editorial Office
MDPI
St. Alban-Anlage 66
4052 Basel, Switzerland

This is a reprint of articles from the Special Issue published online in the open access journal *Applied Sciences* (ISSN 2076-3417) (available at: https://www.mdpi.com/journal/applsci/special_issues/Life_Cycle_Cost_Buildings).

For citation purposes, cite each article independently as indicated on the article page online and as indicated below:

LastName, A.A.; LastName, B.B.; LastName, C.C. Article Title. *Journal Name* **Year**, *Volume Number*, Page Range.

ISBN 978-3-0365-2295-1 (Hbk)
ISBN 978-3-0365-2296-8 (PDF)

Contents

About the Editor

Edyta Plebankiewicz is a Professor of Civil Engineering. She is the head of Construction Management at the Faculty of Civil Engineering at Cracow University of Technology in Poland. Her research interests include tendering and bidding in construction, planning methods in construction projects, cost calculation in the investment process, building life cycle costing, and fuzzy logic. She is the author and coauthor of more than 120 articles in Polish and foreign journals, 2 books, 40 articles in reviewed conference materials, and 2 textbooks.

Preface to "Probabilistic and Fuzzy Approaches for Estimating the Life Cycle Costs of Buildings"

The construction sector is a major consumer of natural resources and incurs high costs. Life cycle cost (LCC) makes it possible for the whole life performance of buildings and other structures to be optimized. The introduction of the idea of thinking in terms of a building life cycle resulted in the need to use appropriate tools and techniques to assess and analyze costs throughout the life cycle of a building. Traditionally, estimates of LCC have been calculated based on the historical analysis of data and have used deterministic models. The concepts of probability theory can also be applied to life cycle costing, treating the costs and timings as a stochastic process. If any subjectivity is introduced to the estimates, then the uncertainty cannot be handled using probability theory alone. The theory of fuzzy sets is a valuable tool for handling such uncertainties.

In this Special Issue, a collection of 11 contributions provide an updated overview of the approaches for estimating the life cycle cost of buildings. In the first paper the importance of information and communication technology use in life cycle cost management are considered. The research assumes that the most critical implementation factor is the investment cost. The second paper uses multiple criteria analysis to define the factors influencing the sustainable construction industries in the EU member states, the UK and Norway. Construction development of Commercial and Recreational Complex Building Projects (CRCBPs) is one of the community needs, but the implementation of these projects is usually very costly. The results and findings of the third article can be considered by CRCBPs in both the private and public sectors for properly effective risk identification, evaluation, and mitigation. The next four papers (4–7) consider the broadly understood maintenance phase of the building, with particular emphasis on the costs incurred at this stage. The eighth article deals with the partial outputs of large-scale infrastructure project risk assessment, specifically in the field of road and motorway construction. A partial section of the research was focused on the analysis of the probability distribution of the input variables, especially "the investment costs". The research topic of the ninth article addresses a part of the evaluation of railway infrastructure project efficiency within its life cycle by using the cost–benefit analysis method. The contractor selection problem, with emphasis on the life cycle costing method as the criterion of choosing the most appropriate company, is discussed in the tenth paper. In the last article, the parameters of tuned mass dampers are optimized to improve the performance level of steel structures during earthquakes.

Edyta Plebankiewicz

Editor

Article

The Implementation Factors of Information and Communication Technology in the Life Cycle Costs of Buildings

Peter Mésároš [1], Tomáš Mandičák [1,*], Marcela Spišáková [1], Annamária Behúnová [2] and Marcel Behún [2]

[1] Institute of Technology, Economics and Management in Construction, Faculty of Civil Engineering, Technical University of Košice, 042 00 Košice, Slovakia; peter.mesaros@tuke.sk (P.M.); marcela.spisakova@tuke.sk (M.S.)

[2] Institute of Earth Sources, Faculty of Mining, Ecology, Process Control and Geotechno-Logy, Technical University of Košice, 040 01 Košice, Slovakia; annamaria.behunova@tuke.sk (A.B.); marcel.behun@tuke.sk (M.B.)

* Correspondence: tomas.mandicak@tuke.sk; Tel.: +421-55-602-4378

Abstract: Life cycle cost management is an integral part of buildings construction. The life cycle cost approach can be considered an objective approach because it considers all life cycles of buildings. Information and communication technology is one of the critical factors for the success of construction projects. Several studies point to the importance of information and communication technology use in life cycle cost management. Generally, information and communication technology can be helpful in the cost management process of buildings. However, few implementation factors of information and communication technology are used in the life cycle cost management of buildings. The research assumes that the most critical implementation factor is the investment cost for information and communication technologies used in cost management during the life cycle. The relative importance index method was used to evaluate and quantify the final rank of implementation factors. The Kruskal–Wallis test was used to confirm or reject research results that were statistically significant.

Keywords: implementation factors; information and communication technology; life cycle costs; buildings

Citation: Mésároš, P.; Mandičák, T.; Spišáková, M.; Behúnová, A.; Behún, M. The Implementation Factors of Information and Communication Technology in the Life Cycle Costs of Buildings. *Appl. Sci.* **2021**, *11*, 2934. https://doi.org/10.3390/app11072934

Academic Editors: Alberto Benato and Edyta Plebankiewicz

Received: 4 February 2021
Accepted: 23 March 2021
Published: 25 March 2021

Publisher's Note: MDPI stays neutral with regard to jurisdictional claims in published maps and institutional affiliations.

1. Introduction

The issue of cost optimization is topical, particularly when participants in construction projects strive to reduce costs from the life cycle cost perspective. The construction project should consider cost management approaches from the buildings' whole life cycle perspective. Life cycle cost management plays an important role that focuses on cost optimization [1]. However, this approach has more potential for use than is currently utilized. This is because the relevant databases of information on the expected lifetime of buildings, the time and extent to which they require repairs, and the structures' maintenance costs are not available. According to Biolek and Hanák [2], these data should be processed in future building information modeling (BIM) systems. Several authors have specified the so-called life cycle cost (LCC) [3]. This is mainly the sum of costs during the construction project's individual stages, such as ownership, implementation, maintenance, and liquidation of the building. Budgetary constraints, environmental conditions, lack of communication, and skilled labor availability affect costs and time, even during the maintenance phase. These factors can also significantly affect the cost-effectiveness and efficiency of design management and the construction phase. This means that there is a close link between the maintenance phase and the design and construction phase. Therefore, if the building fabric's maintenance can be related to the initial stage of the design and construction phase, textile maintenance plans can be planned, and compelling predictions of uncertainties can minimize textile maintenance costs [4].

1

Knezovic et al. [5] noted the application of artificial neural networks, and the specific advantages and disadvantages that characterize econometric models. Further research indicates that life cycle management (LCM) is a concept that is often seen as an aggregation of life cycle tools and methods, and focuses on minimizing environmental impacts throughout their life cycle. Overall, the life cycle costs of a given project have also been plotted [6–8]. Kambanou notes that this method is still not widespread and has more potential [9]. Perceiving a construction project as a business plan to optimize the life cycle cost is one way to achieve efficiency [10,11]. In addition, cost management has been examined via information and communication technology [11]. Other authors agree that the building project should be assessed in terms of its entire life cycle, including the project's cost [2]. A construction project's business success depends largely on accurate estimates, such as the initial investment costs from the design phase to the construction phase, the operating costs required for the operation and maintenance phases, and the profits accumulated during the operation phase [12]. In many cases, relevant socio-economic benefits and costs also affect the construction project's economic efficiency; the influence of these factors cannot be neglected [12]. According to other studies, operating costs exceed implementation [13]. This also applies to the assumption of energy utilization [14].

In connection with cost management, several authors have mentioned information technologies used for the needs of cost management. The use of information and digital technologies increases when more cost-effective applications are found [15]. Generally, it can be said that information technology is gradually expanding in the field of construction. Several scientists have specified the relationship between information and communication technology in the context of successful construction project management or cost management. Building costs commonly occur under various market and legal conditions, which, unfortunately, often negatively influence construction project aims. Numerous research results indicate the scale of this problem. It is possible to define different construction investments that can be specified in various stages of their implementation [16,17]. Investment projects are complex and require appropriate management at all stages. The importance of procurement is due to the main criteria that affect the project's success: cost, quality, time, safety, and how the project meets its envisaged purpose. For this reason, one of the crucial success factors of construction projects is to allow bidding for the contract only by contractors who are sufficiently qualified for the proper performance of that contract [18]. The different life cycles of construction projects can specify various types of investments. These are characterized by different technological, organizational, and economic specifications [19]. There are specific costs associated with repairing defects. Knowledge about implementation factors and defects occurring in residential buildings can be used to better plan the investment budget [20].

Several studies have already partially addressed factors in implementing information and communication technology (ICT) for cost management of construction. Several studies suggest that these are investment costs [21–28]. A detailed overview of studies on this issue and the identification of factors is given in Table 1.

Table 1. Literature review of information and communication technology (ICT) implementation factors in the construction industry [21–32].

Year	Implementation Factors	Relevant Literature
2020	<ul><li>mimetic pressure,</li><li>strategic value judgment,</li><li>behavioral control capability</li></ul>	[24]
2019	<ul><li>communication and work relationship,</li><li>distraction and waste of time,</li><li>better information management on-site,</li><li>better management of construction defects,</li><li>improved work planning</li></ul>	[22]

Table 1. *Cont.*

Year	Implementation Factors	Relevant Literature
2019	• ICT human capital skills, • firm's decision-making process and support of visionary leaders, • inter-organizational research and development collaboration	[29]
2018	• ICT safety, • investment costs, • people acceptance • support management	[21]
2017	• lack of understanding about the process of sensing technology adoption, • its purpose of utilization in construction industry	[30]
2012	• investment costs, • lack of finance, • maintenance costs, • lack of management support, • low level and experience of users, • rejection of changes, • compatibility problems, • law framework, • transparency and ICT safety,	[23]
2012	• costs, • experiences, • IT safety,	[25]
2011	• high investment costs for ICT, • virus infiltration and degradation give, • security and privacy, • continuous need to create system upgrade, • increased IT staff costs, • incompatibility of product solutions, • poor return on investment, • personal abuse of ICT employees • weak management support	[31]
2009	• investments, • IT safety,	[28]
2008	• investment costs, • human resources, • IT equipment	[26]
2007	• diversity of construction industry, • construction project participants cooperation	[27]
2007	• hardware and software costs, • concerns about virus infiltration, • ICT equipment, • return on investment, • fear of dismissing redundant workers, • IT security.	[32]

Information and communication technology includes, in particular, software applications designed for communication, working with data, and information sharing. A lack of confidence is observed among project stakeholders in the documented data's authenticity and integrity [33]. Cost savings, revenues, and improvements in the quality of construction projects increase the practice's credibility and convince potential project participants of these technologies' benefits [34]. Several studies point to some of the benefits of using information technology in cost management [35]. Cost savings were seen as the most significant benefit by Marsh and Flanagan [36]. Increased efficiency and increased transparency, and greater convenience in the procurement process were determined in the research by Khayyat [37]. ICT functionalities mainly relate to construction management, so it can be argued that the integration of a lean management approach with the technical capabilities of ICT will bring benefits to the overall productivity and efficiency of construction projects [38]. Another study examined the impact of ICT on the so-called operational benefits [39]. In this group of benefits, the authors included flexibility in systems to meet clients' needs; strengthening the relationship with suppliers; competitive advantage in economies of scale; shortening the production phase; and flexibility of response to the client.

In contrast, information technologies provide little or no benefit according to previous research [27]. However, this study argues that there are areas where information technology implementation can also be beneficial. Improved monitoring and control have also been identified as crucial in implementing ICT in construction due to the impact on cost management [40,41]. Other authors discussed the methods of measuring the benefits of ICT and BIM technologies in construction [42,43].

Concerning research on the life cycle cost of building and cost management in general, costs and investments can be highlighted as factors mentioned. An analysis of the implementation factors of ICT and subsequent quantification has not yet been carried out in any research found and included in the review. In particular, in life cycle cost management, it is appropriate to examine and verify the relationship of implementation factors in terms of cost optimization, and whether it is a dimension in the management of construction projects or life cycle cost management of construction projects. The research's basic scientific questions were determined based on theoretical overviews in the given area and a summary of this research. What are the implementation factors of ICT in life cycle cost management? The fundamental research problem is that investment costs are the most critical ICT implementation factor in building life cycle cost management.

Based on the literature, interviews with experts (project managers, cost managers, etc.), and formulation of the research problem, four categories of implementation factors were determined, in which the expected implementation factors were defined to be examined in the survey. These implementation factors were first discussed with selected practitioners. Their comments and expert advice were taken into account for the final formulation of implementation factors. This pre-research ensured a strict selection and formulation of implementation factors. Cronbach's alpha was also used to test these questions for the selection of these factors. Implementation factors that were not considered appropriate by the experts were not further investigated. Table 2 shows all research implementation factors. The most important of the these is the first factor (Investment costs of ICTs) from the research hypothesis's point of view.

Table 2. Research ICT implementation factors in the construction industry (based on literature review and expert statements).

Group of IF	Implementation Factors (IF)	Description of Factors and Impact on LCC
Economic factors	1. Investment costs for ICTs 2. System maintenance costs during its lifetime 3. The need to recruit IT staff to manage ICT	1. Investment costs represent all costs related to the implementation of information and communication technologies, infrastructure modification and all installation costs. Their impact represents an increased cost of acquiring the system in the first year, and ICT should have a lower cost in the later period. 2. System maintenance costs represent all costs for technology maintenance, upgrades, improvements, and additional equipment management and service costs. The level of these costs should be lower than the cost savings resulting from ICT implementation in each life phase of a construction project. 3. Wage-related costs for new staff needed to manage ICT. This factor represents the cost burden during the entire construction period or each life cycle stage of the construction project.
Technical factors	4. Compatibility of software solutions 5. Functional possibilities of the system 6. Knowledge of the use of ICT in the field 7. System maintenance and service and the need to upgrade the system (administrative burden, inspections, repairs)	4. Ensuring the compatibility of technologies is challenging, especially in the construction project's design stage, where it is necessary to combine all technologies to ensure a smooth flow of data between devices. This can have a significant positive effect on other life cycle costs. 5. The functions and possibilities of technologies can be a motivation for implementation. It should have a positive impact on costs at each stage, especially concerning increasing productivity. 6. Knowledge is one of the prerequisites for the successful implementation of ICT. Their impact on LCC depends on the value of the people who have this knowledge and their ability to work with new ICTs, which reduces costs at every stage of the project. 7. The need to deal with service and constant upgrade is associated with increased costs and loss of time and energy of employees, which again has a negative impact on LCC.
Personnel factors	8. User qualification (training and certificates) 9. User experience (practical experience) 10. Readiness and disinterest of users 11. Ability to embrace innovation and change 12. Management support	8. From the LCC's point of view, the education and training of employees is a cost. The highest rate is at the beginning of the project, when this level is the highest. 9. Practical user experience can have a positive impact on LCC. Experience and the necessary qualifications represent a lower precondition for the need for training costs at each stage. 10. User lack of interest can have a serious negative impact on LCC. Their reluctance to accept change and innovation can lead to ever-increasing costs. 11. The reluctance to accept changes is equally negatively transmitted to the LCC. 12. Management support should be one of the keys in motivating new technologies to be adopted. The attitude of management can influence the opinion of employees on new technologies. This can have a positive effect on LCC of buildings. Management support can represent a high degree of ICT implementation and thus lead to cost savings at each stage.

Table 2. *Cont.*

Group of IF	Implementation Factors (IF)	Description of Factors and Impact on LCC
Industry factors	13. Fragmentation of the sector and integration among participants in construction projects 14. Legislative framework 15. Level of competition in the use of ICT 16. Level of use of ICT by other participants in the construction project.	13. Fragmentation of the sector and integration between participants in construction projects means a hard way of communication between participants and increases misunderstandings. In LCC, it is reflected as a negative phenomenon, with a large number of sub-suppliers increasing costs and expanding the supply chain. From the LCC point of view, it is primarily the risk of increased costs in the design and construction stage. On the contrary, the use phase does not pose this risk. 14. The legislative framework may also affect the implementation of ICT. If the legislation is simple and fixed, it can lead to the facilitation of the whole implementation process. On the contrary, if the legislative framework is set incorrectly, a number of restrictions, etc. this leads to a negative impact. Legislation can also directly affect the regulation of the use of specific ICT (such as BIM technology) in the procurement process and in selected projects. This may delay earlier implementation of ICT. This can have a positive impact on LCC. Thus, costs can decrease over time. 15. The level of use of ICT by competitors may impact the decision of other construction companies to use technology. To minimize LCC and increase competitiveness, it also has this impact. 16. Other participants can pressure the use of selected ICTs, which can be a motivator for rapid implementation. To call other participants can have a significant positive impact on the LCC.

2. Materials and Methods

2.1. Research Methods and Steps

This research consisted of two phases, the pre-research and the research. The pre-research included determining a basic research question based on a detailed theoretical analysis of previous research. This analysis also provided the basis for identifying implementation factors and grouping. These compiled implementation factors were discussed by relevant experts. Four project managers from large international construction companies discussed the proposed research implementation factors in an interview. Based on the agreement of all, the final implementation factors were determined, and were the subject of the investigation.

The selection of the research sample was based on the structure of the industry. The respondents' selection was from the building industry database (The Statistical Office of the Slovak Republic). The total number of entities in the construction industry in Slovakia is 83,560,000. More than 1200 (sample file size) construction companies were approached to participate in construction projects and final buildings. Respondents were selected as a percentage composition reflecting the number of market participants. The statistical set of respondents included various participants in construction projects. The ratio of real business entities was maintained. Companies were contacted (investors 11.20%, suppliers 52%, sub-contractors 16.80%, and designers 20%), and 125 respondents took part in the survey. The return rate was 10.42%.

Cronbach's alpha verified the suitability of the questions. This ensured an adequate distribution of the research sample. Data processing was based on the relative importance index and five critical levels method. Based on this, the ranking was determined, and the Kruskal–Wallis test for statistical significance was used to verify the results.

Subsequently, for quantification purposes, the selected group's arithmetic means and the specific factor were determined. The Kruskal–Wallis test was also used to verify the influence of a given factor. A detailed overview of the research steps and methods used is given in Table 3.

Table 3. Research methods and steps.

Pre-Research Stage

Research Steps	Data Source	Methods	Results
1. Analysis of previous research and creation of an overview	Web of Science database	Analysis and synthesis	Implementation factors of ICT in LCC overview
2. Determination of research question and hypothesis	Web of Science database	Analysis and synthesis	Problem statement and hypothesis
3. Analysis of implementation factors—assessment of the correctness of implementation factors by experts (construction project managers by large enterprises)	Primary data	Interview	Selection of researched implementation factors of ICT in LCC—Final implementation factors of ICT in LCC overview

Research steps

4. Selection of research sample	Statistical Office database	Research sample was performed at random. However, the research groups and their representation (number) were established according to the number in the market. Thus, no research group was disadvantaged, and the research sample reflected the real market presence.	Research sample
5. Data collection	Questionnaire data	Likert scale	Research data
6. Data processing	Questionnaire data		
(a) Evaluation of the suitability of the questions asked	Questionnaire data	Cronbach's alpha	
(b) Ranking of implementation factors	Questionnaire data	Relative importance index and five critical levels	Ranking
(c) Verification of ranking results by statistical significance	Questionnaire data—Statistica software	Kruskal–Wallis test	Statistical significance
(d) Quantification of the ICT implementation factors on LCC	Questionnaire data	Arithmetic mean and impact rate	Proposal of the influence of ICT factors on implementation and LCC
(e) Verification of implementation factors impact rate results by statistical significance	Questionnaire data—Statistica software	Kruskal–Wallis test	Statistical significance
7. Hypothesis evaluation			

2.2. Research Aim and Hypothesis

The research focused on analyzing the implementation factors of ICT in the life cycle cost of buildings. Monitoring and analyzing construction costs in each phase of a construction project is the first step to managing and attempting to optimize these costs. A thorough analysis of research and studies has focused on the impact and implementation factors of using ICTs.

The research aim is to analyze the implementation factors of ICT in the life cycle costs of buildings and verify that investment costs represent the most critical implementation factor.

The primary research claims that information and communication technology play an important role in construction life cycle cost management based on study knowledge and research. One of the most significant advantages of using information and communication technology is a positive impact on reducing costs at every stage of a construction project's life cycle. In addition, key factors impact the implementation of information and communication technology in life cycle cost management. Based on this, the main hypothesis adopts this claim.

Hypothesis statement: Investment costs are the most important ICT implementation factor in managing the buildings' life cycle costs.

This statement means that, in analyzing the impact ranking of factors, the factor investment costs will achieve the highest value and be ranked first. Based on this, a null hypothesis can be postulated and verified. This means that no implementation factor has a higher impact rate than investment costs. The investigated main or null hypothesis and its alternative have the following form:

Hypothesis 1. *Investment costs are ranked first as the most important ICT implementation factor in managing the life cycle costs of buildings.*

Hypothesis 0. *Investment costs are not ranked first as the most important ICT implementation factor in managing the life cycle costs of buildings.*

2.3. Data Collection and Research Sample

Data collection was carried out through an online questionnaire. The questionnaire was divided into several parts that were content related. The first part of the questionnaire focused on the characteristics of the respondents. This included information about the participant in the construction project; the size of the construction company (How many employees does your construction company have?); the work experience of the project manager (How long have you been working as a project manager?); the participation of foreign capital and know-how (Do you use only domestic capital and know-how?; Do you use foreign capital or know-how of another parent company?); construction project size and characteristic of buildings (How big is the construction project, based on which you assess the level of impact of information and communication technology in the context of the life cycle cost issue?).

The second part of the questionnaire dealt with direct questions on the perception of the implementation factors of using information and communication technology (Specify the information systems used in project management and planning life cycle cost management; Specify the extent and frequency of use of these technologies based on the scale provided). The next part focused on issues related to the impact on construction life cycle management and implementation factors of information and communication technology (based on the Likert scale, respondents defined their perception of selected implementation factors of using ICT for cost management in individual stages of a construction project; these data were based on real results of construction project costs (1—change up to 5%, 2—change from 6% to 10%, 3—change from 11% to 15%, 4—change from 16% to 20%, 5—change over 21%).

The third part of the questionnaire survey also included questions focused on using information and communication technologies and quantifying the impacts of the implementation of information and communication technologies on the life cycle cost. Respondents who stated that they use selected information technologies should also quantify the impact on the life cycle cost (as a percentage) and the degree of improvement in communication between participants in the construction project.

The list of implementation factors resulting from the use of ICT was compiled based on a thorough theoretical analysis of resources and research dealing with the implementation factors of ICT in the cost management of buildings [22–32]. Experts reviewed the long list of implementation factors in the field by interview. These were mainly project managers and financial managers in the field of cost management in the construction industry. These interviewed managers came from Slovak construction companies and investors. Based on a theoretical long list of ICT implementation factors, and consultation with project managers, a researched list of implementation factors was established.

Respondents answered the questions using an evaluation based on a 5-point Likert scale, where the value of 5 represented very significant and 1 represented not significant. Questions on some cost issues were filled in as nominal or relative indicators. The questionnaire also contained a detailed explanation of the interpretation of the Likert scale's values, as mentioned in the previous paragraph.

The research involved respondents who represented the main participants in the construction project. These were most often significant contractors and sub-contractors, but also included investors and architects. Regarding job positions were concerned, among the suppliers were project managers and cost managers. Among architects, participants were often designers from design studios. The investor was represented through the finance department and financial managers. A more detailed specification of the research sample is shown in Table 4.

Table 4. Structure of research participants.

Type of Enterprises (n = 125)	Frequency	% Share
Participant of construction project		
Contractor	65	52.00
Sub-contractor	21	16.80
Investor	14	11.20
Designer	25	20.00
Enterprise size		
Micro enterprises	40	32.00
Small enterprises	42	52.5
Medium-sized enterprises	38	47.5
Large enterprises	5	4.00
Business funding origin		
International private	13	10.40
National private	112	89.60
Working experience in work position		
0–5 years	35	28.00
6–10 years	57	45.60
11–20 years	23	18.40
21 years and more	10	10.00

2.4. Data Processing

Data processing was performed based on verified statistical methods. This processing took place in the software environment of MS Excel and Statistica. The obtained data combined quantitative and qualitative data collection methods, using values of 1 to 5 based on the distribution of the Likert scale, were the basis for subsequent statistical processing. Based on a random selection of a research sample representing the real situation of the construction market, the statistical extremes were adjusted to ensure they did not distort the results.

The relative importance index (RII) was used to determine the quantification and significance of individual implementation factors resulting from the impact of the implementation process of information and communication technology on buildings' life cycle costs. This is a scientific method that is commonly used to determine rankings, often in construction surveys. RII is calculated using the following expression:

$$\text{RII} = \frac{\sum_{i=1}^{5} w \, Xi}{A \times N} \tag{1}$$

RII = relative importance index

w = weighting given to each benefit by respondents and it ranges from 1 to 5

x = frequency of the i-th response given for each cause

A = highest weight (i.e., 5 in this case)

N = total number of participants

The ranking for each area is considered important to discuss construction life cycle cost management. Comparing and focusing on cost parameters is highly important in research. The implementation factors for each area did not reach the same ranking. It is essential to look at the key values and compare them in life cycle cost management. This means their significance in terms of individual stages of the construction project. The areas related to the research area of construction life cycle management were specified for total costs and costs associated with the construction project's management.

The significance of individual implementation factors can be assessed based on interval values, including specific measured values. According to Akadiri [44], five critical levels are transformed from RII values:

- high (H) $(0.8 \leq \mathrm{RII} \leq 1)$,
- high-medium (HM) $(0.6 \leq \mathrm{RII} \leq 0.8)$,
- medium (M) $(0.4 \leq \mathrm{RII} \leq 0.6)$,
- medium-low (ML) $(0.2 \leq \mathrm{RII} \leq 0.4)$ and
- low (L) $(0 \leq \mathrm{RII} \leq 0.2)$.

This is crucial because these values can be clearly specified and directly classified based on a proven scientific method. The results in the form of rankings were compared with the intervals, and preliminary conclusions were drawn. In this case, it is necessary to also verify these results by statistical significance. Based on the research sample distribution, it was evaluated that it is best to perform this distribution using Kruskal–Wallis tests for a given type of data. The Kruskal–Wallis test was chosen for statistical testing. This test was chosen because the researchers worked with an ordinal variable. As the dependent variable was ordinally scaled, the Kruskal–Wallis test was required. Applying the given tests allowed determination of whether statements and assumptions examined by the current research were statistically significant.

The threshold for the use and impact of information technology (IT) and information systems (IS) was set at 3.5. This value was determined based on several sources, however, values above 3.5 are considered significant [45].

2.5. Limitations of Research

The implementation of research activities related to the examined issue uncovered several limitations. It should be considered whether these limitations could reduce the value of the results or change the research conclusions. At the outset, the current study focused on the perception of the research problems identified in the project manager's specific questions (i.e., one person evaluated the success of the project and answered the research questions for the whole project). These facts may raise questions about the subjective evaluation of this respondent. However, this was prevented by a detailed description using percentages for each research area and question. Based on real accounting data, the respondent (project manager) clearly defined the percentage to which his answer belongs in the Likert scale.

Another issue from the point of view of the correctness of the interpretation of the results and the comparison was the size of the companies participating in the research and establishing a condition to compare these results. The responses in the form of a Likert scale with a description of the values (relative indicator in percent) were determined for this purpose. Therefore, in cost perception in projects of different sizes, comparing absolute values was not possible, and a relative indicator appears to be the most relevant type of research data acquisition.

The research also took into account the number of forms information and communication technology used, but only at intervals. The use of information and communication technology was considered in all construction projects that pointed to 3 or more IS and IT. For comparison, however, the results may have been skewed according to the number of forms of information and communication technology used.

3. Results and Discussion

The use of ICTs has several advantages, which was also the statement of several respondents. However, as several also stated, the expectations and benefits of implementing

ICTs should be greater than the factors (in many cases, the concerns) that hinder the implementation of the decision to implement ICT. This research sought to quantify and analyze these implementation factors focusing on cost optimization throughout the life cycle of the construction project and a positive impact on cost management during the life cycle of buildings. Experts in the field answered questions about these implementation factors. They tried to quantify the impact level of implementation factors on the Likert scale based on managing construction projects.

Based on the answers, the RII index was used to obtain the ranking of implementation factors resulting from thinking about ICT implementation in construction projects' cost management. The final ranking, which considers the entire life cycle of the construction project and buildings (i.e., pre-design and design stage, project stage, construction phase, and maintenance and use stage of the building, up to its liquidation) is given in detail in Table 5.

Table 5. Relative importance index of ICT implementation factors in life cycle cost management.

Group of IFs	Implementation Factors (IFs)		RII	Ranking
Economic factors	1.	Investment costs for ICTs	0.879	1
	2.	System maintenance costs during its lifetime	0.758	5
	3.	The need to recruit IT staff to manage ICTs	0.735	7
Technical factors	4.	Compatibility of software solutions		
	5.	Functional possibilities of the system	0.648	9
	6.	Knowledge of the use of ICT in the field	0.357	16
	7.	System maintenance and service and the need to	0.489	14
		upgrade the system (administrative burden,	0.737	6
		inspections, repairs)		
Personnel factors	8.	User qualification (training and certificates)	0.648	9
	9.	User experience (practical experience)	0.567	13
	10.	Readiness and disinterest of users	0.639	11
	11.	Ability to embrace innovation and change	0.706	8
	12.	Management support	0.778	4
Industry factors	13.	Fragmentation of the sector and integration among participants in construction projects	0.583	12
	14.	Legislative framework	0.837	2
	15.	Level of competition in the use of ICT	0.435	15
	16.	Level of use of ICT by other participants in the construction project.	0.784	3

Based on the ranking results, it can be stated that the most extensive relative importance index was achieved for the implementation factor "Investment Costs for ICTs", with a value of 0.879, which falls in the interval (H) ($0.8 \leq$ RII ≤ 1), i.e., a high value. The "legislative framework" achieved a similar value and the same order in the ranking. These two implementation factors achieved a relative importance index higher than 0.8 and are therefore considered significant. These results suggest a research statement and consideration of the most critical implementation factor. However, for a correct evaluation and conclusion, this result must be evaluated by statistical tests that confirm its statistical significance. In addition, however, the research also yielded other quantitative and qualitative results that need to be discussed within the topic.

Significant results, according to the ranking, also included the following implementation factors (factors that reached 0.6 and higher): level of use of ICT by other participants in the construction project; management support; system maintenance costs during its lifetime; system maintenance and service and the need to upgrade the system; the need to recruit IT staff to manage ICTs; ability to embrace innovation and change; compatibility of software solutions; user qualification (training and certificates); and readiness and disinterest of users. Of a total of 16 examined implementation factors, up to 12 (75%) can be considered significant based on ranking. All examined economic factors were identified as significant implementation factors. These fall in the interval of ($0.6 \leq$ RII ≤ 0.8), which can be considered as medium–high significance. This again points to the trend that investment costs and other types of costs associated with ICT implementation represent a serious implementation factor.

It is essential that these findings and the main research hypothesis are confirmed using the Kruskal–Wallis test. An overview of groups of factors shows that the most important are economic factors. Personal factors are also significant. Overall, the setting of people (employees) to accept change, and accept innovation and their skills and knowledge in the IT field, significantly affect the implementation of ICT in the cycle cost management of construction and construction projects. Based on the Likert scale and Kruskal–Wallis testing, an infographic was constructed that highlights the importance of individual factors for specific groups and provides information about the arithmetic mean for individual implementation groups (Figure 1).

It is possible to discuss why experts and managers of construction projects have quantified and determined such a ranking of these implementation factors. These results should be compared with the research already carried out in the field of ICT implementation factors in the context of managing life cycle costs of buildings. The research results also confirm the research conclusions of [22–29], in which the investment cost was a significant implementation factor. This result and comparison point to one of the phases of a construction project in the context of cost management. The research results point to a relatively high degree of implementation factors in life cycle cost management.

An important view of the results is presented in Figure 1, in which the degree of importance of the implementation factor is quantified. The investment costs reached the highest value, and the given rate was 4.395, which represents very high importance from the point of view of ICT implementation. Regarding the degree of saving resulting from the use of ICT, which this research investigates, it can be stated that the turning point at which ICT will cover investment costs, should be in the 6th to 10th construction project. This clearly depends on other factors such as the size of the projects. However, consideration must also be given to operating costs, which extend this period. In terms of the most significant impact on implementing ICT for life cycle cost management, it is possible to define a significant level (3.954) of economic factors. Thus, in addition to the investment costs, the operating costs and the costs for the employees who will manage the selected technologies must be included. The research also included questions focused on using information and communication technologies and their impact on LCC. This impact was mentioned by several respondents, based on which the Kruskal–Wallis test was also carried out to determine if results were statistically significant. Due to the implementation and use of ICT, the value of cost savings was 10% to 15%.

The use of ICTs can increase efficiency, which has a limited impact on achieving optimization or a reduction in costs, and costs in terms of the entire life cycle of the construction. Monitoring the impacts on life cycle cost management is the more complex subject of this research. During interviews or in response to additional questions, several experts indicated that ICT leads to better cost management if communication and sharing of necessary information is faster. This also proved to be significant at every stage of the life cycle of a construction project. Here, however, it must be noted that although this rate was different in the individual phases of the construction cycle, it always reached the level of at least medium or medium–high based on the empirical method. The details of these differences discussed in more detail later. Based on the RII, the ranking and intervals were determined, and indicated the degrees of importance. However, from a statistical point of view, these values should be verified. Therefore, based on the Likert scale and data on the frequency of responses with specific values, a table of importance or the so-called importance rate (IR), with values from 1 to 5, was constructed. This is the average value of all respondents values, which is the important frequency of responses. This importance index shows the frequency of utilization in similar cases and studies globally. The results are mentioned in more detail in Table 6, which shows the value of the Kruskal–Wallis test for the statistical significance level (Table 5).

Figure 1. Implementation factor groups and impact rates.

Table 6. Hypothesis results and Kruskal–Wallis ANOVA results.

Hypothesis	Parameter	K-W Anova (p)	Rejection
H_1: Investment costs are ranked first in the most important ICT implementation factors in managing the life cycle costs of buildings.	RII Ranking	0.0476	accepted
H_0: Investment costs are not ranked first in the most important ICT implementation factors in managing the life cycle costs of buildings	RII Ranking		rejected

From the Kruskal–Wallis ANOVA based on ranking, the variable "ICT implementation factors in life cycle cost management" achieved a *p*-value of 0.0476, which has significant statistical significance. The number of valid responses was 125 for each factor (see Table 5).

Table 5 describes the Kruskal–Wallis test to examine the statistical significance of selected factors' impact on ICT implementation. Table 5 also describes the decision and evaluation of the scientific hypothesis based on the Kruskal–Wallis ANOVA. Thus, the achieved *p*-value was 0.0476. This indicates that the statistical implementation factors confirmed the statistical significance. That is, at the level of probability $\alpha = 0.05$, we can reject H_0: investment costs are not ranked first in the most important ICT implementation factors in managing the life cycle costs of building, and thus accept the hypothesis H_1: investment costs are ranked first in the most important ICT implementation factors in managing the life cycle costs of buildings. Thus, it follows that investment costs represent the most important implementation factor influencing the implementation and use of ICT in the life cycle cost management of buildings.

4. Conclusions

It is necessary to examine the issue of implementation factors of ICT adoption in life cycle cost management and, in particular, to quantify and manage the actual cost of a construction project during its entire life cycle. Several studies point to factors that may influence the implementation of ICT in life cycle cost management. This research sought to verify these claims. The assumption was that investment costs represent the most critical implementation factor of ICT in life cycle cost management. This statement can be accepted based on the empirical methods performed. Based on a selected sample of respondents and projects, it was determined that investment costs are the most critical implementation factor. Several studies have confirmed the benefits of using ICT in life cycle cost management of buildings. Therefore, examining the factors influencing this implementation was highly important, and the research findings are essential for practitioners. Simultaneously, the research noted other essential implementation factors of ICT in life cycle cost management. These are the level of use of ICT by other participants in the construction project; management support; system maintenance costs during its lifetime; system maintenance and service and the need to upgrade the system; the need to recruit IT staff to manage ICTs; ability to embrace innovation and change; compatibility of software solutions; user qualifications (training and certificates); and readiness and disinterest of users and legislative framework. The research also highlighted the importance of all economic factors related to the implementation of ICT in life cycle cost management. Maintenance costs and the costs associated with recruiting new staff to the IT department greatly influence the decision to implement ICT for life cycle cost management.

This issue is closely linked to the issue of implementation benefits. This represents another research gap that this research should address. The extension of this scientific and practical issue should be addressed using the same research sample, which would allow a confrontation with the results of this study. The direct benefits of implementing ICT in life cycle cost management can contribute to the growth of ICT implementation in construction.

This research confirms the claims of previous studies [12,16]. These previous studies note the importance of implementing information and communication technologies for life cycle cost management. The positive impact on cost financing was also confirmed. As also mentioned in a previous study [18], one of the main factors is investment costs. Another study [18] also found that this research has not substantiated that operating costs exceed investment costs, and that this is a larger implementation factor.

In contrast, research has also examined the effects of information and communication technologies, and a positive impact of the use of information and communication technologies was found, as stated in the study [20].

The most important research results can be summarized as:

- high investment costs are the most critical implementation factor,
- operating costs are also a critical implementation factor for the adoption of ICT, but this is not the most important, as some studies claim;
- the survey showed that the use of information and communication technologies has the effect of reducing life cycle cost management costs, and this result has also been quantified at 10% to 15% of costs;
- implementation has improved communication between research participants;
- research quantified the importance of specific implementation factors for adopting ICT in life cycle cost management (this important information for practice has not yet been mentioned in any research).

Knowledge of implementation factors in practice also means focusing on specific processes that can contribute to better implementation of ICT in construction. It can also challenge the views and support of management, which can positively affect the industry's development.

Author Contributions: Conceptualization, P.M. and T.M.; methodology, T.M.; software, T.M.; validation, T.M., A.B. and M.B.; formal analysis, M.S.; investigation, T.M., A.B. and M.B.; resources, T.M., A.B. and M.B.; data curation, T.M., A.B.; writing—original draft preparation, T.M., M.S., P.M.; writing—review and editing, P.M.; visualization, T.M.; supervision, P.M.; project administration, T.M.; funding acquisition, P.M. All authors have read and agreed to the published version of the manuscript.

Funding: This work was supported by the Slovak Research and Development Agency under the contract no. APVV-17-0549. The paper presents partial research results of project VEGA 1/0828/17 "Research and application of knowledge-based systems for modelling cost and economic parameters in Building Information Modelling". The paper also presents partial research results of project KEGA 059TUKE-4/2019 "M-learning tool for intelligent modeling of building site parameters in a mixed reality environment".

Institutional Review Board Statement: Not applicable.

Informed Consent Statement: Not applicable.

Data Availability Statement: All research activities have been carried out in accordance with the MDPI Ethics and there is no obstacle on our part.

Acknowledgments: This work was supported by the Slovak Research and Development Agency under the contract no. APVV-17-0549. Paper presents a partial research results of project VEGA 1/0828/17 "Research and application of knowledge-based systems for modelling cost and economic parameters in Building Information Modelling". The paper presents a partial research results of project KEGA 059TUKE-4/2019 "M-learning tool for intelligent modeling of building site parameters in a mixed reality environment".

Conflicts of Interest: The authors declare no conflict of interest.

References

1. Hromádka, V.; Korytárová, J.; Vítková, E.; Seelman, H.; Funk, T. New aspects of socioeconomic assessment of the railway infrastructure project life cycle. *Appl. Sci.* **2020**, *10*, 7355. [CrossRef]
2. Biolek, V.; Hanák, T. LCC Estimation Model: A Construction Material Perspective. *Buildings* **2019**, *9*, 182. [CrossRef]

3. Zabielski, J.; Zabielska, I. Life Cycle of a Building (LCC) in the Investment Process-Case Study. In Proceedings of the Baltic Geodetic Congress, BGC-Geomatics, Olsztyn, Poland, 21–23 June 2018; pp. 254–259.
4. Chen, C.; Tang, L. BIM-based integrated management workflow design for schedule and cost T planning of building fabric maintenance. *Autom. Constr.* **2020**, *107*, 1–12. [CrossRef]
5. Knezevic, M.; Cvetkovska, M.; Hanák, T.; Braganca, L.; Soltesz, A. Artificial Neural Networks and Fuzzy Neural Networks for Solving Civil Engineering Problems. *Complexity* **2018**, 1–2–2. [CrossRef]
6. De Haes, H.A.U.; Rooijen, M.V. *Life Cycle Approaches: The Road from Analysis to Practice*; UNEP/SETAC Life Cycle Initiative: Paris, France, 2005.
7. Westkämper, E.; Alting, A. Life Cycle Management and Assessment: Approaches and Visions Towards. *Sustain. Manuf. Cirp Ann.* **2009**, *49*, 501–526. [CrossRef]
8. Bey, N. Life Cycle Management. In *Life Cycle Assessment: Theory and Practice*; Springer: Berlin/Heidelberg, Germany, 2018.
9. Kambanou, M.L. Life Cycle Costing: Understanding How It Is Practised and Its Relationship to Life Cycle Management—A Case Study. *Sustainability* **2020**, *12*, 3252. [CrossRef]
10. Grieves, M. CIM data, Product lifecycle management—Empowering the future of business. *Technic. Repet.* **2002**, *2*, 1–21.
11. Corallo, A.; Latino, M.E.; Lazoi, M.; Verardi, S. Defining Product Lifecycle Management: A Journey across Features, Definitions, and Concepts. *Int. Sch. Res. Not.* **2013**, *2013*, 1–10. [CrossRef]
12. Kim, G.T.; KIM, K.T.; Lee, D.H.; Han, C.H.; Kim, H.B.; Jun, J.T. Development of a life cycle cost estimate system for structures of light rail transit infrastructure. *Autom. Constr.* **2010**, *19*, 308–325. [CrossRef]
13. Hanák, T.; Marović, I.; Aigel, P. Perception of Residential Environment in Cities: A Comparative Study. *Procedia Eng.* **2015**, *117*, 495–501. [CrossRef]
14. Fantozzi, F.; Gargari, C.; Rovai, M.; Salvadori, G. Energy Upgrading of Residential Building Stock: Use of Life Cycle Cost Analysis to Assess Interventions on Social Housing in Italy. *Sustainability* **2019**, *11*, 1452. [CrossRef]
15. De Soto, B.G.; Augustí-Juan, I.; Hunhevicz, J.; Joss, S.; Graser, K.; Habert, G.; Adey, B.T. Productivity of digital fabrication in construction: Cost and time analysis of a robotically built wall. *Autom. Constr.* **2018**, *92*, 297–311. [CrossRef]
16. Plebankiewicz, E.; Wieczorek, D. Prediction of cost overrun risk in construction projects. *Sustanability* **2020**, *12*, 9431. [CrossRef]
17. Korytarova, J.; Hromadka, V. Building life cycle economic impacts. In Proceedings of the International Conference on Management and Service Science, Wuhan, China, 24–26 August 2010.
18. Korytárová, J.; Hanák, T.; Kozik, R.; Radziszewska-Zielina, E. Exploring the contractors' qualification process in public works contracts. *Procedia Eng.* **2015**, *123*, 276–283. [CrossRef]
19. Plebankiewicz, E.; Wieczorek, D. Adaptation of a cost overrun risk prediction model to the type of construction facility. *Symmetry* **2020**, *12*, 1739. [CrossRef]
20. Plebankiewicz, E.; Malara, J. Analysis of defects in residential buildings reported during the warranty period. *Appl. Sci.* **2020**, *10*, 6123. [CrossRef]
21. Vasista, T.G.; Abone, A. Benefits, Barriers and Applications of Information Communication Technology in Construction Industry: A Contemporary Study. *Int. J. Eng. Technol.* **2018**, *7*, 492–499. [CrossRef]
22. Hasan, A.; Ahn, S.; Rameezdeen, R.; Baroudi, B. Empirical Study on Implications of Mobile ICT Use for Construction Project Management. *J. Manag.Eng.* **2019**, *35*, 04019029. [CrossRef]
23. Sargent, K.; Hyland, P.; Sawang, S. Factors influencing the adoption of information technology in a construction business. *Australas. J. Constr. Econ. Build.* **2012**, *12*, 72–86. [CrossRef]
24. Wang, G.; Lu, H.; Gao, X. Understanding Behavioral Logic of Information and Communication Technology Adoption in Small- and Medium-Sized Construction Enterprises: Empirical Study from China. *J. Manag. Eng.* **2020**, *36*, 1–16. [CrossRef]
25. Sekou, A.E. Promoting the Use of ICT in the Construction Industry, Assessing the Factors Hindering Usage by Building Constructors in Ghana. Ph.D. Thesis, Kwame Nkrumah University, Kumasi, Ghana, September 2012.
26. Linderoth, H.C.J.; Jacobsson, M. Understanding adoption and use of ICT in construction projects through the lens of context, actors and technology. In Proceedings of the International Conference on Information Technology in Construction, CIB W78, Santiago, Chile, 15–17 July 2008.
27. Wigforss, O.; Lofgren, A. Rethinking communication in construction. *ITcon* **2007**, *12*, 337–345.
28. Woksepp, S.; Olofsson, T. An evaluation model for ICT investments in constriction project. *ITconstruction* **2009**, *13*, 343–361.
29. Lu, H.; Pishdad-Bozorgi, P.; Wang, G.; Xue, Y.; Tan, D. ICT Implementation of Small- and Medium-Sized Construction Enterprises: Organizational Characteristics, Driving Forces, and Value Perceptions. *Sustainability* **2019**, *11*, 3441. [CrossRef]
30. Sepasgozaar, S.M.E.; Shirowzhan, S.; Wang, C. A scanner technology acceptance model for construction projects. *Procedia Eng.* **2017**, *180*, 1237–1246. [CrossRef]
31. Wang, G.B.; Cao, D.P.; Chong, D. Investigation to factors affecting IT implementation in Chinese construction projects. In Proceedings of the 4th Conference on Systems Science, Management Science and Systems Dynamics, New York, NY, USA, 10–11 April 2011; Volume 3, pp. 1–13.
32. Mahamadu, A.M.; Mahdjoubi, L.; Booth, C.A. Determinants of building information modelling (BIM) acceptance for supplier integration: A conceptual model. In *Proceedings of the 30th Annual ARCOM Conference*; Portsmouth, UK, 1–3 September 2014, Association of Researchers in Construction Management; pp. 723–732.

33. Soetanto, R.; Dainty, A.R.J.; Price, A.D.F.; Glass, J. A framework for investigating human issues associated with the implementation of new ICT systems in construction organizations. *Innov. Dev. Archit. Eng. Constr.* **2003**, *2*, 121–130.

34. Tetik, M.; Peltokorpi, A.; Seppänen, O.; Holmstrom, J. Direct digital construction: Technology-based operations management T practice for continuous improvement of construction industry performance. *Autom. Constr.* **2019**, *107*, 1–13. [CrossRef]

35. Love, P.E.D.; Matthews, J. The how of benefits management for digital technology: From engineering to asset management. *Autom. Constr.* **2019**, *107*, 1–15. [CrossRef]

36. Flanagan, R.; Marsh, L. Measuring the costs and benefits of information technology in construction. *Eng. Constr. Archit. Manag.* **2000**, *7*, 423–435. [CrossRef]

37. Khayyat, N.T. Effects of information technology on cost, quality and efficiency in provision of public services. In *Information and Communication Technologies*; Nova Science Publishers, Inc.: Hauppauge, NY, USA, 2010.

38. Pérez-López, R.J.; Olguín-Tiznado, J.E.; García-Alcaraz, J.L.; Camargo-Wilson, C.; López-Barreras, J.A. The Role of Planning and Implementation of ICT in Operational Benefits. *Sustainability* **2018**, *10*, 2261. [CrossRef]

39. Koskela, L.J.; Kazi, A.S. *Information Technology in Construction: How to Realise the Benefits?* Information Science Publishing: Manchester, UK, 2003.

40. Lambrecht, J.F.; Vestergaard, F.; Karlshøj, J.; Hauch, P.; Mouritsen, J. Measuring the effects of using ICT/BIM in construction projects. In Proceedings of the 33rd CIB W78 Conference 2016, Brisbane, Australia, 31 October–2 November 2016.

41. Sheng, D.; Ding, L.; Zhong, B.; Love, P.E.D.; Luo, H.; Chen, J. Construction quality information management with blockchains. *Autom. Constr.* **2020**, *120*, 1–16. [CrossRef]

42. Heigermoser, D.; Garcia de Soto, B.; Sidney Abbott, E.L.; Chua, D.K.H. BIM-based Last Planner System tool for improving construction project management. *Autom. Constr.* **2019**, *104*, 246–254. [CrossRef]

43. Chi, C.; Yang, L. Measurement of ICT impact. *J. Adv. Commun. Syst. Technol.* **2011**, *3*, 1–11.

44. Akadiri, P.O.; Olomolaiye; Chinyio, E.A. Multi-criteria evaluation model for the selection of sustainable materials for building projects. *Autom. Constr.* **2013**, *30*, 113–125. [CrossRef]

45. Mutesi, E.T.; Kyakula, M. Application of ICT in the construction industry in Kampala. In Proceedings of the Second International Conference on Advances in Engineering and Technology, Kampala, Uganda, 31 January–2 February 2011; pp. 263–269.

Article

Multiple Criteria Evaluation of the EU Country Sustainable Construction Industry Lifecycles

Arturas Kaklauskas [1,*], Edmundas Kazimieras Zavadskas [2,*], Arune Binkyte-Veliene [2], Agne Kuzminske [3], Justas Cerkauskas [2], Alma Cerkauskiene [2] and Rita Valaitiene [4]

[1] Department of Construction Management and Real Estate, Faculty of Civil Engineering, Vilnius Gediminas Technical University, Sauletekio av. 11, LT-10223 Vilnius, Lithuania

[2] Institute of Sustainable Construction, Faculty of Civil Engineering, Vilnius Gediminas Technical University, Sauletekio av. 11, LT-10223 Vilnius, Lithuania; arune.binkyte-veliene@vgtu.lt (A.B.-V.); justas.cerkauskas@vgtu.lt (J.C.); alma.cerkauskiene@vgtu.lt (A.C.)

[3] Lithuanian Business Support Agency, Savanorių pr. 28, LT-03116 Vilnius, Lithuania; a.kuzminske@lvpa.lt

[4] Faculty of Civil Engineering, Vilnius Gediminas Technical University, Sauletekio av. 11, LT-10223 Vilnius, Lithuania; rita.valaitiene@vgtu.lt

* Correspondence: arturas.kaklauskas@vgtu.lt (A.K.); edmundas.zavadskas@vgtu.lt (E.K.Z.); Tel.: +370-5-274-5234 (A.K.)

Received: 16 April 2020; Accepted: 26 May 2020; Published: 28 May 2020

Abstract: This article looks at the trends and success of the sustainable construction industries in the EU member states, the UK and Norway. The research, covering the past three decades, revealed that different quality of life, macroeconomic, human development, construction and well-being factors define the sustainable construction industries in the EU member states, the UK and Norway. A multiple criteria decision matrix was created and analysed to look at the EU member countries, the UK and Norway from the perspective of their macro level environment and construction industries. Assessments of the sustainable construction industries were completed by using the COmplex PRoportional Assessment (COPRAS) and Degree of Project Utility and Investment Value Assessments (INVAR), two analysis methods. A look was taken at the dependencies linking the indicators related to the construction industries and macro level in the EU member countries, the UK and Norway. Then, the multiple criteria analysis of the construction industry's utility degree and performances were completed, and recommendations were generated. A country's perceived image and success can influence the economic behaviour of consumers. By and large, advanced and successful countries rarely become associated with a negative national image and their products and services rarely suffer negative consequences due to such association. This research, then, offers findings that can assist potential buyers in more rational decision-making when choosing of products and services based on a country of origin.

Keywords: sustainable construction industry; lifecycles; European Union Member States; complex evaluation; multiple criteria analysis; COPRAS and INVAR methods; success and image of a country; marketing

---

1. Introduction

World scientists have studied the construction industry [1–7], energy and buildings [8–11], building information modelling [12–14] and building and projects lifecycle [15–20]. Each stage of construction has certain environmental impacts associated with it and life-cycle assessment (analysis) can be applied to analyse construction throughout its lifecycle comprising all these stages [21]. Studies suggest that working fewer hours could improve sustainability as the scale of economic output would drop along with the severity of environmental pressures related to consumption patterns [22].

The necessity of having consumption and production systems in synchronisation with society and the environment first called for identification. The word sustainability was a response; thereby, it has presently been broadly inserted in policy and research aligned with such concepts as "circular economy" and "inclusive growth" [23,24]. The amount of construction waste produced annually by the construction industry in the UK alone is 100 million tonnes, which contain around 13 million tonnes of unused materials. However, the capacity for recycling such waste materials is merely 20% of the volume. Most of it gets dumped in landfills, which further adds to polluting the biosphere. There are numerous reasons for such negative impacts, according to the literature in the field. Among others, the reasons probably consist of poor management, embedded cultural values, obsolete technologies and inappropriate logistics [25]. Previously, environmental quality would be substituted for economic growth and vice versa in discussions regarding development. Now, amendments have been included to such discourses. Currently, talks on growth, environmental sustainability and societal development more and more frequently regard identifying simultaneous targets [26]. The concern of the construction industry now more than ever before points to defining needed improvements to sustainability in the spheres of society, the economy and the environment. The foundation of sustainability and building and construction improvements consists of applying lifecycle assessment (LCA). These must be understood by SMEs for their industrial activities. It is a necessity for increasing green construction market productivity and competitiveness as well as for satisfying consumers who now call for environmentally friendly products [27].

Therefore, only consumption habits require change for sustainability, without reductions in the present-day life quality, to foresee continuous development. Being sustainable in this development also relates to universal solidarity and democratic and fair allocations. In other words, via a sustainable development model, the suggestion is that a full understanding of development aims to reach environmental management as well as cover social responsibility and economic solutions by abandoning the existence of a consumer society. Thus, it can be stated that sustainable construction has three main dimensions/components called environmental, economic and societal. Interactions between ecological protection, economic progression and social fairness are significant parameters of sustainability [28]. Sustainability in the construction industry involves various interest groups with different demands, awareness, knowledge, communication skills, implementation skills and commitments. However, all such interest groups orient to the same tasks: climate adaptation, procurement, carbon and energy, environmental management, waste, water, materials, biodiversity, the community and the economy for developing a sustainable construction industry.

LCA enjoys widespread international acceptance as one way to improve environmental processes and services, and this is the reason Ortiz et al. [27] decided to examine it. Additionally, they wanted ways to evade negative environmental impacts, thus they needed to develop appropriate aims. The result was bound to generate a healthy environment for people's lives and an overall enhancement in the quality of life. The building sector must turn to governmental administrations along with environmental agencies to improve sustainability in the industry by generating appropriate construction codes and other environmental policies. Meanwhile the construction industry itself must pay attention to its involved individual players encouraging them to be proactive in developing the sorts of environmental, social and economic guideposts that would achieve sustainability within the industry [27]. Roads leading to sustainable development must insure an efficient metrics system for measuring an adequate transition to a greener accomplishment. Such an effort will require inclusion of performances, which distinguish not only recent achievements but also the matters that need improvement. Thereby, the performance is bound to result in a policy that is better informed [29]. The metrics gap is a focal point in the investigation by Doyle et al. [29]. They measured the "global competitiveness" of environmental and social sustainability by estimating the cross-country influences on economic achievements. The purpose of the research presented here was to develop an effective system consisting of environmental, social and economic criteria as well and to include an instrument for analysis, which would support an evolving lifecycle of a sustainable construction industry.

Knight et al. [22] performed a panel analysis of 29 high-income the Organisation for Economic Co-operation and Development (OECD) countries looking at carbon dioxide emissions, carbon footprint and ecological footprint, three environmental indicators, to see the effects of working hours. Their research, based on data for 1970–2007, suggests a significant link between working hours and increased environmental strains; policies intended to boost environmental sustainability, thus, could target this aspect. Hayden and Shandra [30] used the STochastic Impacts by Regression on Population, Affluence and Technology (STIRPAT) design to validate the hypothesis that shows a positive link between hours of work and ecological footprint (EF). Developed countries (Sweden, Australia and others) have recently started to reduce the number of working hours per week after reaching high levels of labour productivity per person. Several researchers have analysed the relationship between Gross Domestic Product (GDP) and EF [31–34], EPI [35,36] and environmental quality [37,38]. These and other studies [39] confirm that construction, macroeconomic, quality of life, human development and well-being factors impact the environment. The United Nations [40] emphasised the need to ensure the green economy development, when human development, social equity and economic growth go alongside environmental security. In today's world increasingly concerned about resource draining, both developed and developing countries can improve their construction industries' sustainability and address environmental issues by basing their decisions on lifecycle assessment.

Sustainable development has been an internationally recognised aim since the UN Conference on Environment and Development in Rio de Janeiro in 1992. Its central challenges are the maintenance of social security and justice, sustainable economic development and the preservation and creation of an intact environment [41]. In the 21st century, sustainability has become the most important issue concerning the construction industry lifecycle [42]. Looking at industrial sectors, the construction sector is of particular importance. On the one hand, it makes a vital contribution to the social and economic development of every country by providing housing and infrastructure; on the other hand, this sector is an important consumer of non-renewable resources, a substantial source of waste, a polluter of air and water and an important contributor to land dereliction [41].

The construction industry is a large and critical sector within the world economy, having a significant impact on the environment [43]. It is considered one of the main contributors to global warming. To mitigate global warming effects, the construction industry has been exploring various approaches to mitigate the impacts of carbon dioxide emissions over the entire lifecycle of buildings [1]. The need to minimise the negative impacts of construction lifecycle activities is increasing pressure on construction organisations to adopt proactive, environmentally sustainable strategies and actions in the design and construction process [43].

Construction firms are increasingly faced with sustainable development-based requirements that are influencing many facets of their activities, ranging from proactive environmentally conscious design and construction through to sustainable procurement, project efficiency and effectiveness and investment management. A lot of literature suggests that the implementation of sustainable construction lifecycle practices is influenced by environmentally sustainable development-based requirements in the form of government regulations, as well as stakeholders pressures—from clients, environmental groups, financial institutions and top management commitment [43]. The construction of buildings brings about a substantial ecological load: about 40% of energy consumption and about 25% of material moved by our economy is due to the construction of buildings. New construction lifecycle technologies and new building components would allow us to reduce the ecological load of buildings to a fraction of its present value [44].

Europeans spend over 90% of their time indoors (homes, workplaces, cars and public transport means, etc.) and are exposed to a complex mixture of pollutants at concentration levels that are often several times higher than outdoors [45]. Resource-efficient Europe is an effort to make economic growth less dependent on the use of resources, promote the transition to a low carbon economy, increase the share of renewable sources in the energy sector and promote energy efficiency. By 2020, energy efficiency must go up by 20% [46]. The flagship initiative "Innovation union" aims to change the focus of R&D and innovation policy by shifting it to the challenges facing our society, including climate change, energy efficiency and lower resource use.

Kotler and Gertner [47] analysed the influence of established country images on attitudes towards the services and products a country offers and the country's ability to attract tourists, businesses and investment. Kotler and Gertner [47] also investigated strategic marketing management and its role in improving a national image, making a country more attractive and its products more popular. A country's economy highly depends on its brand image and identity. The literature review revealed the key factors that influence a country's brand image and its impact on the economy through the intentions of consumers to buy the country's brands and products [48].

Studies suggest that stereotypes associated with a country can influence the way consumers see a brand irrespective of their intention [49]. Herz and Diamantopoulos [49] carried out three experiments that complement each other. They aimed to examine how country cues can lead to various country stereotypes (emotional vs. functional), which, in turn, make an automatic impact on brand-related behaviour, as well as affective and cognitive brand evaluations by consumers [49].

Saridakis and Baltas [50] examined the impact the country of origin has on new car prices. They applied hedonic price analysis to an extensive dataset. Their models demonstrate that, in addition to implicit prices related to technological characteristics and performance, prices of new cars offered by a certain brand reflect price distortions stemming from the heterogeneity related to the country of origin. Universally seen as a source of high-quality products, Japan may have lower demand in some other Eastern Asian countries due to their historical animosity [51]. Less economically developed countries are usually associated with a negative country image and the products they supply suffer the related negative effect [52].

Companies apply many different strategies to present their country of origin and make their customers more aware of what that country represents [53]. They use flags and symbols of the country of origin; label their products with the phrase "Made in..."; incorporate imagery of famous buildings or typical landscapes of their country, as well as famous or stereotypical people; attach origin and quality labels; make the country of origin part of the company name; and use the language of the country of origin.

Pappu et al. [54] applied canonical correlation analysis to examine the relationships that link the perception consumers have of a country at the macro-level (the country itself) and micro-level (the products associated with the country), and the equity a brand from that country has in the eyes of consumers. They interviewed residents of an Australian state capital city in mall-intercept surveys. The results show a significant impact of both the micro and macro images of the brand's country of origin on the equity of a brand perceived by consumers. The two sets of constructs were linked by a positive relationship that depended on the product category. The product category also influenced the type of contribution each dimension of the brand equity perceived by consumers made to the relationship. The contribution of both macro and micro country image dimensions depended on the product category as well. An interesting finding is that, among product categories, the country image has a bigger impact on cars than on TV sets. The results of this research can give international marketers important direct insights [54].

Roth et al. [55] measure the added value the name of a country can give to a brand or a product in the eyes of an individual consumer. They applied the construct of brand equity in a country context and their results suggest that product preferences are influenced by country brand equity in a positive way [55].

Elliot and Papadopoulos [56] explored the multifaceted nature of the image of a place and how it impacts buyer behaviour. Their interdisciplinary approach combines tourism, country and product variables. The authors selected two countries for empirical tests of their integrated model with four target countries analysed in each case. The eight model tests showed the relationships linking the subcomponents of the image of a place. Affective country image made the biggest impact on destination evaluations, cognitive country image made the biggest impact on product evaluations and beliefs associated with a product made an impact on tourism, receptivity [56].

The subject of the current study is the EU sustainable construction industry lifecycle and the construction, macroeconomic, quality of life, human development and well-being factors context as a whole.

To achieve the purpose of the research, following objectives were identified:

- Develop integrated numerical and qualitative indicators that define the sustainable construction industry and the macro context affecting it.
- Calculate and analyse the correlations linking the sustainable construction industry indicators;
- Calculate and analyse the correlations across states.
- Create and analyse a decision matrix for the multiple criteria analysis of the macro-level and the sustainable construction industry in EU countries.
- Analyse the dependencies linking the indicators describing the macro level and the sustainable construction industry.
- Make a multiple criteria analysis of the sustainable construction industry in EU countries and offer recommendations.

An all-encompassing analysis of the EU sustainable construction industry required the application of techniques of multiple criteria assessment that allow the user to take a comprehensive look at many aspects, including the construction, macroeconomic, quality of life, human development and well-being. The variety of investigated factors matches the different forms of data required in multiple criteria decision making. The analysis makes use of statistical, decision-making and biometric techniques, as well as big data analytics.

The structure of the rest of this paper is as follows. Section 2 presents a multiple criteria assessment of the sustainable construction industry in the European Union Member States. Section 3 shows a comparison of the sustainable construction industry indicators in the EU countries, the UK and Norway over the past 29 years. Section 4 presents an analysis of the interdependencies between the macro-level indicators and the indicators describing the construction industry in the EU member countries, the UK and Norway. Section 5 presents recommendations for EU macro-level and construction sectors. Section 6 concludes the paper and lays the groundwork for future research.

2. Multiple Criteria Assessment of the Sustainable Construction Industry in European Union Member States

Degree of Project Utility and Investment Value Assessments (INVAR), a new multiple criteria decision analysis method (Degree of Project Utility and Investment Value Assessment with recommendations by Kaklauskas [57]) (see Figure 1), applied in this research to analyse countries, shares the first five stages with the COmplex PRoportional Assessment (COPRAS) method [58]. The rankings and weights of the countries depend, directly and proportionally, on an appropriate system of specific decision criteria, as well as the weights and values of the criteria. At the start, experts develop the system of decision criteria and then determine their weights and values.

The basis for the exhaustive subsystem of criteria describing the sustainable construction industry of the countries considered, which is characterised herein, consists of studies from around the world [59–70].

The values for the indicators were obtained from the human development index [71], GDP growth data (annual %) [72–75], GDP per capita in PPP terms [75–77], inflation growth data, consumer prices (annual %) [78], unemployment rates (annual %) [79,80], the ease of doing business ranking [81], the labour productivity per person employed in 1991 USD (converted at Geary Khamis PPPs) [82], public debt (% of GDP) [83–85], the education index [71], the happiness index [86], the social progress index [87–92], the construction cost index (residential buildings, except for community housing) in national currency (index, 2015 = 100) [93], the building permits index (the amount of new residential construction, except for community housing) (index, 2015 = 100) [94], the production in construction (production volume index) (index, 2015 = 100) [95] and the labour input in construction (number of persons employed) (index, 2015 = 100) [96] (see Table 1).

For some countries (Spain and Malta), certain data (the production in construction, 2018 (production volume index) and the ease of doing business ranking, 2020) were not available; hence, the sustainable construction industry multiple criteria evaluation decision matrix (see Table 2) includes only 27 countries (25 EU member states, the UK and Norway) out of the 29 (27 EU member states, the UK, Norway) considered.

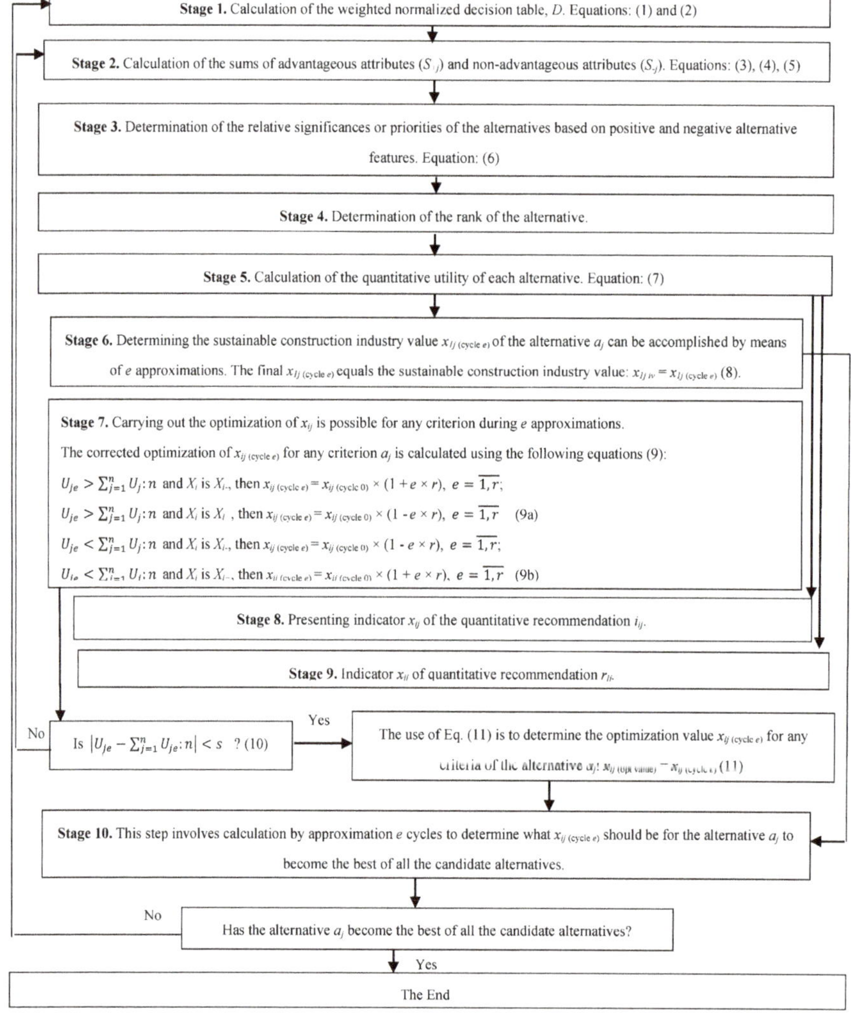

Figure 1. Degree of Project Utility and Investment Value Assessments (INVAR) method.

Table 1. Data from different databases.

Data	Unit	Source
X_1: Average GDP growth (annual %) from 1990 to 2019 [1]	Percentage	[73–76]
X_2: GDP per capita in USD, 2018	USD	
X_3: Average GDP per capita in PPP terms, 1995–2019 [1]	USD	[76–78]
X_4: GDP per capita in PPP terms, 2019	USD	
X_5: Average inflation, consumer prices (annual %), 1995–2019 [1]	Percentage	[79]
X_6: Inflation growth, consumer prices (annual %), 2019	Percentage	
X_7: Average unemployment rate (annual %), 1999–2019 [1]	Percentage	[80,81]
X_8: Unemployment rate (annual %), 2019	Percentage	
X_9: Average labour productivity per person employed in 1990 USD (converted at Geary Khamis PPPs), 1990–2018 [1]	USD	[83]
X_{10}: Labour productivity per person employed in 1990 USD (converted at Geary Khamis PPPs), 2018	USD	
X_{11}: Average public debt (% of GDP), 2000–2019 [1]	Percentage	[84–86]
X_{12}: Public debt (% of GDP), 2019	Percentage	
X_{13}: Average ease of doing business ranking, 2006–2020 [1]	Position	[82]
X_{14}: Ease of doing business ranking, 2020	Position	
X_{15}: Average human development index, 1990–2018 [1]	Index	[72]
X_{16}: Human development index, 2018	Index	
X_{17}: Average social progress index, 2014–2019 [1]	Index	[88–93]
X_{18}: Social progress index, 2019	Index	
X_{19}: Average education index, 1990–2018 [1]	Index	[72]
X_{20}: Education index, 2018	Index	
X_{21}: Average happiness index, 2013–2019 [1]	Index	[87]
X_{22}: Happiness index, 2019	Index	
X_{23}: Average construction cost index (residential buildings, except for community housing) in national currency (index, 2015 = 100), 2000–2018 [1]	Index	[94]
X_{24}: Construction cost index ((residential buildings, except for community housing) in national currency (index, 2015 = 100), 2018	Index	
X_{25}: Average building permits index (the amount of new residential construction, except for community housing) (index, 2015 = 100), 2000–2018 [1]	Index	[95]
X_{26}: Building permits index (the amount of new residential construction, except for community housing) (index, 2015 = 100), 2018	Index	
X_{27}: Average production in construction (production volume index) (index, 2015 = 100), 2000–2018 [1]	Index	[96]
X_{28}: Production in construction (production volume index) (index, 2015 = 100), 2018	Index	
X_{29}: Average labour input in construction (number of persons employed) (index, 2015 = 100), 2000–2018 [1]	Index	[97]
X_{30}: Labour input in construction (number of persons employed) (index, 2015 = 100) 2018	Index	

[1] Average was calculated using primary data.

All criteria were grouped into two categories. The first category covers the average value spanning the entire analysis period and the second category covers the value for the latest available year (see Tables 1 and A1). Fifteen experts in the area of well-being, macroeconomics, construction, human development and quality of life assigned the same significance (equal to 1) to all 30 well-being, macroeconomic, construction, human development and values-based decision factors, which means the significances of the 30 criteria add up to a total of 30. The units of the criteria were determined, and their values and significances were calculated.

When the criteria values and significances are available, and multiple criteria decision methods applied, the success and utility degree of the construction industry is rather easy to determine and the industry's priority (efficiency) easy to establish.

Table 2. Outcomes of the multiple criteria evaluation of the construction industries of 25 EU countries, the UK and Norway.

Compared Countries (a_j)	Success of a Country (Q_j)	Country's Priority Rank (P_j)	Country's Utility Degree (U_j), %
Germany	1.3039	4	86.93%
France	1.0668	15	71.12%
United Kingdom	1.2027	9	80.18%
Denmark	1.4228	2	94.86%
Norway	1.5	1	100%
Sweden	1.2803	6	85.36%
Finland	1.2552	7	83.68%
Lithuania	1.0863	13	72.42%
Estonia	1.1553	11	77.02%
Latvia	1.0034	19	66.90%
Belgium	1.1107	12	74.05%
Bulgaria	0.7662	27	51.08%
Greece	0.891	25	59.40%
Austria	1.1899	10	79.33%
Poland	0.9892	20	65.95%
Portugal	1.0111	18	67.41%
Romania	0.8416	26	56.11%
Slovenia	1.0549	17	70.33%
Slovakia	0.9692	22	64.62%
Cyprus	1.085	14	72.34%
Czech Republic	1.0556	16	70.38%
Ireland	1.301	5	86.74%
Hungary	0.9824	21	65.50%
Croatia	0.9035	24	60.24%
Luxembourg	1.3771	3	91.81%
Italy	0.9682	23	64.55%
Netherlands	1.2261	8	81.75%

Stages 1–5 of the INVAR technique [57] were applied to perform multiple criteria assessment of the sustainable construction industries in the EU member states (see Figure 1).

Stage 1. Calculate the weighted normalised decision table, D. Applied equations:

$$d_{ij} = \frac{x_{ij} \cdot q_i}{\sum_{j=1}^{n} x_{ij}}, i = \overline{1, m}, j = \overline{1, n} \tag{1}$$

$$\sum_{j=1}^{n} d_{ij} = q_i \tag{2}$$

Calculations:

$$d_{11} = 1 * 4.53/(4.53 + 2.19 + 1.96 + 3.03 + 8.4 + 1.64 + 5.83 + 4.22 + 4.14 + 2.92 + 4.2 + 3.88$$
$$+ 2.57 + 4.15 + 3.67 + 4.16 + 4.31 + 3.76 + 3.15 + 4.38 + 2.02 + 1.71 + 4.98 + 2.82 + 4.62$$
$$+ 3.27 + 4.4) = 1 * 4.53/100.91 = 0.0449;$$

$$d_{61} = 1 * 1.5/(1.5 + 1.2 + 1.8 + 1.3 + 2.3 + 1.7 + 1.2 + 2.3 + 2.5 + 3 + 1.5 + 2.5 + 0.6 + 1.5$$
$$+ 2.4 + 0.9 + 4.2 + 1.8 + 2.6 + 0.7 + 2.6 + 1.2 + 3.4 + 1 + 1.7 + 0.7 + 2.5) = 1 * 1.5/50.6 = 0.0296$$

Stage 2. Calculate the sums of advantageous attributes (S_{+j}) and non-advantageous attributes (S_{-j}). Applied equations:

$$S_{+j} = \sum_{i=1}^{m} d_{+ij}, \ S_{-j} = \sum_{i=1}^{m} d_{-ij}, \ i = \overline{1, m}, \ j = \overline{1, n} \tag{3}$$

$$S_+ = \sum_{j=1}^{n} S_{+j} = \sum_{i=1}^{m} \sum_{j=1}^{n} d_{+ij}, = \overline{1, m}, \; j = \overline{1, n} \tag{4}$$

$$S_- = \sum_{j=1}^{n} S_{-j} = \sum_{i=1}^{m} \sum_{j=1}^{n} d_{-ij}, = \overline{1, m}, \; j = \overline{1, n} \tag{5}$$

Calculations:

$$S_{-1} = 0.0301 + 0.0073 + 0.0103 + 0.0202 + 0.0083 + 0.0344 + 0.0296 + 0.0447 + 0.0171 = 0.202$$

$$S_{+1} = 0.0449 + 0.0432 + 0.0392 + 0.047 + 0.0444 + 0.0402 + 0.0391 + 0.0453 + 0.0376$$
$$+ 0.0375 + 0.0312 + 0.0286 + 0.0338 + 0.0355 + 0.039 + 0.0409 + 0.0404 + 0.0398 + 0.0357$$
$$+ 0.0389 + 0.0404 = 0.8226$$

Stage 3. Determine the relative significances or priorities of the alternatives based on positive and negative alternative features. Applied equation:

$$Q_j = S_{+j} + \frac{S_{-min} \cdot \sum_{j=1}^{n} S_{-j}}{S_{-j} \cdot \sum_{j=1}^{n} \frac{S_{-min}}{S_{-j}}}, \; j = \overline{1, n} \tag{6}$$

Calculations:

$$\sum_{j=1}^{n} \frac{S_{-min}}{S_{-j}} = \frac{0.1661}{0.202} + \frac{0.1661}{0.3277} + \frac{0.1661}{0.2251} + \frac{0.1661}{0.1661} + \frac{0.1661}{0.1844} + \frac{0.1661}{0.2025} + \frac{0.1661}{0.2331} +$$
$$\frac{0.1661}{0.2437} + \frac{0.1661}{0.2164} + \frac{0.1661}{0.2914} + \frac{0.1661}{0.339} + \frac{0.1661}{0.68} + \frac{0.1661}{0.7724} + \frac{0.1661}{0.2617} + \frac{0.1661}{0.3231} +$$
$$\frac{0.1661}{0.4192} + \frac{0.1661}{0.4601} + \frac{0.1661}{0.3123} + \frac{0.1661}{0.3393} + \frac{0.1661}{0.3764} + \frac{0.1661}{0.2751} + \frac{0.1661}{0.2914} +$$
$$\frac{0.1661}{0.4076} + \frac{0.1661}{0.4256} + \frac{0.1661}{0.278} + \frac{0.1661}{0.4793} + \frac{0.1661}{0.267} = 15.37543039$$

$$\sum_{J=1}^{N} S_{-J} = 0.202 + 0.3277 + 0.2251 + 0.1661 + 0.1844 + 0.2025 + 0.2331$$
$$+0.2437 + 0.2164 + 0.2914 + 0.339 + 0.68 + 0.7724 + 0.2617 + 0.3231 + 0.4192 + 0.4601$$
$$+0.3123 + 0.3393 + 0.3764 + 0.2751 + 0.2914 + 0.4076 + 0.4256 + 0.278 + 0.4793 + 0.267 = 8.9999$$

$$Q_1 = 0.8226 * \frac{0.1661*8.9999}{0.202*15.37543039} = 0.8226 * \frac{1.49488339}{3.105836} = 0.8226 + 0.4813143 = 1.3039$$
$$Q_2 = 1.0668$$

Stage 4. Determine the rank of the alternative. The greater is the significance Q_j, the higher is the rank of the alternative: $Q_1 < Q_2$. The success Q_j of a country a_j shows to what degree the country has fulfilled the requirements and achieved its needs in the sustainable construction industry. Each country is assigned a success with the most efficient country always taking the top spot with the success Q_{max}. Other countries that are below the best country in terms of their achievements related to the sustainable construction industry are, accordingly, assigned lower success.

Stage 5. Calculate the quantitative utility of each alternative. A higher success of a country also means that the country's utility degree is higher and a lower success then means lower utility degree. National utility degrees are determined by comparing countries with their most efficient counterpart in terms of the performance related to the sustainable construction industry. The countries considered will, therefore, have their utility degrees between 0% (worst case) and 100% (best case). Such ranking offers easier visual assessment of the efficiency of the countries. The utility degree U_j of a country a_j shows the country's performance in terms of the requirements. Higher utility degree shows that a bigger number of more important requirements were achieved. Applied equation:

$$U_j = \left(Q_j : Q_{max} \right) \cdot 100\% \tag{7}$$

Calculations:

$$U_1 = (1.3039/1.5) * 100\% = 86.93\%$$

$$U_2 = (1.0668/1.5) * 100\% = 71.12\%$$

The multiple criteria evaluation outcomes for the countries considered (see Table 2) show that Norway (a_5) scored best (significance $Q_{max} = Q_5 = 1.5$ and utility degree $U_{max} = U_5 = 100\%$) in terms of the criteria considered here.

Lithuania (a_8) ranked eleventh with its significance $Q_8 = 1.0863$, well below the top performer. The country's utility degree was $U_8 = 72.42\%$ (see Table 2).

Lithuania and Portugal in Table A1 can be examples for discussion. The evaluation of the 1990–2019 period shows that, at 4.22%, Lithuania recorded a similar average annual GDP growth to Portugal with 4.16%. The 2019 data show that the GDP per capita in PPP terms was lower in Portugal (30,487.7 USD) than in Lithuania (32,378.6 USD). At 8.81, the average 1999–2019 unemployment rate in Portugal was about 21% lower than that in Lithuania (10.89), and the 2019 unemployment rate in Portugal (6.10) was equal to that in Lithuania (6.10). Between 1995 and 2019, the average inflation growth in Lithuania (3.65%) was 1.825 times as high as in Portugal (2.00%). During 1990–2018, the average labour productivity per person employed was greater in Portugal (58,505.00 USD) than it was in Lithuania (45,530.00 USD) in 1990 USD terms (converted at Geary Khamis PPPs). However, the average public debt levels between 2000 and 2019 show that Portugal's debt burden, at 92.42%, was higher than that of Lithuania (29.12%). The conditions of the business environment better in Lithuania than in Portugal. Lithuania ranked 11th and Portugal 39th in the 2020 ease of doing business index; the average for 2006–2020, meanwhile, shows that Lithuania (21st) offered better conditions in terms of the ease of doing business than Portugal (34th) had. The human development index in 2018 is similar in both countries, in Portugal 0.85 and Lithuania 0.87. The education index in 2018 in Portugal (0.76) is less than in Lithuania (0.89). Conversely, appreciate the social progress index 2019 index is bigger in Portugal (87.12) than in Lithuania (81.30). It is clear from the summary of all 16 indicators under evaluation that, in four instances, Lithuania performed better than Portugal. The remaining 13 indicators, however, show Portugal performing better than Lithuania. The combination of these comparative data reflects the results of the multiple criteria assessment (see Table 2), showing that, in terms of the criteria considered here, Lithuania (a_8) scored better (priority $P_8 = 13$ and utility degree $N_8 = 72.42\%$) [52]. Portugal (a_{16}) ranked 18th with its utility degree $N_{16} = 67.41\%$ (see Table 2).

3. Comparison of the Sustainable Construction Industry Indicators in the EU Countries, the UK and Norway over the Past 29 Years

Over the past 29 years, many EU member countries under evaluation purposed considerable economic gains compared with the global level. Next, indicators such as GDP per capita (USD), the ease of doing business ranking and a few others are considered as examples (see Table A1 and Figure 2).

The leading countries among EU members, Norway and the UK in terms of GDP per capita (USD) in 2018 were Luxembourg, followed by Norway, Ireland, Denmark, Sweden, Netherlands, Austria, Finland, Germany, Belgium, the UK amd France. Italy is not far behind, followed by Cyprus, Slovenia, Portugal, Czech Republic, Estonia, Greece, Slovakia and Lithuania. The following are countries with GDP per capita (USD) 1.9 times or more below Italy's: Latvia, Hungary, Poland and Croatia. The weakest economies are in Romania and Bulgaria (see Figure 2). Spain (30,523.86 USD) and Malta (30,074.74 USD), which are not included in Table A1, have similar economies to Italy.

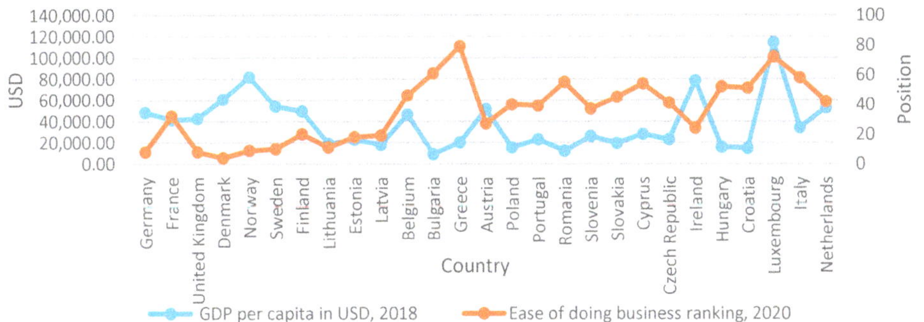

Figure 2. GDP per capita in USD, 2018, and ease of doing business ranking, 2020, values in analysed countries.

A look at the 2006–2020 ease of doing business ranking shows Romania moving up from 78 in 2006 to 55 in 2020, Slovenia improving from 63 in 2006 to 37 in 2020, Poland from 54 in 2006 to 40 in 2020 and Croatia, from 118 in 2006 to 51 in 2020, as the top improvers of their business environments among the EU member countries (see Figure 2).

The human development index (HDI) measures a country's performance in terms of its social living standards. The index looks at literacy, educational achievement, average life expectancy and the standard of living in every country around the world. All 25 EU member countries, the UK and Norway under evaluation fall into two HDI groups. Lithuania, Latvia, Estonia, Bulgaria, Slovakia, Hungary and Croatia have the smallest human development average range and fall in the high human development range (HDI = 0.71–0.80). All the other countries under evaluation (Germany, France, the UK, Denmark, Norway, Sweden, Finland, Belgium, Greece, Austria, Poland, Portugal, Romania, Slovenia, Cyprus, Czech Republic, Ireland, Luxembourg, Italy and Netherlands) fall in the very high human development range (HDI = 0.81–0.99) based on the 1990–2018 average. Spain (0.893) and Malta (0.885), which are not included in Table A1, fall in the very high human development range too. The same situation is shown for the average human development index, 1990–2018 (see Figure 3). Countries that are not under evaluation include Spain (0.84) and Malta (0.81).

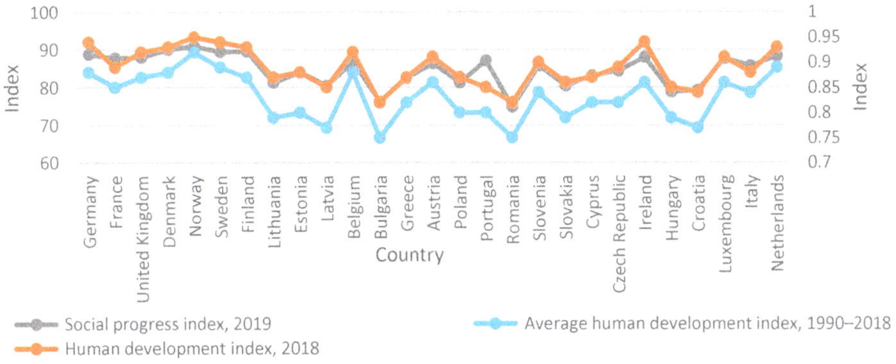

Figure 3. Social progress index, 2019, and HDI, 2018, and average HDI, 1990–2018, values in analysed countries.

A look at the change in social creation under the HDI shows that the EU countries, the UK and Norway considered in the study made remarkable progress over 28 years. A comparison of the 1990 and 2018 data shows that the highest positive HDI change recorded among the 25 EU member countries,

Norway, the UK under evaluation occurred in Ireland (0.18); Croatia (0.17); Czech Republic, Latvia and Poland (0.16); Estonia and the UK (0.15); Germany, Finland, Lithuania, Portugal, Hungary and Cyprus (0.14); Denmark (0.13); Sweden, Luxembourg, Bulgaria, Greece, Austria, Romania and Slovakia (0.12); France, Belgium and Italy (0.11); and Norway and Netherlands (0.10). The smallest recorded, positive HDI change among the 25 EU member countries, the UK and Norway under evaluation occurred in Slovenia (0.07) (see Figure 3). Countries that are not under evaluation include Spain (0.14) and Malta (0.14).

For the social progress index in 2019, the leading countries are Norway (90.95) and Denmark (90.09), followed by Finland (89.56), Sweden (89.45), Germany (88.84), Netherlands (88.31), the UK (87.98), Ireland (87.97), France (87.79), Luxembourg (87.66), Portugal (87.12), Belgium (86.77), Austria (86.4), Slovenia (85.8), Italy (85.69), Czech Republic (84.36), Estonia (83.98), Cyprus (83.14), Greece (82.48), Lithuania (81.3), Poland (81.25), Slovakia (80.43) and Latvia (80.42). The following are countries with social progress index 12.90% or more below that of Norway: Croatia (79.21), Hungary (78.77), Bulgaria (76.17) and Romania (74.81) (see Figure 3). Countries that are not under evaluation include Spain (87.47) and Malta (82.63), which have similar social progress index values as the UK and Greece.

For the average GDP growth during 1990–2019 (annual %), the leading country is Norway (8.40), followed by Finland (5.83), Hungary (4.98), Luxemburg (4.62), Germany (4.53), Netherlands (4.40), Cyprus (4.38), Romania (4.31) and Lithuania (4.22). The following are countries have average growth during 1990–2019 (annual %) two times or more below that of Norway: Belgium (4.20), Portugal (4.16), Austria (4.15), Estonia (4.14), Bulgaria (3.88), Slovenia (3.76), Poland (3.67), Italy (3.27), Slovakia (3.15), Denmark (3.03), Latvia (2.92), Croatia (2.82), Greece (2.57), France (2.19) and Czech Republic (2.02). The lowest average GDP growth in 1990–2019 (annual %) is in the UK (1.96), Ireland (1.71) and Sweden (1.64) (see Figure 4). Countries that are not under evaluation include Spain (2.29) and Malta (4.80), which have similar average GDP growth during 1990–2019 (annual %) as France and Hungary.

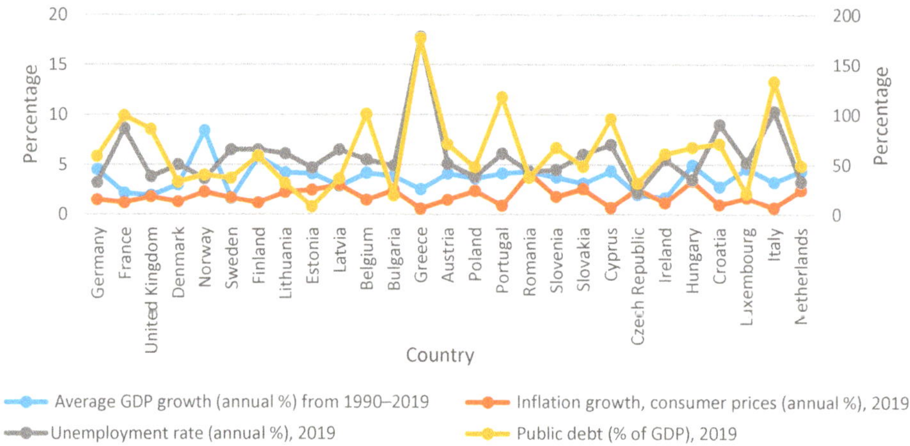

Figure 4. Average GDP growth, 1990–2019, unemployment rate, 2019, inflation growth, 2019, and public debt, 2019, values in analysed countries.

For inflation growth, consumer prices (annual %) in 2019, the leading EU member country is Greece (0.60%), followed by Cyprus (0.70%), Italy (0.70%), Portugal (0.90%) and Croatia (1.00%). The following countries have twice or more inflation growth than Greece: France (1.20%), Finland (1.20%) and Ireland (1.20%), followed by Denmark (1.30%), Germany (1.50%), Belgium (1.50%), Austria (1.50%), Sweden (1.70%), Luxemburg (1.70%), the UK (1.80%), Slovenia (1.80%), Norway (2.30%), Lithuania (2.30%), Poland (2.40%), Estonia (2.50%), Bulgaria (2.50%), Netherlands (2.50%), Slovakia (2.60%) and Czech Republic (2.60%). The highest inflation growth is in Latvia (3.00%), Hungary (3.40%)

and Romania (4.20%) (see Figure 4). Countries that are not under evaluation include Spain—(0.70%), which has the lowest indicator from all comparing countries, and Malta (1.70%), which has the same inflation growth as in Sweden or Luxemburg.

For unemployment rate (annual %) in 2019, the leading EU member country is the Czech Republic (2.2%), followed by Germany (3.2%), Netherlands (3.3%), Hungary (3.5%), Norway (3.6%), the UK (3.8%), Poland (3.8%) and Romania (4.3%). The following countries have at least twice the unemployment rate of Czech Republic: Slovenia (4.5%), Estonia (4.7%), Bulgaria (4.9%), Denmark (5.00%), Austria (5.1%), Luxembourg (5.2%), Belgium (5.5%), Ireland (5.5%), Slovakia (6.00%), Portugal (6.1%), Lithuania (6.1%), Sweden (6.5%), Finland (6.5%), Latvia (6.5%) and Cyprus (7.00%). The following countries have high or very high inflation rates: France (8.6%), Croatia (9.00%), Italy (10.3%) and Greece (17.8%) (see Figure 4). Countries that are not under evaluation include Spain (13.9%), which has a very high unemployment rate, and Malta (3.8%), which has an unemployment rate similar to the UK or Poland.

For public debt (% of GDP) in 2019, the leading EU member country is Estonia (8.2%), followed by Bulgaria (19.2%), Luxembourg (21.3%), Czech Republic (31.6%), Lithuania (31.8%), Denmark (33.0%), Latvia (36.3%), Sweden (36.9%), Romania (37.4%), Norway (40.0%), Poland (47.8%), Slovakia (48.4%), Netherlands (49.2%), Germany (58.6%), Finland (58.9%), Ireland (60.9%), Slovenia (67.1%), Hungary (67.5%), Austria (70.7%), Croatia (71.1%), the UK (85.6%), Cyprus (96.1%) and France (99.3%). Four EU member states have more than 100% public debt: Belgium (101%), Portugal (117.6%), Italy (133.2%) and Greece (176.6%) (see Figure 4). Countries that are not under evaluation include Spain (96.4%) and Malta (42.3%), which have similar public debt as Cyprus and Poland, respectively.

Comparison of indicators in EU member states show that countries that have a high economy, fall in the high human development range, have a high social progress index, have a low unemployment rate and take a higher position on ease of doing business ranking. However, they have a relatively high inflation rate and public debt. The comparison also showed that lower economy performance countries during 29 years boosted their economic, social and environmental performance. In conclusion, it seems that economic and social gains encourage boosting the sustainable construction industry performance of various EU member countries.

4. Analysis of the Interdependencies between the Indicators of the Macro Level and the Indicators Describing the Construction Industry in the EU Countries, the UK and Norway

This study identified correlational relationships among the 25 EU countries, the UK and the Norwegian construction industry (construction cost index, year-over-year change; production in construction in terms of the production volume index, year-over-year change; building permits index in terms of the number of dwellings, year-over-year change; and labour input in construction in terms of the number of persons employed, year-over-year change) and the macro-level indicators of the countries (GDP annual growth rate, GDP per capita current USD, unemployment rate, public debt, human development index, education index, gender inequality index and life expectancy at birth (total years)).

The outcomes of the correlation analysis are discussed with France taken as an example. The following indicators show strong correlations:

- GDP per capita current USD and construction cost index 2000–2018 ($r = 0.86$, linear dependence) (Figure 5) and production in construction 2000–2018 ($r = 0.81$, linear dependence) (Figure 5).
- Construction cost index and public debt 2000–2018 ($r = 0.92$, linear dependence), human development index 2000–2018 ($r = 0.97$, linear dependence), education index 2000–2018 ($r = 0.95$, linear dependence) and gender inequality index 2000–2018 ($r = -0.94$, inverse dependence).
- Production in construction and unemployment rate 2000–2018 ($r = -0.84$, inverse dependence) and public debt 2000–2018 ($r = -0.74$, inverse dependence).
- Unemployment rate and public debt 2000–2018 ($r = 0.76$, linear dependence).

- Public debt and human development index 2000–2018 (r = 0.94, linear dependence), education index 2000–2018 (r = 0.94, linear dependence) and gender inequality index 2000–2018 (r = −0.96, inverse dependence).
- Human development index and education index 2000–2018 (r = 0.99, linear dependence) and gender inequality index 2000–2018 (r = −0.96, inverse dependence).
- Education index and gender inequality index 2000–2018 (r = −0.96, inverse dependence).
- Life expectancy at birth (total years) and construction cost index 2000–2017 (r = 0.98, linear dependence), public debt 2000–2017 (r = 0.92, linear dependence), human development index 2000–2017 (r = 0.98, linear dependence), education index 2000–2017 (r = 0.97, linear dependence) and gender inequality index 2000–2017 (r = −0.94, inverse dependence).

Figure 5. GDP per capita current USD, 2000–2018; construction cost index, 2000–2018; production in construction, 2000–2018; and labour input in construction, 2000–2018, data and their variation in France.

Average correlations were identified between the following indicators:

- GDP annual growth rate and labour input in construction 2000–2018 (r = −0.42, inverse dependence).
- Construction cost index and production in construction 2000–2018 (r = −0.45, inverse dependence), labour input in construction 2000–2018 (r = 0.50, linear dependence) (Figure 5) and unemployment rate 2000–2018 (r = 0.53, linear dependence).
- Production in construction and labour input in construction 2000–2018 (r = 0.48, linear dependence) (Figure 5), building permits index 2000–2018 (r = 0.51, linear dependence), human development index 2000–2018 (r = −0.54, inverse dependence), education index 2000–2018 (r = −0.58, inverse dependence) and gender inequality index 2000–2018 (r = 0.64, linear dependence).
- Unemployment rate and human development index 2000–2018 (r = 0.62, linear dependence), Education index 2000–2018 (r = 0.67, linear dependence) and gender inequality index 2000–2018 (r = −0.63, inverse dependence).
- Life expectancy at birth (total years) and production in construction 2000–2017 (r = −0.45, inverse dependence), labour input in construction 2000–2017 (r = 0.48, linear dependence) and unemployment rate 2000–2017 (r = 0.63, linear dependence).

Figure 5 shows GDP per capita current USD, 2000–2018; construction cost index, 2000–2018; production in construction, 2000–2018; and labour input in construction, 2000–2018 data, and their variation in France.

The data of the indicators for 25 EU countries, the UK and Norway appear in Figures 6–8. Strong linear correlations were identified between the 2017 life expectancy at birth (total years) [97] and the 2019 social progress index [92], the 2018 labour productivity (USD) [82] and the 2018 human development index [71]. Strong inverse correlations were identified between the 2017 life expectancy at birth (total years) and the 2018 gender inequality index [71], the 2018 distribution of population by tenure status and type of household and income group [98]. Average linear correlations was identified between the 2017 life expectancy at birth (total years) [97] and the 2018 corruption perception index [99], the 2020 quality of life [100] and the 2019 happiness index [86].

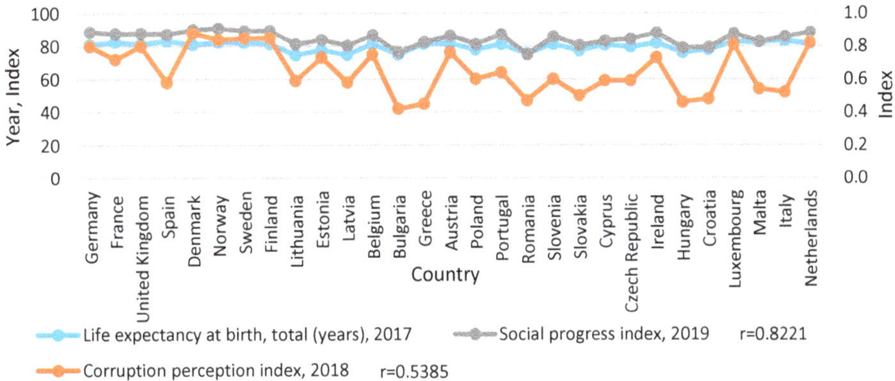

Figure 6. Correlation between the life expectancy at birth, 2017, and corruption perception index, 2018, and social progress index, 2019, of the 25 EU countries, the UK and Norway under evaluation.

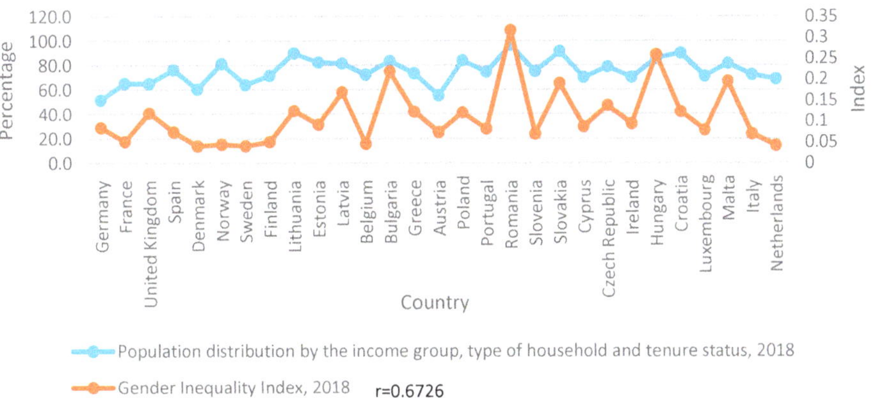

Figure 7. Correlation between the population distribution by the income group, type of household and tenure status, and the gender inequality index of the 25 EU countries, the UK and Norway under evaluation (data for 2018).

Research shows the existence of a strong dependency between the 2018 population distribution by the income group, type of household and tenure status, the 2019 social progress index ($r = -0.7633$, inverse dependence) and the 2018 human development index ($r = -0.7125$, inverse dependence). There are average dependences between the 2018 population distribution by the income group, type of household and tenure status and the 2018 gender inequality index ($r = 0.6743$, linear dependence) (Figure 7), the 2018 corruption perception index ($r = -0.6794$, inverse dependence) (Figure 7), the 2018 labour productivity ($r = -0,5096$, inverse dependence), the 2020 quality of life (except Luxembourg) ($r = -0,6008$, inverse dependence) (Figure 7) and the 2018 happiness index ($r = -0,5804$, inverse dependence) (Figures 7 and 8).

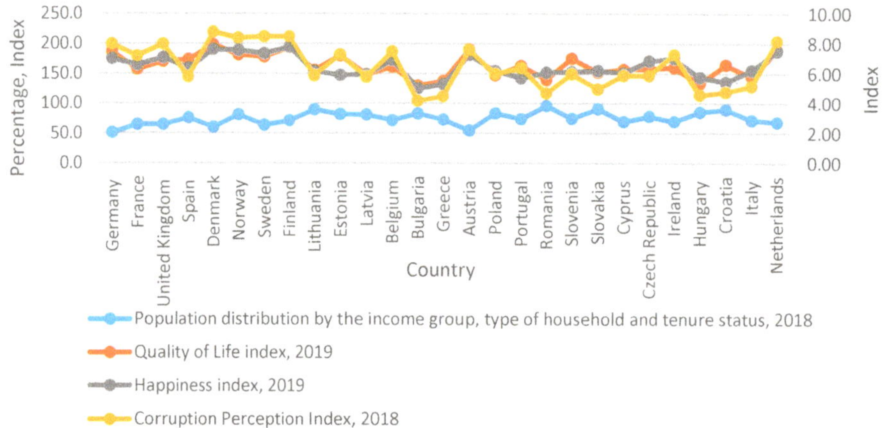

Figure 8. Correlation between the population distribution by the income group, type of household and tenure status, 2018, the quality of life, 2019, the happiness index, 2019, and the corruption perception index, 2018, of the 25 EU countries, the UK and Norway under evaluation.

The basis for the assessment of 25 EU members, the UK and Norway is the input data matrix presented in Table A1. The multiple criteria evaluation ranked 25 EU members, the UK and Norway by priority and determined their utility degrees.

The same assessments can be used in other cases where a best practice needs to be created and implemented.

The end of World War II brought communism to the countries in Eastern Europe, and the state typically promised to provide housing for the people. This meant that the majority of housing stock in such countries were public rental housing, and the rents, unlike in public housing offered in advanced capitalist societies, were very low. Everything changed after 1989 when democracy returned to Communist Europe. The major part of public housing was sold off. The elimination of rent control and privatisation of public housing encouraged the creation of a significant private renting sector in a few places [101].

A correlation analysis comparing the sustainable construction industry and housing price indicators of the EU member states, the UK and Norway was performed (see Table 3). The correlations among the 2018 real house price index (index, 2015 = 100), the 2018 nominal house price indices (index, 2015 = 100), the 2020 price-to-rent ratio (index, 2010 = 100), the 2018 house price indices with the 2018 production in construction (production volume index) (index, 2015 − 100), the 2018 construction cost index in national currencies (index, 2015 = 100) and the 2018 building permits indices (the amount of new residential construction, except for community housing) (index 2015 = 100) were analysed (Table 3).

Table 3. The correlation between the construction and housing price indices of the 25 EU countries, the UK and Norway.

Indicators	Production in Construction (Volume Index of Production), 2018 (index, 2015 = 100)	Construction Cost Index—in National Currency 2018 (index, 2015 = 100)	Real House Price Indices, 2018 (index, 2015 = 100)	Nominal House Price Indices, 2018 (index, 2015 = 100)	Building Permits—Number of Residential Buildings (Except Residences for Communities) Dwellings, (index, 2015 = 100), 2018
Real house price indices, 2018 (index, 2015 = 100) [102]	0.4049	0.4264	1	0.9788	0.6196
Nominal house price indices, 2018 (index, 2015 = 100) [102]	0.4286	0.5350	0.9788	1	0.5611
Price-to-rent ratio, 2020 (index, 2010 = 100) [103]	-	-	-	-	−0.4196
House price indices, 2018 [104]	0.4309	0.5430	0.9713	0.9975	0.5522

The outcomes of the correlation analysis are discussed below. Strong correlations were identified between the following indicators (Table 3):

- Real house price indices in 2018 (index, 2015 = 100) and nominal house price indices in 2018 (index, 2015 = 100) (r = 0.9788, linear dependence).
- House price indices in 2018, and real house price indices in 2018 (index, 2015 = 100) (r = 0.9713, linear dependence), and nominal house price indices in 2018 (index, 2015 = 100) (r = 0.9975, linear dependence).

Average and below average correlations were identified between the following indicators:

- Real house price indices in 2018 (index, 2015 = 100), and production in construction (production volume index) in 2018 (index, 2015 = 100) (r = 0.4049, linear dependence), construction cost index in 2018 (index, 2015 = 100) (r = 0.4264, linear dependence) (Figure 6), and building permits index (the amount of new residential construction, except for community housing) (index, 2015=100) in 2018 (r = 0.6196, linear dependence).
- Nominal house price indices in 2018 (index, 2015 = 100), and production in construction (production volume index) (index, 2015 = 100) in 2018 (r = 0.4286, linear dependence), construction cost index (index, 2015 = 100) in 2018 (r = 0.5350, linear dependence) (Figure 9), and building permits index (the amount of new residential construction, except for community housing) (index, 2015 = 100) in 2018 (r = 0.5611, linear dependence).
- Price to rent ratio in 2020 (index, 2010 = 100), and building permits index (the amount of new residential construction, except for community housing) (index, 2015 = 100) in 2018 (r = −0.4196, inverse dependence).
- House price indices in 2018, and production in construction (production volume index) in 2018 (index, 2015 = 100) (r = 0.4309, linear dependence), construction cost index (index, 2015 = 100) in 2018 (r = 0.5430, linear dependence), and building permits index (the amount of new residential construction, except for community housing) (index, 2015 = 100) in 2018 (r = 0.5522, linear dependence).

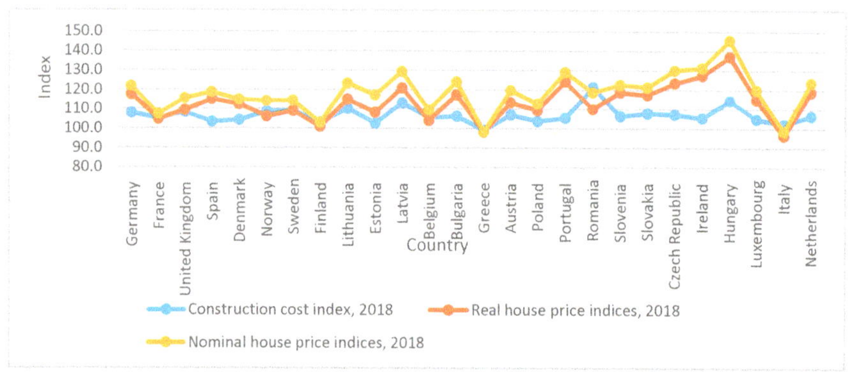

Figure 9. The correlations between the construction cost indices and real house price (nominal house price) indicators of the 25 EU countries, the UK and Norway (data for 2018).

Figure 9 shows the data for the 2018 construction cost index (index, 2015 = 100), real house price indices and nominal house price indices data. The rise of housing prices comes from higher construction prices determined by the growth in the construction sector (see Figure 9).

5. Recommendations for EU Construction Sectors

The next step, where INVAR [57] can be applied, is the investment value of a specific building lifecycle (Stage 6). INVAR can also optimise any selected criterion to make a specific project as competitive in the market as the other projects being compared (Stage 7) and determine the value that would propel the specific project to the top position among all the projects being analysed (Stage 10) (see Figure 1). Stages 6, 7 and 10 were not applied in this paper. For that reason, Equations (8)–(11) were not applied too.

Stage 6. Determining the sustainable construction industry value $x_{1j\,(cycle\,e)}$ of the alternative a_j can be accomplished by means of e approximations. The final $x_{1j\,(cycle\,e)}$ equals the sustainable construction industry value:

$$x_{1j\,iv} = x_{1j\,(cycle\,e)} \tag{8}$$

Stage 7. Carrying out the optimization of x_{ij} is possible for any criterion during e approximations.

The corrected optimization of $x_{ij\,(cycle\,e)}$ for any criterion a_j is calculated using the following Equations (9):

$$U_{je} > \sum_{j=1}^{n} U_j : n \text{ and } X_i \text{is } X_{i-}, \text{ then } x_{ij(cycle\,e)} = X_{ij\,(cycle\,0)} \times (1 + e \times r), e = \overline{1,r},$$
$$U_{je} > \sum_{j=1}^{n} U_j : n \text{ and } X_i \text{is } X_{i+}, \text{ then } x_{ij(cycle\,e)} = X_{X_{ij\,(cycle\,0)}} \times (1 - e \times r), e = \overline{1,r} \tag{9a}$$

$$U_{je} < \sum_{j=1}^{n} U_j : n \text{ and } X_i \text{is } X_{i-}, \text{ then } x_{ij(cycle\,e)} = x_{ij\,(cycle\,0)} \times (1 - e \times r), e = \overline{1,r};$$
$$U_{je} < \sum_{j=1}^{n} U_j : n \text{ and } X_i \text{ is } X_{i+}, \text{ then } x_{ij(cycle\,e)} = x_{ij\,(cycle\,0)} \times (1 + e \times r), e = \overline{1,r} \tag{9b}$$

$$\text{Is } \left| U_{je} - \sum_{j=1}^{n} U_{je} : n \right| < s ? \tag{10}$$

The use of Equation (11) is to determine the optimization value $x_{ij\,(cycle\,e)}$ for any criteria of the alternative a_j:

$$x_{ij\,(opt\,value)} = x_{ij\,(cycle\,e)} \tag{11}$$

When the value has been determined, digital recommendations (Stages 8 and 9) can then be provided on ways to improve projects [57]. For example, in Stages 8 and 9 of the INVAR technique [57], countries are offered digital recommendations on ways to achieve better scores. They also learn the effect of the new scores on their cumulative sustainable construction industry ranking. All recommendations are delivered as a matrix (see Table 4).

Table 4. A sample of digital recommendations on ways to improve certain scores for specific countries and the impact of the new scores on their cumulative position on the sustainable construction industry ranking.

Criteria Describing the Alternatives	Compared Alternatives Possible Improvement of the Analysed Criterion by % Possible Market Value Growth of Alternatives by % as First Impacted by Criterion Value Growth										
	Germany	France	United Kingdom	Denmark	Norway	Sweden	Finland	Lithuania	Estonia	Latvia	⋮
Average GDP growth (by annual %) from 1990–2019	4.53 (85.43%) (2.8477%)	2.19 (283.56%) (9.4521%)	1.96 (328.57%) (10.9524%)	3.03 (177.23%) (5.9076%)	8.4 (0%) (0%)	1.64 412.2% (13.7398%)	5.83 (44.08%) (1.4694%)	4.22 (99.05%) (3.3017%)	4.14 (102.9%) (3.43%)	2.92 (187.67%).. (6.2557%)	
...	...	...	...	...	...	...	...	...	...	...	...
Ease of doing business ranking, 2020	8 (50%) (1.6667%)	32 (87.5%) (2.9167%)	8 (50%) (1.6667%)	4 (0%) (0%)	9 (55.56%) (1.8519%)	10 (60%) (2%)	20 (80%) (2.6667%)	11 (63.64%) (2.1212%)	18 (77.78%) (2.5926%)	19 (78.95%)... (2.6316%)	
Average Human Development Index, 1990–2018	0.88 (4.55%) (0.1515%)	0.85 (8.24%) (0.2745%)	0.87 (5.75%) (0.1916%)	0.88 (4.55%) (0.1515%)	0.92 (0%) (0%)	0.89 (3.37%) (0.1124%)	0.87 (5.75%) (0.1916%)	0.79 (16.46%) (0.5485%)	0.8 (15%) (0.5%)	0.77 (19.48%)... (0.6494%)	
Human Development Index, 2018	0.94 (1.06%) (0.0355%)	0.89 (6.74%) (0.2247%)	0.92 (3.26%) (0.1087%)	0.93 (2.15%) (0.0717%)	0.95 (0%) (0%)	0.94 (1.06%) (0.0355%)	0.93 (2.15%) (0.0717%)	0.87 (9.2%) (0.3065%)	0.88 (7.95%) (0.2652%)	0.85 (11.76%)... (0.3922%)	
Average Social progress index, 2014–2019	88.37 (1.44%) (0.0479%)	87.48 (2.47%) (0.0823%)	87.78 (2.12%) (0.0706%)	89.31 (0.37%) (0.0123%)	89.64 (0%) (0%)	88.75 (1%) (0.0334%)	89.08 (0.63%) (0.021%)	80.79 (10.95%) (0.3651%)	82.8 (8.26%) (0.2754%)	79.41 (12.88%)... (0.4294%)	
...	...	...	...	...	...	...	...	...	...	...	...

Stage 8. Present indicator x_{ij} of the quantitative recommendation i_{ij}. Applied equation:

$$i_{ij} = |x_{ij} - x_{i\,max}|: x_{ij} \times 100\% \tag{12}$$

Calculations:

$$i_{16\,8} = |0.87 - 0.9|: 0.87 \times 100\% = 9.2\%$$

Stage 9. Present indicator x_{ij} of quantitative recommendation r_{ij}. r_{ij} shows the percentage of possible enhancement in U_j of a_j, supposing the value of x_{ij} can be enhanced up to the best value $x_{i\,max}$ of the indicator of criterion X_i. Applied equation:

$$r_{ij} = (q_i \times x_{i\,max}): (S_{-j} + S_{+j}) \times 100\% \tag{13}$$

Application:

Table 4 shows that Norway (a_5), for instance, performed best in key dimensions of human development, i.e., has the highest human development index ($x_{16\,5} = 0.95$) among the countries under evaluation. Meanwhile, Lithuania (a_8) has a human development index of 0.87 ($x_{16\,8} = 0.87$). Indicator $x_{16\,8} = 0.87$ of quantitative recommendation $r_{16\,8}$ shows the percentage of possible improvement of utility degree $U_8 = 72.42$ of Lithuania (a_8) upon presentation of $x_{16\,8} = x_{16\,5}$ (0.87 = 0.95). In other words, $r_{16\,8}$ shows the percentage of possible improvement in the utility degree U_8 of Lithuania (a_8), assuming the value of indicator $x_{16\,8}$ can be improved up to the best value $x_{16\,5}$ of the indicator of the human development index criterion X_{16}.

If Lithuania (a_8) aims to achieve the level of the human development index (X_{16}) achieved by Norway (a_5), the country must boost its performance by 9.2% ($i_{16\,8} = 9.2\%$, calculated in Stage 8 of the INVAR technique (see Table 4 [57]). Lithuania's (a_8) position in the overall country ranking would then improve by 0.3065% ($r_{16\,8} = 0.3065\%$ (see Table 4)).

Stage 10. This step involves calculation by approximation e cycles to determine what $x_{ij\,(cycle\,e)}$ should be for the alternative a_j to become the best of all the candidate alternatives.

Other countries can similarly examine ways to improve their scores on the sustainable construction industry ranking.

6. Conclusions

Among the latest available studies on sustainable construction, this research offers three innovative elements. The first is the use of the INVAR method [57]. This method can be used as the basis for creating efficient construction, macroeconomic, quality of life, human development and well-being factors to ensure improved macro-environments for the lifecycle of sustainable construction. Macro environments are important in attempts to make the lifecycle of the sustainable construction industry efficient. The multiple criteria decision matrix makes an integrated assessment of aspects that characterise the sustainable construction industries in the EU member countries, the UK and Norway. The variety of aspects considered is matched by that of the forms of data required in decision making. The COPRAS and INVAR techniques were applied for the evaluation of the sustainable construction industries.

The second innovation is that this research applied INVAR and its capabilities to expand the analysis of various indicators with extra features. Among them are digital tips for specific construction industries analysed against these criteria, deriving rationalised indicators and determining which values of these criteria would push the rating of a specific construction industry up to the expected level. When the INVAR method is applied, a broader look at the contexts such as well-being, macroeconomic factors, quality of life, construction and human development becomes possible, as well as a more thorough interpretation of the changes and shifts observed in construction industries over recent years. The multiple criteria analysis of the macro environment and construction industries in the EU member countries, the UK and Norway was performed and recommendations offered.

The third innovation is that we eliminated the need to stick to only construction and other traditional measures when the indicators for the 27 construction industries analysed in our research have to be improved. Corruption, happiness, education and social progress are other, less explored areas where improvements are possible. The dependencies that link the indicators describing the macro level and construction industry in the EU member countries, the UK and Norway were analysed.

The perceived image and success of a country can influence economic behaviour. Companies operate in a macro environment comprising a wide range of technological, ecological, legal, economic, success, image, political, health-related and social aspects, visible on a national level, and this environment is relevant to corporate marketing efforts. Due to its intuitive and cost-conscious nature, the emoticon measure can be a handy marketing tool to discover the sentiments elicited by one nation. Marketing managers then can apply these insights to different export markets [105]. A high developed country national image makes a country more attractive and its products more popular, and the impact such image makes on the economy through the intentions of consumers to buy the country's brands and products [18]. The findings of this research, then, can help potential buyers make decisions regarding the best choice in terms of the country of origin.

The future plans include research on the sustainable construction industries developments in American, African and Asian countries and the supply of recommendations. An innovative integrated evaluation of the sustainable construction industries would allow us in the future to investigate the lifecycle of buildings and its phases, the parties involved in the project and the context as a whole. The current models and intelligent systems available worldwide do not offer these functions.

Author Contributions: Conceptualisation, A.K. (Arturas Kaklauskas) and E.K.Z.; methodology, A.K. (Arturas Kaklauskas), E.K.Z., A.B.-V., A.K. (Agne Kuzminske), J.C., A.C. and R.V.; software, A.K. (Arturas Kaklauskas), A.B.-V., A.K. (Agne Kuzminske), J.C., A.C. and R.V.; validation, A.K. (Arturas Kaklauskas), A.B.-V., A.K. (Agne Kuzminske), J.C., A.C. and R.V.; formal analysis, A.K. (Arturas Kaklauskas), A.B.-V., A.K. (Agne Kuzminske), J.C., A.C. and R.V.; investigation, A.K. (Arturas Kaklauskas), A.B.-V., A.K. (Agne Kuzminske), J.C., A.C. and R.V.; resources, A.K. (Arturas Kaklauskas), A.B.-V., A.K. (Agne Kuzminske), J.C., A.C. and R.V.; data curation, A.K. (Arturas Kaklauskas), A.B.-V., A.K. (Agne Kuzminske), J.C., A.C. and R.V.; writing—original draft preparation, A.K. (Arturas Kaklauskas), A.B.-V., A.K. (Agne Kuzminske), J.C., A.C. and R.V.; writing—review and editing, A.K. (Arturas Kaklauskas), A.B.-V., A.K. (Agne Kuzminske), J.C., A.C. and R.V.; visualisation, A.K. (Arturas Kaklauskas), A.B.-V., A.K. (Agne Kuzminske), J.C., A.C. and R.V.; supervision, A.K. (Arturas Kaklauskas) and E.K.Z.; project administration, A.K. (Arturas Kaklauskas) and E.K.Z.; and funding acquisition, A.K. (Arturas Kaklauskas) and E.K.Z. All authors have read and agreed to the published version of the manuscript.

Funding: This project received funding from European Regional Development Fund (project No 01.2.2-LMT-K-718-01-0073) under grant agreement with the Research Council of Lithuania (LMTLT). The sponsor had no involvement in the study design, the collection, analysis and interpretation of data, or the preparation of the article.

Conflicts of Interest: The authors declare no conflict of interest.

Appendix A

Table A1. The sustainable construction industry multiple criteria evaluation for 25 EU member states, the UK and Norway.

Defining Criteria	1	2	3	4	5	6	7	8	9
	Germany	France	United Kingdom	Denmark	Norway	Sweden	Finland	Lithuania	Estonia
X1	4.53	2.19	1.96	3.03	8.4	1.64	5.83	4.22	4.14
X2	48,195.58	41,463.64	42,491.36	60,726.47	81,807.20	54,111.97	49,648.15	19,089.71	22,927.74
X3	36,883.78	33,758.78	32,814.96	37,373.91	55,564.39	36,991.77	33,989.35	17,836.78	19,598.09
X4	50,803.60	44,080.70	44,292.20	50,071.50	72,057.60	51,185.00	44,492.20	32,378.60	31,648.80
X5	1.45	1.5	2	1.63	2.12	1.46	1.59	3.65	4.61
X6	1.5	1.2	1.8	1.3	2.3	1.7	1.2	2.3	2.5
X7	6.92	9.17	5.76	5.5	3.73	7.1	8.39	10.89	9.19
X8	3.2	8.6	3.8	5	3.6	6.5	6.5	6.1	4.7
X9	85,500.00	91,826.00	80,570.00	87,298.00	124,259.00	86,944.00	84,874.00	45,530.00	46,716.00
X10	94,634.36	102,188.24	90,454.28	97,713.58	136,315.45	104,569.24	96,093.9	69,218.06	66,790.43
X11	68.28	79.89	63.16	40.59	37.66	43.1	48.37	29.12	6.95
X12	58.6	99.3	85.6	33	40	36.9	58.9	31.8	8.2
X13	7	32	7	5	8	13	13	21	18
X14	8	32	8	4	9	10	20	11	18
X15	0.88	0.85	0.87	0.88	0.92	0.89	0.87	0.79	0.8
X16	0.94	0.89	0.92	0.93	0.95	0.94	0.93	0.87	0.88
X17	88.37	87.48	87.78	89.31	89.64	88.75	89.08	80.79	82.8
X18	88.84	87.79	87.98	90.09	90.95	89.45	89.56	81.3	83.98
X19	0.85	0.75	0.84	0.84	0.88	0.85	0.83	0.79	0.81
X20	0.95	0.81	0.92	0.92	0.92	0.91	0.92	0.89	0.88
X21	6.89	6.56	6.84	7.57	7.56	7.35	7.51	5.85	5.6
X22	6.99	6.59	7.05	7.6	7.55	7.34	7.77	6.15	5.89
X23	90.3	89.8	86.4	87.7	82.7	85.7	89.5	86.4	86.5
X24	107.9	105	108.3	104.2	108.7	109	103.1	110.2	102.7
X25	90.2	115.9	100.6	94.9	100.6	67.5	106.3	81.3	93.2
X26	115.4	114.2	110.3	147.2	110.3	111.2	134.7	122.5	125.1
X27	98.2	109.8	93.7	97.5	84.7	85.1	87.6	91.2	93.3
X28	108.6	102.2	110.2	112.7	111.6	108	113	112.4	149.2
X29	99.8	103.5	101.9	102	85.7	85.7	100.2	94.7	95.5
X30	105.8	101.5	108.5	111.4	111.2	114.9	116	101.1	110.4

Defining criteria	10	11	12	13	14	15	16	17	18
	Latvia	Belgium	Bulgaria	Greece	Austria	Poland	Portugal	Romania	Slovenia
X1	2.92	4.2	3.88	2.57	4.15	3.67	4.16	4.31	3.76
X2	18,088.93	46,556.10	9,272.63	20,324.25	51,512.91	15,424.05	23,145.73	12,301.19	26,234.02
X3	15,965.26	35,637.57	12,755.57	24,637.00	37.,995.57	17,413.65	23,647.78	14,003.70	24,062.48
X4	27,701.60	46,621.30	21,767.60	27,795.90	50,031.00	29,642.20	30,487.30	24,605.30	34,480.00
X5	4.39	1.87	53.52	2.5	1.76	4.37	2	20.44	3.91
X6	3	1.5	2.5	0.6	1.5	2.4	0.9	4.2	1.8
X7	11.51	7.63	10.64	15.52	5.57	11.29	8.81	6.65	6.89
X8	6.5	5.5	4.9	17.8	5.1	3.8	6.1	4.3	4.5
X9	41,298.00	101,755.00	32,833.00	70,713.00	90,814.00	47,982.00	58,505.00	33,497.00	57,058.00

Table A1. *Cont.*

X10	62,101.79	111,234.55	44,353.13	71,095.93	99,888.18	71,218.16	65,232.77	57,001.93	72,911.29
X11	28.02	101.09	29.12	140.79	73.91	47.97	92.42	29.04	46.65
X12	36.3	101	19.2	176.6	70.7	47.8	117.6	37.4	67.1
X13	23	33	51	84	27	51	34	55	44
X14	19	46	61	79	27	40	39	55	37
X15	0.77	0.88	0.75	0.82	0.86	0.8	0.8	0.75	0.84
X16	0.85	0.92	0.82	0.87	0.91	0.87	0.85	0.82	0.9
X17	79.41	86.48	74.69	81.59	86.43	81.1	85.33	74.33	84.76
X18	80.42	86.77	76.17	82.48	86.4	81.25	87.12	74.81	85.8
X19	0.76	0.84	0.71	0.73	0.77	0.79	0.68	0.7	0.82
X20	0.87	0.89	0.81	0.83	0.87	0.87	0.76	0.76	0.89
X21	5.57	6.93	4.51	5.2	7.18	5.95	5.27	5.59	5.92
X22	5.94	6.92	5.01	5.29	7.25	6.18	5.69	6.07	6.12
X23	81.1	90.7	90.4	96.6	87.3	94.9	91.4	78.7	88.5
X24	113.2	105.5	106.3	99.1	107.1	103.7	105.4	121.4	106.3
X25	138.8	109.2	131.3	1041.9	89.9	90.4	553.6	104.5	191.4
X26	167.6	136.2	205.8	175.9	107	138.2	245.8	109.2	136.2
X27	94.9	101.7	83.4	286.5	97	85.5	200.5	84.7	133
X28	120.6	103	88.5	74.7	115.9	116.3	100.4	86.7	116
X29	96.9	95.3	106.4	134.5	99.7	107	182.3	102.8	120.4
X30	104.1	104.9	92.1	131.2	107.7	105.7	99.9	110.6	107.5

	19	20	21	22	23	24	25	26	27
Defining criteria	Slovakia	Cyprus	Czech Republic	Ireland	Hungary	Croatia	Luxembourg	Italy	Netherlands
X1	3.15	4.38	2.02	1.71	4.98	2.82	4.62	3.27	4.4
X2	19,546.90	28,159.30	23,078.57	78,806.43	15,938.84	14,869.09	114,340.50	34,318.35	53,024.06
X3	19,979.43	29,844.96	23,655.00	41,426.35	19,634.74	17,136.39	79,529.57	32,216.30	39,980.96
X4	33,069.90	37,172.10	35,537.00	73,214.70	29,558.70	24,748.60	105,147.60	38,233.50	53,933.00
X5	4.05	1.87	3.12	1.74	6.35	2.53	2.08	1.88	1.87
X6	2.6	0.7	2.6	1.2	3.4	1	1.7	0.7	2.5
X7	13.52	7.6	6.22	8.1	7.2	16.73	4.72	9.44	4.94
X8	6	7	2.2	5.5	9	9	5.2	10.3	3.3
X9	53,695.00	72,390.00	54,439.00	103,942.00	53,355.00	51,526.00	140,215.00	92,307.00	90,424.00
X10	76,662.91	81,134.22	71,002.58	149,707.06	65,736.31	62,514.81	142,966.76	91,588.91	101,949.93
X11	44.03	74.67	32.6	61.02	68.65	56.51	15.15	115.37	54.85
X12	48.4	96.1	31.6	60.9	67.5	71.1	21.3	133.2	49.2
X13	39	47	52	14	50	80	57	64	29
X14	45	54	41	24	52	51	72	58	42
X15	0.79	0.82	0.82	0.86	0.79	0.77	0.86	0.84	0.89
X16	0.86	0.87	0.89	0.94	0.85	0.84	0.91	0.88	0.93
X17	79.59	81.09	83.46	87.21	79.08	78.76	86.02	85.19	88.32
X18	80.43	83.14	84.36	87.97	78.77	79.21	87.66	85.69	88.31
X19	0.75	0.71	0.78	0.82	0.76	0.68	0.72	0.72	0.84
X20	0.82	0.81	0.89	0.92	0.82	0.8	0.8	0.79	0.91
X21	6.09	5.86	6.59	6.98	5.24	5.49	6.96	6.02	7.42
X22	6.2	6.05	6.85	7.02	5.76	5.43	7.09	6.22	7.49
X23	89.4	91	91.9	97.7	82.1	95.5	88.6	89.8	91.2
X24	107.8	100.6	107.2	105.4	114.4	100.3	104.8	102.2	106.5
X25	92.7	325.6	125.7	371	268.3	226.8	103.1	344.9	122.5
X26	116.6	194	127.2	223.1	302.4	168.9	137.3	127.4	129.3
X27	94.1	188.2	99.1	183	102.2	131.3	102	134.8	108.3
X28	99.8	180.7	106.3	147.2	127.5	110.3	108.7	101.1	125.8
X29	99.6	138.5	103.7	127.8	109	113.7	104.1	116.5	123.2
X30	107.2	138	99	138.6	117.1	107.6	107.1	98	106.4

References

1. Paik, I.; Na, S. Evaluation of Carbon Dioxide Emissions amongst Alternative Slab Systems during the Construction Phase in a Building Project. *Appl. Sci. (Basel)* **2019**, *9*, 4333. [CrossRef]
2. Morkunaite, Z.; Kalibatas, D.; Kalibatiene, D. A bibliometric data analysis of multi-criteria decision making methods in heritage buildings. *J. Civ. Eng. Manag.* **2019**, *25*, 76–99. [CrossRef]
3. Wu, P.; Song, Y.; Zhu, J.; Chang, R. Analyzing the influence factors of the carbon emissions from China's building and construction industry from 2000 to 2015. *J. Clean. Prod.* **2019**, *221*, 552–566. [CrossRef]
4. Tetik, M.; Peltokorpi, A.; Seppänen, O.; Holmström, J. Direct digital construction: Technology-based operations management practice for continuous improvement of construction industry performance. *Autom. Constr.* **2019**, *107*, 102910. [CrossRef]

5. Chowdhury, T.; Adafin, J.; Wilkinson, S. Review of digital technologies to improve productivity of New Zealand construction industry. *J. Inf. Technol. Constr. (ITcon)* **2019**, *24*, 569–587.

6. Xu, X.; Wang, Y.; Tao, L. Comprehensive evaluation of sustainable development of regional construction industry in China. *J. Clean. Prod.* **2019**, *211*, 1078–1087. [CrossRef]

7. Fang, Q.; Chen, L.; Zeng, D.; Zhang, L. Drivers of Professional Service Model Innovation in the Chinese Construction industry. *Sustainability* **2019**, *11*, 941. [CrossRef]

8. Bahramian, M.; Yetilmezsoy, K. Life cycle assessment of the building industry: An overview of two decades of research (1995–2018). *Energy Build.* **2020**, *219*, 109917. [CrossRef]

9. Hong, T.; Wang, Z.; Luo, X.; Zhang, W. State-of-the-art on research and applications of machine learning in the building life cycle. *Energy Build.* **2020**, *212*, 109831. [CrossRef]

10. Mourão, J.; Gomes, R.; Matias, L.; Niza, S. Combining embodied and operational energy in buildings refurbishment assessment. *Energy Build.* **2019**, *197*, 34–46. [CrossRef]

11. Gluch, P.; Gustafsson, M.; Baumann, H.; Lindahl, G. From tool-making to tool-using and back: Rationales for adoption and use of LCC. *Int. J. Strateg. Prop. Manag.* **2018**, *22*, 179–190. [CrossRef]

12. Succar, B.; Poirier, E. Lifecycle information transformation and exchange for delivering and managing digital and physical assets. *Autom. Constr.* **2020**, *112*, 103090. [CrossRef]

13. Ozturk, G.B. Interoperability in building information modeling for AECO/FM industry. *Autom. Constr.* **2020**, *113*, 103122. [CrossRef]

14. Muller, M.F.; Esmanioto, F.; Huber, N.; Loures, E.R.; Junior, O.C. A systematic literature review of interoperability in the green building information modeling lifecycle. *J. Clean. Prod.* **2019**, *223*, 397–412. [CrossRef]

15. Santos, R.; Costa, A.A.; Silvestre, J.D.; Vandenbergh, T.; Pyl, L. BIM-based life cycle assessment and life cycle costing of an office building in Western Europe. *Build. Environ.* **2020**, *169*, 106568. [CrossRef]

16. Orgut, R.E.; Batouli, M.; Zhu, J.; Mostafavi, A.; Jaselskis, E.J. Critical Factors for Improving Reliability of Project Control Metrics throughout Project Life Cycle. *J. Manag. Eng.* **2020**, *36*, 04019033. [CrossRef]

17. Kotb, M.H.; Aly, A.S.; Muhammad, E.K.M.A. Risk Assessment of Time and Cost Overrun Factors throughout Construction Project Lifecycle. *Life Sci. J.* **2019**, *16*, 78–91. [CrossRef]

18. Janjua, S.; Sarker, P.; Biswas, W. Sustainability Assessment of a Residential Building using a Life Cycle Assessment Approach. *Chem. Eng. Trans.* **2019**, *72*, 19–24.

19. Noktehdan, M.; Shahbazpour, M.; Zare, M.R.; Wilkinson, S. Innovation Management and Construction Phases in Infrastructure Projects. *J. Constr. Eng. Manag.* **2019**, *145*, 04018135. [CrossRef]

20. Tupenaite, L.; Kaklauskas, A.; Voitov, I.; Trinkunas, V.; Siniak, N.; Gudauskas, R.; Naimaviciene, J.; Kanapeckiene, L. Multiple criteria assessment of apartment building performance for refurbishment purposes. *Int. J. Strateg. Prop. Manag.* **2018**, *22*, 236–251. [CrossRef]

21. Kaklauskas, A. *Analysis of the Life Cycle of a Built Environment: Monograph*; Nova Science Publishers, Incorporated: Hauppauge, NY, USA, 2016; Available online: http://iti.vgtu.lt/ceneast/Media/Default/Documents/books/ Textbook_Analysis_of_the_life_cycle_of_the_built_environment.pdf (accessed on 15 March 2020).

22. Knight, K.W.; Rosa, E.A.; Schor, J.B. Could working less reduce pressures on the environment? A cross-national panel analysis of OECD countries, 1970–2007. *Glob. Environ. Chang.* **2013**, *23*, 691–700. [CrossRef]

23. Corrigan, G.; Crotti, R.; Hanouz, M.D.; Serin, C.; Drzeniek Hanouz, M.; Serin, C. Assessing progress toward sustainable competitiveness. In *The Global Competitiveness Report 2014–2015*; World Economic Forum: Cologny, Switzerland, 2014; Chapter 1.2; pp. 53–83.

24. Piketty, T.; Goldhammer, A. *Capital in the Twenty-First Century*; The Belknap Press of Harvard University Press: Cambridge MA, USA, 2014.

25. Alwan, Z.; Jones, P.; Holgate, P. Strategic sustainable development in the UK construction industry, through the framework for strategic sustainable development, using Building Information Modelling. *J. Clean. Prod.* **2017**, *140*, 349–358. [CrossRef]

26. Ambec, S.; Cohen, M.A.; Elgie, S.; Lanoie, P.; Canada, B.; Chabot, R.; Thornton, G. The porter hypothesis at 20: Can environmental regulation enhance innovation and competitiveness? *Rev. Environ. Econ. Policy* **2013**, *7*, 2–22. [CrossRef]

27. Ortiz, O.; Castells, F.; Sonnemann, G. Sustainability in the construction industry: A review of recent developments based on LCA. *Constr. Build. Mater.* **2009**, *23*, 28–39. [CrossRef]

28. Yılmaz, M.; Bakış, A. Sustainability in Construction Sector. *Procedia Soc. Behav. Sci.* **2015**, *195*, 2253–2262. [CrossRef]

29. Doyle, E.; Alaniz, M.P. Dichotomous impacts on social and environmental sustainability: Competitiveness and development levels matter. *Compet. Rev. Int. Bus. J.* **2020**, in press. [CrossRef]

30. Hayden, A.; Shandra, J.M. Hours of work and the ecological footprint of nations: An exploratory analysis. *Local Environ.* **2009**, *14*, 575–600. [CrossRef]

31. Liu, L.; Lei, Y. Dynamic changes of the ecological footprint in the Beijing-Tianjin-Hebei region from 1996 to 2020. *Ecol. Indic.* **2020**, *112*, 106142. [CrossRef]

32. Shujah-ur-Rahman; Chen, S.S.; Saud, S.; Saleem, N.; Bari, M.W. Nexus between financial development, energy consumption, income level, and ecological footprint in CEE countries: Do human capital and biocapacity matter? *Environ. Sci. Pollut. Res.* **2019**, *26*, 31856–31872. [CrossRef]

33. Toth, G.; Szigeti, C. The historical ecological footprint: From over-population to over-consumption. *Ecol. Indic.* **2016**, *60*, 283–291. [CrossRef]

34. Al-Mulali, U.; Weng-Wai, C.; Sheau-Ting, L.; Mohammed, A.H. Investigating the environmental Kuznets curve (EKC) hypothesis by utilizing the ecological footprint as an indicator of environmental degradation. *Ecol. Indic.* **2015**, *48*, 315–323. [CrossRef]

35. Tektüfekçi, F.; Kutay, N. The Relationship between EPI and GDP Growth: An Examination on Developed and Emerging Countries. *J. Mod. Account. Audit.* **2016**, *12*, 268–276. [CrossRef]

36. Hsu, A.; Lloyd, A.; Emerson, J.W. What progress have we made since Rio? Results from the 2012 Environmental Performance Index (EPI) and Pilot Trend EPI. *Environ. Sci. Policy* **2013**, *33*, 171–185. [CrossRef]

37. Sineviciene, L.; Kubatko, O.; Derykolenko, O.; Kubatko, O. The impact of economic performance on environmental quality in developing countries. *Int. J. Environ. Technol. Manag.* **2018**, *21*, 222–237. [CrossRef]

38. Bernard, J.; Mandal, S.K. The impact of trade openness on environmental quality: An empirical analysis of emerging and developing economies. *WIT Trans. Ecol. Environ.* **2016**, *203*, 195–208.

39. Kaklauskas, A.; Herrera-Viedma, E.; Echenique, V.; Zavadskas, E.K.; Ubarte, I.; Mostert, A.; Podvezko, V.; Binkyte, A.; Podviezko, A. Multiple criteria analysis of environmental sustainability and quality of life in post-Soviet states. *Ecol. Indic.* **2018**, *89*, 781–807. [CrossRef]

40. United Nations. Industry and environment. In *Sustainable Building and Construction Initiative: Information Note*; UNEP, Division of Technology, Industry and Economics: Paris, France, 2003; Volume 26, Available online: http://www.uneptie.org/pc/pc/SBCI/SBCI_2006_InformationNote.pdf (accessed on 1 April 2020).

41. Wallbaum, H.; Buerkin, C. Concepts and instruments for a sustainable construction sector. *Ind. Environ.* **2003**, *26*, 53–57.

42. Dobson, D.W.; Sourani, A.; Sertyesilisik, B.; Tunstall, A. Sustainable construction: Analysis of its costs and benefits. *Am. J. Civ. Eng. Archit.* **2013**, *1*, 32–38.

43. Akadiri, P.O.; Fadiya, O.O. Empirical analysis of the determinants of environmentally sustainable practices in the UK construction industry. *Constr. Innov.* **2013**, *13*, 352–373. [CrossRef]

44. Rohracher, H. Managing the technological transition to sustainable construction of buildings: A socio-technical perspective. *Technol. Anal. Strateg. Manag.* **2001**, *13*, 137–150. [CrossRef]

45. de Bruin, Y.B.; Koistinen, K.; Kephalopoulos, S.; Geiss, O.; Tirendi, S.; Kotzias, D. Characterisation of urban inhalation exposures to benzene, formaldehyde and acetaldehyde in the European Union. *Environ. Sci. Pollut. Res.* **2008**, *15*, 417–430. [CrossRef] [PubMed]

46. EU 2020 Target for Energy Efficiency. 2014. Available online: https://ec.europa.eu/energy/en/topics/energy-efficiency/targets-directive-and-rules/eu-targets-energy-efficiency (accessed on 5 January 2020).

47. Kotler, P.; Gertner, D. Country as brand, product, and beyond: A place marketing and brand management perspective. *J. Brand Manag.* **2002**, *9*, 249–261. [CrossRef]

48. Saeed, S.A. Creating a country brand identity: A Review of Literature. *Pak. Bus. Rev.* **2018**, *20*, 91–98.

49. Herz, M.F.; Diamantopoulos, A. Activation of country stereotypes: Automaticity, consonance, and impact. *J. Acad. Mark. Sci.* **2013**, *41*, 400–417. [CrossRef]

50. Saridakis, C.; Baltas, G. Modeling price-related consequences of the brand origin cue: An empirical examination of the automobile market. *Mark. Lett.* **2016**, *27*, 77–87. [CrossRef]

51. Maheswaran, D.; Chen, C.Y. Nation equity: Country-of-origin effects and globalization. *Handb. Int. Mark.* **2009**, 91–113. [CrossRef]

52. Lumb, R.; Lall, V. Perception of the Quality of Products Made in India by Consumers from the United States: A Longitudinal Analysis. *Indian J. Econ. Bus.* **2006**, *47*.

53. Aichner, T. Country-of-origin marketing: A list of typical strategies with examples. *J. Brand Manag.* **2014**, *21*, 81–93. [CrossRef]

54. Pappu, R.; Quester, P.G.; Cooksey, R.W. Country image and consumer-based brand equity: Relationships and implications for international marketing. *J. Int. Bus. Stud.* **2007**, *38*, 726–745. [CrossRef]

55. Roth, K.P.Z.; Diamantopoulos, A.; Montesinos, M.Á. Home country image, country brand equity and consumers' product preferences: An empirical study. *Manag. Int. Rev.* **2008**, *48*, 577–602. [CrossRef]

56. Elliot, S.; Papadopoulos, N. Of products and tourism destinations: An integrative, cross-national study of place image. *J. Bus. Res.* **2016**, *69*, 1157–1165. [CrossRef]

57. Kaklauskas, A. Degree of Project Utility and Investment Value Assessments. *Int. J. Comput. Commun. Control* **2016**, *11*, 666–683. [CrossRef]

58. Zavadskas, E.K.; Kaklauskas, A.; Sarka, V. The new method of multicriteria complex proportional assessment of projects. *Technol. Econ. Dev. Econ.* **1994**, *1*, 131–139.

59. Marx, H.J. Outcomes of the 2014 Survey of the CIDB Construction Industry Indicators. 2014. Available online: http://www.cidb.org.za/papers/Documents/Construction%20Industry%20Indicators%20-%202014%20Survey%20Outcomes%20-%20Full%20Report.pdf (accessed on 1 April 2020).

60. Zhang, P.; London, K. A Comparative Evaluation of Construction Industry International Performance between China and United States Using the International Advanced Index. In *Management of Construction: Analysis to Practice, Proceedings of the Joint CIB International Symposium of W055, W065, W089, W118, TG76, TG78, TG81 and TG84 (CIB 2012), Montreal, QC, Canada, 26–29 June 2012*; In-House Publishing: Brisbane, Australia, 2012; pp. 25–36.

61. Zhou, Q. *Advances in Applied Economics, Business and Creation: International Symposium, Proceedings of the ISAEBD 2011, Dalian, China, 6–7 August 2011*; Part 2; Springer Science & Business Media: Berlin, Germany, 2011.

62. Meikle, J.L.; Grilli, M.T. Measuring European Construction Output: Problems and Possible Solutions. 1999. Available online: http://citeseerx.ist.psu.edu/viewdoc/download?doi=10.1.1.201.-4682&rep=rep1&type=pdf (accessed on 1 April 2020).

63. Šaparauskas, J. The key aspects of sustainability evaluation in construction. In Proceedings of the 9th International Conference on Modern Building Materials, Structures and Techniques, Vilnius, Lithuania, 16–18 May 2007.

64. Sultan, B.; Kajewski, S. The behaviour of construction costs and affordability in developing countries: A yemen case study. In Proceedings of the Joint International Symposium on Knowledge Construction of CIB Working Commissions W55, W65 and W107, Singapore, 22–24 October 2003.

65. Deloitte Real Estate. European Construction Monitor, 2015–2016: PE Firms Aim for European Construction Market. 2016. Available online: https://www2.deloitte.com/content/dam/Deloitte/glo-bal/Documents/About-Deloitte/gx-real-est-construction-monitor.pdf (accessed on 11 March 2020).

66. Building Radar. Construction Industry in Europe. 2016. Available online: https://buildingradar.com/construction-blog/construction-industry-europe/ (accessed on 11 March 2020).

67. European Foundation for the Improvement of Living and Working Conditions. Trends and Drivers of Change in the European Construction Sector: Mapping Report. 2005. Available online: http://www.certificazione.unimore.it/site/home/documento124000637.html (accessed on 1 April 2020).

68. European Commission. *European Construction Sector Observatory, Country Profile France*; European Commision: Brussels, Belgium, 2018.

69. Olanrewaju, A.L.; Abdul-Aziz, A.R. Building Maintenance Processes, Principles, Procedures, Practices and Strategies. In *Building Maintenance Processes and Practices*; Springer: Singapore, 2015; pp. 79–129.

70. Nazarko, J.; Chodakowskaa, E. Measuring productivity of construction industry in Europe with Data Envelopment Evaluation. *Procedia Eng.* **2015**, *122*, 204–212. [CrossRef]

71. United Nations Development Programme. Human Development Reports, Education Index, Gender Inequeality. 2019. Available online: http://hdr.undp.org/en/data (accessed on 5 January 2020).

72. The World Bank Group. GDP Growth (Annual %). 2019. Available online: https://data.worldbank.org/indicator/ny.gdp.mktp.kd.zg (accessed on 10 February 2020).

73. The World Bank Group. GDP Per Capita (Current US$). 2019. Available online: https://data.worldbank.org/indicator/NY.GDP.PCAP.CD (accessed on 10 February 2020).

74. CountryEconomy.com. Gross Domestic Product. 2020. Available online: https://countryeconomy.com/gdp (accessed on 5 January 2017).
75. Trading Economics. GDP Growth Rate, GDP Per Capita, GDP per Capita PPP. 2020. Available online: https://tradingeconomics.com/ (accessed on 5 January 2020).
76. The World Bank Group. GDP per Capita, PPP (Current International $). 2019. Available online: https://data.worldbank.org/indicator/NY.GDP.PCAP.PP.CD (accessed on 10 February 2020).
77. StatisticsTimes.com. List of Countries by GDP (PPP) per Capita. 2020. Available online: http://statisticstimes.com/economy/countries-by-gdp-capita-ppp.php (accessed on 10 February 2020).
78. The World Bank Group. Inflation, Consumer Prices (Annual %). 2019. Available online: https://data.worldbank.org/indicator/FP.CPI.TOTL.ZG?locations=AT (accessed on 5 March 2020).
79. The World Bank Group. Unemployment, Total. 2019. Available online: https://data.worldbank.org/indicator/sl.uem.totl.zs (accessed on 10 February 2020).
80. Global Economy. Unemployment Rate—Country Rankings. 2020. Available online: http://www.theglobaleconomy.com/rankings/Unemployment_rate/ (accessed on 5 January 2020).
81. The World Bank Group. Doing Business Reports. 2019. Available online: https://www.doingbusiness.org/en/reports/global-reports/doing-business-2020 (accessed on 5 April 2020).
82. Conference Board. The Conference Board Total Economy Database™. November 2018. Available online: https://www.conference-board.org/data/economydatabase/index.cfm?id=30565 (accessed on 5 January 2020).
83. CountryEconomy.com. General Government Gross Debt. 2020. Available online: http://countryeconomy.com/national-debt (accessed on 10 February 2020).
84. IMF. Debt % of GDP. 2019. Available online: https://www.imf.org/external/datamapper/DEBT1@DEBT/OEMDC/ADVEC/WEOWORLD (accessed on 10 February 2020).
85. Central Intelligent Agency. Public Debt. 2020. Available online: https://www.cia.gov/library/publications/the-world-factbook/rankorder/2186rank.html (accessed on 10 February 2020).
86. World Happiness Report. 2020. Available online: https://worldhappiness.report/ (accessed on 5 January 2020).
87. Social Progress Index. 2014. Available online: https://www2.deloitte.com/content/dam/Deloitte/cr/Documents/public-sector/2014-Social-Progress-IndexRepIMP.pdf (accessed on 10 February 2020).
88. Social Progress Index. 2015. Available online: https://progresosocial.org.gt/wp-content/uploads/2015/04/2015-Methodology-Report.pdf (accessed on 10 February 2020).
89. Social Progress Index. 2016. Available online: https://www2.deloitte.com/content/dam/Deloitte/mx/Documents/about-deloitte/Social-Progress-Index-2016-Report.pdf (accessed on 10 February 2020).
90. Social Progress Index. 2017. Available online: https://www2.deloitte.com/content/dam/Deloitte/de/Documents/public-sector/Social-Progress-Index-Findings-Report-SPI-2017.pdf (accessed on 10 February 2020).
91. Social Progress Index. 2018. Available online: https://www.socialprogress.org/assets/downloads/resources/2018/2018-Social-Progress-Index-Exec-Summary.pdf (accessed on 10 February 2020).
92. Social Progress Index. 2019. Available online: https://www.socialprogress.org/assets/downloads/resources/2019/2019-Social-Progress-Index-executive-summary-v2.0.pdf (accessed on 10 February 2020).
93. Eurostat.eu. Construction Cost (or Producer Prices), New Residential Buildings—Annual Data. 2020. Available online: http://appsso.eurostat.ec.europa.eu/nui/show.do?dataset=sts_copi_a&lang=en (accessed on 5 January 2020).
94. Eurostat.eu. Building Permits—Annual Data. 2020. Available online: http://appsso.eurostat.ec.europa.eu/nui/show.do?dataset=sts_cobp_a&lang=en (accessed on 5 January 2020).
95. Eurostat.eu. Production in Construction—Annual Data. 2020. Available online: https://appsso.eurostat.ec.europa.eu/nui/show.do?dataset=sts_copr_a&lang=en (accessed on 5 January 2020).
96. Eurostat.eu. Labour Input in Construction—Annual Data. 2020. Available online: https://appsso.eurostat.ec.europa.eu/nui/show.do?dataset=sts_colb_a&lang=en (accessed on 5 January 2020).
97. The World Bank Group. Life Expectancy at Birth. 2017. Available online: https://data.worldbank.org/indicator/SP.DYN.LE00.IN?end=2017&start=1990 (accessed on 5 January 2020).
98. Eurostat.eu. Distribution of Population by Tenure Status, Type of Household and Income Group—EU-SILC Survey. 2018. Available online: https://appsso.eurostat.ec.europa.eu/nui/submitViewTableAction.do (accessed on 5 January 2020).

99. Transparency.org. Corruption Perceptions Index. 2020. Available online: https://www.transparency.org/cpi2019?/news/feature/cpi-2019 (accessed on 5 January 2020).
100. Numbeo.com. Quality of Life. 2019. Available online: https://www.numbeo.com/quality-of-life/rankings_by_country.jsp?title=2019 (accessed on 5 January 2020).
101. Gilbert, A. Rental housing: The international experience. *Habitat Int.* **2016**, *54*, 173–181. [CrossRef]
102. Oecd.org. Real House Prices, Nominal House Prices. 2019. Available online: https://data.oecd.org/price/housing-prices.htm (accessed on 5 January 2020).
103. Numbeo.com. Price to Rent Ratio. 2020. Available online: https://www.numbeo.com/property-investment/rankings.jsp (accessed on 5 January 2020).
104. Eurostat.eu. House Price Index—Annual Data. 2020. Available online: https://ec.europa.eu/eurostat/databrowser/view/tipsho20/default/table?lang=en (accessed on 5 January 2020).
105. Kock, F.; Josiassen, A.; Assaf, A.G. Toward a Universal Account of Country-Induced Predispositions: Integrative Framework and Measurement of Country-of-Origin Images and Country Emotions. *J. Int. Mark.* **2019**, *27*, 43–59. [CrossRef]

applied
sciences

Article

Identification and Prioritization of Critical Risk Factors of Commercial and Recreational Complex Building Projects: A Delphi Study Using the TOPSIS Method

Jolanta Tamošaitienė [1], **Mojtaba Khosravi** [2], **Matteo Cristofaro** [3], **Daniel W. M. Chan** [4] and **Hadi Sarvari** [5,*]

[1] Faculty of Civil Engineering, Institute of Sustainable Construction, Vilnius Gediminas Technical University, LT-10223 Vilnius, Lithuania; jolanta.tamosaitiene@vilniustech.lt

[2] Department of Civil Engineering, Najafabad Branch, Islamic Azad University, Najafabad 8514143131, Iran; m.khosravi@sci.iaun.ac.ir

[3] Department of Management and Law, University of Rome "Tor Vergata", 00133 Rome, Italy; matteo.cristofaro@uniroma2.it

[4] Department of Building and Real Estate, The Hong Kong Polytechnic University, Hung Hom, Kowloon, Hong Kong, China; daniel.w.m.chan@polyu.edu.hk

[5] Department of Civil Engineering, Isfahan (Khorasgan) Branch, Islamic Azad University, Isfahan 8155139998, Iran

* Correspondence: h.sarvari@khuisf.ac.ir

Citation: Tamošaitienė, J.; Khosravi, M.; Cristofaro, M.; Chan, D.W.M.; Sarvari, H. Identification and Prioritization of Critical Risk Factors of Commercial and Recreational Complex Building Projects: A Delphi Study Using the TOPSIS Method. *Appl. Sci.* **2021**, *11*, 7906. https://doi.org/10.3390/app11177906

Academic Editor: Edyta Plebankiewicz

Received: 12 July 2021
Accepted: 23 August 2021
Published: 27 August 2021

Publisher's Note: MDPI stays neutral with regard to jurisdictional claims in published maps and institutional affiliations.

Abstract: Construction development of Commercial and Recreational Complex Building Projects (CRCBPs) is one of the community needs of many developing countries. Since the implementation of these projects is usually very costly, identifying and evaluating their Critical Risk Factors (CRFs) are of significant importance. Therefore, the current study aims to identify and prioritize CRFs of CRCBPs in the Iranian context. A descriptive-survey method was used in this research; the statistical population, selected based on the purposive sampling method, includes 30 construction experts with hands-on experience in CRCBPs. A questionnaire related to the risk identification stage was developed based on a detailed study of the research literature and also using the Delphi survey method; 82 various risks were finally identified. In order to confirm the opinions of experts in identifying the potential risks, Kendall's coefficient of concordance was used. In the first stage of data analysis, qualitative evaluation was performed by calculating the severity of risk effect and determining the cumulative risk index, based on which 25 CRFs of CRCBPs were identified for more accurate evaluation. At this stage, the identified CRFs were evaluated based on multi-criteria decision-making techniques and using the TOPSIS technique. Results show that the ten CRFs of CRCBPs are external threats from international relations, exchange rate changes, bank interest rate fluctuations, traffic licenses, access to skilled labor, changes in regional regulations, the condition of adjacent buildings, fluctuations and changes in inflation, failure to select a suitable and qualified consultant, and employer's previous experiences and records. Obviously, the current study's results and findings can be considered by CRCBPs in both the private and public sectors for proper effective risk identification, evaluation, and mitigation.

Keywords: risk identification; risk assessment; MCDM; critical risk factors; commercial and recreational complex building projects; construction

---

1. Introduction

"Risk is an uncertain event or conditions that, if it occurs, has a positive or negative effect on a project objective" [1]; from that, construction projects are massively pervaded by risks due to the fact that they are planned and managed on the basis of uncertain forecasts [2,3]. These uncertainties come from the 'variability' and 'ambiguity' in relation to performance measures like cost, duration, or quality and, according to [4], they can be grouped into five areas in relation to construction projects: The variability associated with

---

estimates of project parameters, the basis of estimates of project parameters, design and logistics, objectives and priorities, and relationships between project parties. From the cited categories of uncertainty, some risks are inherently related to the project operating organizations and only they are responsible for managing them, while others are related to the economic, social, political, and technological environments [5].

Hence, identifying and evaluating risks in projects is necessary and can play a very important role in achieving project objectives. Project management literature identified several techniques for identifying and assessing risks involved in the construction industry. For instance, it has been found that the main reference in risk identification is historical data, past experience, and judgement [6]. According to Chapman [7], risk identification methods can be grouped into three general categories: Identifying the risks by the risk analyst; risk identification by interviewing key members of the project team; and risk identification through brainstorming meetings. In this regard, research has shown that the questionnaire survey is the most frequently used technique for risk identification in the construction sector [8,9]. However, despite the literature produced (see also [10–14]), few contributions (e.g., Tamošaitienė et al. [12]) have been dedicated to the risks in Commercial and Recreational Complex Building Projects (CRCBPs)—which comprise shops connected to each other with sidewalks that are designed and built alongside recreational, residential, office, hotel, restaurant, and cinema spaces [15,16]. Among them, it is worth mentioning the work of Comu et al. [17] that, despite classifying risks for CRCBPs, has the drawback of diverging from a very extensive list of risk factors (i.e., 21; much less than the 82 included in this study). This is the gap addressed in this work and that can be synthesized in the following research question: *What are the most severe risks in commercial and recreational complex building projects (CRCBPs)?*

To answer the above research question, the Delphi method is used to identify common risks of CRCBPs. Then, the qualitative method of probability of occurrence and the effect of risk on project objectives is adopted to identify CRFs of CRCBPs. Finally, a multi-criteria decision making (MCDM) method is employed, i.e., Technique for Order of Preference by Similarity to Ideal Solution (TOPSIS). This is another great advancement with respect to prior works that have been interested in CRFs of CRCBPs (i.e., [5]). As a result, the findings of this research are an unseen contribution to the original body of CRCBPs and the construction industry. Such an outcome would enable decision makers to make more informed decisions with regard to, for example, proper risk allocation, bid pricing, selection of the optimum procurement route, and evaluation of different construction projects.

2. Commercial and Recreational Complex Building Projects (CRCBPs)

The increase of the urban population in large metropolises, followed by the increase in demand for better building infrastructure on a global scale, paints a positive outlook for improving the construction market's situation in developing countries [18]. Over the past two decades, the construction of CRCBPs, to develop the welfare of citizens and increase socio-economic development indicators and subsequent sustainable development in Iran, has been greatly increased. Stemming from that, realizing projects faster and cheaper has been prioritized by policymakers and, as a consequence, the amount of construction has exceeded its quality [19].

Projects are divided into different types depending on the purpose and type of expected operation. In the field of construction projects, we are faced with different types that the 2015 United Nations' product classification (see Section 5) categorizes as Constructions Buildings and Civil Engineering Works [20]. A subclass of Construction Buildings is Commercial Buildings and among them there are Commercial and Recreational Complex Buildings. CRCBPs comprise a series of interconnected sidewalk shops designed and constructed in conjunction with entertainment, residential, office, hotel, restaurant, and cinema spaces [5]. In order to achieve economic growth, developing countries are forced to increase investment in infrastructures [21]. In fact, development of an infrastructure, such as construction of CRCBPs, can positively affect economic development [22]. Some examples of CRCBPs include Siam Paragon in Bangkok, Berjaya Times in Kuala Lumpur, Jewelry in Istanbul, West Edmonton in Edmonton, Canada, Dubai Mall in the UAE, Aventra Mall in Miami, and Harrods in London.

However, some CRCBPs construction projects have been unsuccessful due to various marketing, financial and investment issues. A clear example of such projects is the CRCBP of Arge Jahannam in Isfahan, which has failed due to a lack of detailed market studies and non-compliance with social rules and conditions [23]. In general, the existence of risk and uncertainty in the project reduces the accuracy and proper estimation of objectives and the efficiency of the project itself; in sum, the need to recognize and manage the risk of full CRCBPs is being increasingly perceived.

3. Perceived Risks of CRCBPs

Risk is an inherent component of all projects, and it is not possible to eliminate it completely. Therefore, identifying and analyzing risks can play an important role in the success of the project. In identifying the critical risk factors (CRFs) of any project, it is not enough to identify the risks that occur gradually. However, with a well-defined risk statement, weighing up not only what might happen, but also all the characteristics of the time of occurrence, probability of occurrence, and its impacts must also be considered. Determining a process for identifying, evaluating, and responding to CRFs will cause improvements in the mechanism, increase the accuracy and quality of work, and have a direct impact on time and cost [24]. Marle and Gidel [25] also stated that uncertainty and risk in projects affect project objectives more than anything else. Therefore, risk management plays an important role in the quality and reliability of decisions during a project.

The first step of risk management is to identify and record the characteristics of the risks that may affect the project (i.e., listing any potential risk to a project's cost, schedule, or any other critical success factor [1]). Risk identification is an iterative process, as new risks may be identified and discovered as the project progresses through its lifespan [16]. In order to identify the risks, several methods have been proposed, including interviews, hypothesis analysis, document review, the Delphi technique, brainstorming, graphing methods [26], Artificial Neural Network (ANN) [27], Failure Mode and Effects Analysis (FMEA) [28], Cross Analytical-Machine Learning models [29], and the Integration Definition for Function Modeling (IDEF0) process [30]. In the risk identification phase, the risks affecting the objectives should be prioritized and the impact of risks that do not affect the objectives should be avoided [12]. In this regard, the methods such as the Delphi technique are useful due to their simplicity, flexibility, and ease of access to experts [10,31–34]. With regard to CRCBPs, Chen and Khumpaisal [35] adopted an Analytic Network Process model based on 29 defined risk assessment criteria associated with commercial real estate development, then classified the data under four risk clusters: Social risks, economic risks, environmental risks, and technological risks. However, no prioritization of these risks was offered. The same applies for the work by Tamošaitienė et al. [12], who identified 19 types of risks for CRCBPs divided into macro- (i.e., country, industry), meso- (i.e., project, enterprise), and micro- (i.e., management, organization) categories.

The second step is risk analysis (i.e., the scope of the risk must be determined [1]), considered as a key factor in risk management that greatly aids the process. Risk classification is an important part of risk management issues with significant effects on the risk management process. General classifications can include cost, financing, demand, and political risks [36]. Risk classification should always be done with regard to the project's objectives [37]. There are different classification methods for risks that can be used for different purposes. Based on this, the risks can be divided into *main* and *subsidiary* risks [38]. They can also be categorized according to their impact on project objectives for project status reporting purposes [39]. However, according to PMI [1], the most appropriate approach for identifying and responding to risks is to determine risk groups based on their origin (instead of their impact), e.g., external risks, internal risks, technical risks, and legal

works [5,53] because it has been proved to work satisfactorily across different application areas and industrial sectors with varying terms and subjects [54], and "although several techniques have been combined or integrated with the classical TOPSIS, many other techniques have not been investigated. These techniques make the classical TOPSIS more representative and workable in handling practical and theoretical problems". Stemming from that recognized value of the TOPSIS method, in a series of studies. it has been empirically found to be better performing than other techniques, such as the Analytical Hierarchy Process (e.g., [55]), under some contextual circumstances. With regard to the proposed application, TOPSIS has been successfully used in works concerning the assessment of risks in construction projects and it is the preferred method, rather than simple/probit/logit regressions in risk analysis works [54]; this is due to its ability to fully use attribute information, providing a cardinal ranking of alternatives, and not requiring attribute preferences to be independent. Indeed, Gebrehiwet and Luo [56] recently adopted it for risk level evaluation on a construction project lifecycle and found that the construction stage is the most influenced by risk factors. Yet, Dandage et al. [57] used TOPSIS for reviewing the risk categories that are predominant in international projects and ranked them according to their effect on project success; they found political risks, technical risks, and design-related risks as the most important.

Investing in construction projects in Iran is one of the most lucrative decisions, but the lack of regular supervision in this sector has caused the people in the community to be exposed to human and financial damage due to the quality of construction of buildings [58]. Iran's economic problems, along with the country's situation in the international arena, are issues that foster project-related risks. In fact, political issues related to nuclear energy and subsequent sanctions against Iran have led to an increase in Iran's economic risk index in recent years [11]. The growth of the economic risk index has undoubtedly reduced Iran's economic interactions with other parts of the world, which can increase the likelihood of occurrence and severity of the impact of various other internal and external risks of projects [59]. At the same time, building construction is one of the main problems in developing countries today and because of rapid population growth, lack of financial resources, land problems, lack of skilled manpower, and, most importantly, lack of proper policy and planning, this issue has become critical [60]. In light of the above, construction projects in Iran, as a developing country, are always associated with many risks and uncertainties. Therefore, Iran was selected for this study to identify and evaluate the CRFs of large and complex CRCBPs.

4.1. Delphi Survey Technique

The Delphi technique is often used for risk determination and screening before the application of a MCDM method. The Delphi technique's main goal is to obtain the most reliable experts' opinions through a series of structured questionnaires with controlled feedback. For the selection of experts that were asked to respond to the Delphi questionnaires, one important rule is to prioritize the quality of experts over their quantity [50]. From that, participants of the Delphi survey are experts with solid knowledge and experience in the same subject, with time to participate in the research, and with effective communication skills [61,62]. Regarding the number of involved experts, this is usually less than 50, and often from 10 to 20 [62,63]. The survey also depends on a series of factors, such as desired sample homogeneity, the Delphi goal, difficulty range, quality of decision, ability of the research team, internal and external validity, time of data collection, available resources, and the scope of the problem under study [63]. The study adopted a purposive sampling technique in the selection of respondents, as done by other scholars for similar research [51,62]. In this regard, 30 experts were selected among practitioners of CRCBPs in Iran based on their level of knowledge and expertise in the field. In this regard, all the experts in this study have experience in the construction of CRCBPs. The survey was launched in May and June 2019. Table 1 shows the demographic data of the experts participating in the Delphi process.

Table 1. Specifications of interviewed experts.

Sample Features	Code	No. (%)
Age	<30 years	6 (20.0)
	30–45 years	15 (50.0)
	>45 years	9 (30.0)
Education	Bachelor's degree	14 (46.7)
	Master's degree	11 (36.7)
	PhD degree	5 (16.7)
Tenure in the construction sector	<10 years	9 (30.0)
	10–20 years	8 (26.7)
	>20 years	13 (43.3)
Field of activity	Public	2 (6.7)
	Private	21 (70.0)
	Both	7 (23.3)
Role	Client	8 (26.7)
	Consultant	16 (53.3)
	Contractor	6 (20.0)
Job Position	Architect	5 (16.7)
	Director	3 (13.3)
	Engineer—Civil, Electrical and Mechanical	7 (23.3)
	General Manger—Procurement and Contracts	3 (10.0)
	Project Manager	3 (10.0)
	Senior Project Manager	4 (13.3)
	Technical Director	4 (13.3)

In this study, 53 construction risks—classified into 14 categories—of CRCBPs were identified based on a detailed and comprehensive literature review (e.g., [5,16,24,31,64]). Table 2 outlines the risks affecting the objectives of CRCBPs and categorizes them into internal and external risks as well as grouping them into 14 clusters at the second level (i.e., social, economic, political, legal, natural, technical, work force, investment, management, safety, design, contract, market and environmental) and 53 risks at the third level. Because of different uses in previous studies and because the risks of each project vary widely depending on the environmental and social conditions, the present study uses past records and library studies as well as interviews with reporters to design a comprehensive RBS for CRCBPs.

The first stage of the Delphi questionnaire was developed based on the risks identified from the literature. By collecting the Delphi first-round questionnaires and statistically analyzing them, a small number of risks were eliminated, and new ones, such as tax and toll risk, site access risk, and traffic permits risk, were added. In the second stage, a questionnaire containing 67 risks of CRCBPs was sent to the experts. By reviewing the results of the second round, a number of risks were removed, and a number of new ones were added; 15 new risks were finally added. As a result, 82 risks were identified as relevant for CRCBPs and were classified into 16 different groups. By distributing the questionnaire based on the risks categorized in the third round, it was found that according to the Delphi panel experts, all 82 identified risks can be considered as relevant for CRCBPs. As it can be seen, 29 new risk factors were identified by the experts in three rounds of the Delphi survey. The authors believe that the identification of this volume of new risks could be due to several different reasons, including (i) the high volume of construction risks of CRCBPs compared to the construction of other urban projects and (ii) the high volume of construction project risks in developing countries compared to developed countries. Yet, this huge addition of risk factors is in line with other similar works [16,24]. In each Delphi round, the questionnaires were confirmed in terms of reliability and validity. Cronbach's alpha coefficient was used to evaluate reliability, while the content validity of Kendall's Coefficient of Concordance (W) was used to examine the degree of agreement (similarly to Sarvari et al. [51] and Khosravi et al. [5]). Kendall's coefficient of concordance shows (*i*) whether people who sorted items according to their importance used similar criteria for their judgment with regard to these items and (*ii*) whether these people agree with each other [65,66]. Kendall's coefficient of concordance is calculated using the following Formula (1).

$$W = \frac{S}{\frac{1}{12} k^2 \left(N^3 - N\right)},\tag{1}$$

Within this formula, K is the sum of all rankers (number of judges); N is the number of ranked items; $\frac{1}{12} k^2 \left(N^3 - N\right)$ is the maximum value of the sum of squares of variations from average Rj (which is equal to S in case of complete agreement between K judges); S is the sum of squares of Rj variations minus the mean Rj (i.e., all ranks for an item). From that, S is, therefore, calculated as follows: $S = \sum \left[R_j - \frac{\sum R_j}{N}\right]^2$. However, due to the complexity and time-consuming calculations of the value of S, Kendall's coefficient of concordance is computed by the Statistical Packages for Social Sciences (SPSS) computer software (similarly to other scholars [67–69]).

Table 2. Identified risks affecting the objectives of CRCBPs based on the review of the literature.

No.	Chapter RBS Level 1	Group RBS Level 2	Risk RBS Level 3	References
R_1	External	Social	Dissatisfaction	[2,24,70]
R_2			Sabotage	[2,24,37,70,71]
R_3		Economical	Exchange rate fluctuation	[10,35,36,70]
R_4			Inflation	[8,37,56,64,70]
R_5			Government economic policies	[11,70,72]
R_6		Political	Government policies	[37,41,70,73]
R_7			Foreign threats	[8,11,49,70,74]
R_8			Political events	[12,70,71]
R_9		Legal	Changes in law	[8,11,70]
R_{10}			Standards and requirements	[3,35,48,70,71]
R_{11}			Regional standards	[10,33,64,70]
R_{12}			Changing point view of government organization	[12,36,70,72]
R_{13}		Natural	Earthquake	[12,33,64,70]
R_{14}			Storm	[11,35,49,70]
R_{15}			Flood	[8,36,70]
R_{16}			Fire	[37,41,70,73]
R_{17}		Technical	Lack of documentation on the changes in project	[3,36,70–72]
R_{18}			Lack of acceptance changes control	[11,24,56,70,74]

Table 2. *Cont.*

No.	Chapter RBS Level 1	Group RBS Level 2	Risk RBS Level 3	References
R_{19}		Work force	Availability of skilled worker	[3,36,56,70]
R_{20}			Salary amount	[10,24,64,70]
R_{21}			Work standards and behavior	[8,64,70]
R_{22}			Skill efficiency	[35,56,70,73]
R_{23}		Investment	Unrealistic primary estimation	[12,33,49,70]
R_{24}			Lack of finance	[3,11,64,70,71]
R_{25}			Bankruptcy	[35,36,56,70]
R_{26}			Mismatch between demand and available resources	[24,33,70,72,74]
R_{27}		Management	Client records and experience	[12,33,49,70]
R_{28}			Delay in land hand over	[11,41,70]
R_{29}			Poor coordination and management	[3,10,24,33,70,73]
R_{30}			Lack of using management methods	[8,48,70]
R_{31}		Safety	Building site safety	[36,70,72]
R_{32}			Hygiene	[12,35,49,56,70]
R_{33}	Internal		Environment	[3,64,70–72]
R_{34}		Design	Technical ability and authority of counselor	[8,48,70]
R_{35}			Inadequate geotechnical studies	[24,33,35,70,73]
R_{36}			Failure to identify underground factors	[3,11,56,70,72]
R_{37}			Workshop supervision	[41,64,70]
R_{38}			Incomplete plans	[35,37,49,70]
R_{39}			Poor technical characteristics	[36,56,70,73,74]
R_{40}		Contract	Contractor contract (listed, fixed)	[8,36,70]
R_{41}			Contractor policies to enter biddings	[33,35,48,56,70,73]
R_{42}			Incomplete duties, agreements, and contracts	[10,41,70]
R_{43}			Contractor claims	[11,56,70,71,74]
R_{44}			Legal claims	[3,12,24,64,70,73]
R_{45}		Market	Increasing work competition	[32,36,70,72]
R_{46}			Change in demand purchases	[10,11,70]
R_{47}			Facilitating sales and commercial marketing	[35,70,71]
R_{48}		Environmental	Adjacent building condition	[64,70,74]
R_{49}			Smoke, pollution, noise	[37,48,70]
R_{50}			Building workshop security	[35,70,71]
R_{51}			Historical condition	[36,41,48,70]
R_{52}			Historical buildings' privacy space	[8,11,70,73]
R_{53}			Geographic and climatic condition	[10,24,33,70]

4.2. Qualitative Risk Assessment

To identify CRFs of CRCBPs, a qualitative method of probability of occurrence and impact of risks on project objectives (i.e., cost, time, and quality) was used. To do this, a questionnaire, concerning the 82 identified risks, was developed. Based on this questionnaire, experts were asked to comment on the probability of occurrence and impact of each of the risks of CRCBPs based on a 5-point Likert scale measurement (very low, low, medium, high and very high). The questionnaire used at this stage, like the Delphi stage, was evaluated and approved in terms of content reliability and validity. Risk refers to the number of expectations for that event to occur; in cases where the probability of occurrence is random, it is only possible to rely on the opinion of experts [12]. The magnitude of the impact of risk and the probability of occurrence are expressed by using descriptive or numerical expressions. Unlike the probability of occurrence, which is one, the impact of risk can have more than one effect; that is, it affects more than one project goal. In preparing the questionnaire, an attempt was made to obtain more valid results by inserting the structure of risk failure and determining the group of internal and external origins of the risks. Yet, with the aim of prioritizing risks by using the risk failure structure and calculating the effect of each risk, the score of each risk in the set of risk failure structure can be determined. The intensity of the impact of each risk is obtained by multiplying the probability of occurrence of each risk by the impact of the same risk on the project objectives. Thus, at first an initial risk index is defined based on the criteria of probability of occurrence and the effect of risk on project time, cost, and quality (Formula (2)).

$$PIR = \sum (PIt) + (PIc) + (PIq), \tag{2}$$

In this equation, PIR represents the initial risk index for each risk. Furthermore, 'P' is the probability of occurrence of risk; 'It' is the impact of risk on project time; 'Ic' is the impact of risk on project cost; and 'Iq' is the impact of risk on project quality. These indicators are separately measured based on each expert's opinion. To this end, according to Formula (3), the arithmetic mean method is used to aggregate indicators and the aggregated initial risk index is calculated for each of the risks.

$$APIR = \frac{\sum_{i=1}^{30} (PIR_i)}{N} \tag{3}$$

In this formula, the APIR represents the cumulative primary risk index for each of the risks. 'PIR$_i$' means initial risk index per risk for each specialist and 'N' is the total number of experts; in this study, there are 30. Finally, by using this index, it is possible to rank the risks qualitatively based on the severity of the impact of each risk.

4.3. Quantitative Risk Assessment

After evaluating and qualitatively prioritizing the risks, a quantitative evaluation is performed for CRFs of CRCBPs. To do this, the TOPSIS technique was used as one of the MCDM methods (Taylan et al., 2014). There are eight steps of the TOPSIS technique, based on Hwang and Yoon [75]: (i) setting risk assessment criteria; (ii) adjusting the decision matrix based on the prepared questionnaires; (iii) converting the decision matrix into a scaleless matrix; (iv) creating a weightless scale matrix; (v) identifying positive and negative ideal solutions; (vi) calculating the relative distance through the ideal positive and negative solutions; (vii) determining the relative proximity to each alternative; and (viii) determining the most important and least significant risks based on the proximity obtained. In this study, quantitative evaluation of CRFs has been done based on the same steps. In addition, the calculations related to the TOPSIS method were done using Microsoft Excel Office software.

In particular, to identify the risk assessment criteria, the results of the research of Draji Jahromi et al. [48]—which was performed to evaluate the risk assessment criteria—were used. Thus, seven risk assessment criteria were selected to evaluate CRFs of CRCBPs. These criteria are (i) risk response criteria, (ii) risk management, (iii) influencing the occurrence of other risks, (iv) accepting threat, (v) risk detection, (vi) risk probability, and (vii) vulnerability. Risks iii, iv, and vii were considered as criteria with a negative effect on risk assessment, while risks i, ii, v, and vi were considered as criteria with a positive effect on risk assessment.

The questionnaire at this stage was developed based on seven criteria and 25 CRFs of CRCBPs, based on which the results of the decision matrix will be formed. In this questionnaire, the importance of each risk was measured based on the criteria evaluated using a 9-point Likert scale measurement, so that 1 indicates very low importance and 9 indicates extremely high importance.

5. Calculation Results

5.1. Results of Delphi Survey

The risk identification step aimed to record the details of the uncertainties before the occurrence of risks. In the present study, after identifying the various risks of CRCBPs based on the study of research literature (e.g., [5,16,24,31,64,70,76]), the relevance of the identified risks was evaluated and monitored. Finally, based on three rounds of the Delphi technique, 82 risks of CRCBPs were identified and recorded; see Tables 3 and 4, respectively, for internal and external risks.

The agreement of experts in the Delphi method was investigated using Kendall's coefficient of concordance (W); the experts rank several categories based on their importance in a similar manner by using, essentially, the same judgment criteria of importance for each category. This Kendall's coefficient of concordance has a range from zero to one, indicating the degree of consensus between individuals (with W > 0.9 indicating very strong consensus; W > 0.7 strong consensus; W = 0.5 average consensus; W = 0.3 weak consensus and W = 0.1 very weak consensus). Furthermore, the significance of W is not enough for stopping the Delphi panel since for panels with more than 10 members, even small values of W are sometimes significant [65,66]. In this study, Kendall's coefficient of concordance was computed using SPSS computer software. According to the calculations, Kendall's concordance coefficient of the current study was equal to W = 0.734, which indicates a strong consensus and favorable agreement between respondents in identifying risks. The results of calculating the Kendall's coefficient of concordance are reported in Table 5. In addition, based on Formula (1) the value of W has been provided.

$$W = \frac{S}{\frac{1}{12}\,k^2\left(N^3 - N\right)} = \frac{\sum\left[R_j - \frac{\sum R_j}{N}\right]^2}{\frac{1}{12}\,k^2\left(N^3 - N\right)} = \frac{11092347}{\frac{1}{12}\,(30)^2\left((82)^3 - 82\right)} = \frac{30182908}{41121128} = 0.734$$

Table 3. The identified risks of CRCBPs based on literature review and the Delphi survey technique—internal risks.

Code	Social Risks
R_1	General dissatisfaction with the project's location
R_2	Sabotage
R_3	Cultural difference between people
R_4	Regional and ethnic limitations

Code	Economic Risks
R_5	Exchange rate fluctuation
R_6	Inflation fluctuation
R_7	Bank interest fluctuation
R_8	Change in duties of imported equipment
R_9	Law changes and economic policies of materials

Code	Political Risks
R_{10}	Government internal policies contradiction
R_{11}	Foreign threats
R_{12}	Inappropriate work relation of government organizations
R_{13}	Government instability

Table 3. *Cont.*

Code	Legal Risks
R_{14}	Changes in law
R_{15}	Changes in binding legal obligations in contracts
R_{16}	Regional standard changes (firefighting-master plans, etc.)

Code	Accident Risks
R_{17}	Natural disasters (flood—earthquake, etc.)
R_{18}	Sewage and water network unexpected accidents
R_{19}	Annual change in weather
R_{20}	Electrical distribution network unexpected accident

Code	Market Risks
R_{21}	Mismatching spaces with customer needs
R_{22}	Public's lack of interest
R_{23}	Increased work competition around project area
R_{24}	Changes in demand for the purchase of spaces with different uses
R_{25}	Facilitate sales and marketing conditions for specific user spaces

Code	Work force Risks
R_{26}	Access to skilled worker
R_{27}	Changes in the legal obligations of contracts
R_{28}	Behaviour, standards, work commitment
R_{29}	Mismatch of job referrals to personnel with related specialized skills

Code	Investment Risks
R_{30}	Unrealistic primary estimation
R_{31}	Inappropriate finance
R_{32}	Lack of on time finance
R_{33}	Bankruptcy
R_{34}	Mismatch between demand and available resources

Table 4. The identified risks of CRCBPs based on literature review and the Delphi survey technique—external risks.

Code	Management Risks
R_{35}	Previous employer-related experience and background
R_{36}	Site unavailability and delay in delivery of land to the presenter
R_{37}	Unauthorized allocation of funds at various stages
R_{38}	Lack of realistic goals
R_{39}	Poor coordination and management

Code	Project communication Risks
R_{40}	Lack of using appropriate methods in workshop management
R_{41}	Lack of proper organizational coordination
R_{42}	Project staff crisis in different units
R_{43}	Assigning responsibility of units to a third party
R_{44}	Lack of qualified consultant
R_{45}	Incomplete plan

Code	Design Risks
R_{46}	Poor technical specifications
R_{47}	Mismatch of layout with site location
R_{48}	Inaccuracies in realistic calculations and estimates
R_{49}	Non-compliance with design codes
R_{50}	Lack of maintenance period in designing process
R_{51}	Lack of a specific contract with contractors
R_{52}	Contractor's claim
R_{53}	Lack of coordination between the design process and manufacturing technology

Table 4. *Cont.*

Code	Construction Risks
R_{54}	Claims
R_{55}	Lack of timely completion of geotechnical studies and identification of underground factors
R_{56}	Delays in construction
R_{57}	Poor quality of workshop supervision
R_{58}	Incomplete description of tasks in contracts
R_{59}	Failure to complete work items in anticipated times

Code	Timetable Risks
R_{60}	Mismatching physical progress with the comprehensive project schedule
R_{61}	Delay in project duration due to lack of parallel work
R_{62}	Delay in completion of the project
R_{63}	Increase in exploitation costs
R_{64}	Increase in maintenance cost
R_{65}	Inappropriate pricing of saleable spaces
R_{66}	Lack of proper internal zoning of spaces in the business centre

Code	Exploitation Risks
R_{67}	Luxury businesses in the vicinity of ordinary businesses
R_{68}	Poor wide advertising
R_{69}	Ignorance of security and safety protocol
R_{70}	Lack of crisis management in CRCBPs
R_{71}	Lack of specific instructions in case of unexpected events
R_{72}	Lack of maintenance team stationed in the CRCBs
R_{73}	Adjacent building condition
R_{74}	Historical conditions

Code	Environmental Risks
R_{75}	Traffic permits
R_{76}	Privacy of monuments in the area
R_{77}	Workshop security in terms of side access

Code	Logistics Risks
R_{78}	Timely supply of materials
R_{79}	Supply of materials according to technical specifications
R_{80}	Predicting spare parts for emergency repairs and installations
R_{81}	Lack of instructions for ordering goods and services
R_{82}	Lack of instructions for ordering items in project warehouse

Table 5. Results of Kendall's coefficient of concordance analysis using SPSS software version 25.

N	Kendall's (W) [a]	Chi-Square	df	Asymp. Sig.	Result
30	0.734	1784.082	81	0	strong consensus

[a] Kendall's Coefficient of Concordance.

5.2. Calculation and Results of Qualitative Risk Assessment

The importance and severity of each risk depend on the probability of occurrence and the effect of that risk. Hence, a qualitative evaluation is adopted based on these two dimensions. In particular, qualitative ranking is determined according to the source of risk by using the risk failure structure [12]. The severity of each risk's impact is calculated by using the probability of occurrence for each risk and its impact on the objectives. After calculating the PI values, the importance of each area of the risk failure structure can be calculated as the sum of the PI values. In the present study, in order to qualitatively rank the risks—by calculating the probability of occurrence and the effect of risk on the main objectives of the project (i.e., time, cost, and quality) with the same weight—qualitative risk rating has been performed. In order to achieve the desired result, the Primary Risk Index (PIR) is calculated after determining the probability of occurrence of each risk and each risk's impact on the time, cost, and quality of the project. It is worth mentioning that the above index is calculated separately based on the presentation of each expert. PIR1 to PIR30 is then determined for each of the 82 risks. The aggregated risk index will then be calculated for each of the risks. By calculating the PIR of all risks, the APIR value is calculated, and the final ranking is provided. Table 6 shows the overall PIR and APIR results of the risks and the degree of rating of each risk based on the probability and effect method. According to the results of the qualitative evaluation stage, risks with an APIR value of 0.6 and above were identified as CRFs. Thus, a total of 25 risks were identified as CRFs of CRCBPs that will be quantitatively analyzed.

Table 6. PRI and APRI results of qualitative risk evaluation.

Code	$\sum PRI$	Sample Size	APRI	Risk Ranking	Code	$\sum PRI$	Sample Size	APRI	Risk Ranking
R_1	6.30	28	0.225	67	R_{42}	10.22	28	0.365	49
R_2	8.26	28	0.295	58	R_{43}	8.32	28	0.297	55
R_3	3.48	28	0.124	81	R_{44}	23 44	28	0.837	7
R_4	4.18	28	0.149	79	R_{45}	17.22	28	0.611	23
R_5	46.36	28	1.655	1	R_{46}	17.04	28	0.608	24
R_6	25.23	28	1.615	2	R_{47}	12.62	28	0.455	39
R_7	14.44	28	0.872	6	R_{48}	14.90	28	0.532	30
R_8	18.92	28	0.675	18	R_{49}	7.84	28	0.280	60
R_9	17.76	28	0.634	21	R_{50}	8.64	28	0.308	53
R_{10}	13.67	28	0.488	35	R_{51}	15.14	28	0.540	29
R_{11}	24.49	28	0.874	5	R_{52}	24.98	28	0.892	4
R_{12}	11.68	28	0.417	46	R_{53}	13.44	28	0.480	37
R_{13}	10.20	28	0.346	50	R_{54}	12.50	28	0.446	42
R_{14}	12.38	28	0.442	43	R_{55}	11.02	28	0.393	48
R_{15}	8.28	28	0.295	57	R_{56}	20.26	28	0.723	13
R_{16}	19.58	28	0.699	14	R_{57}	16.36	28	0.580	26
R_{17}	13.50	28	0.482	36	R_{58}	9.10	28	0.325	52
R_{18}	2.28	28	0.081	82	R_{59}	16.90	28	0.603	25
R_{19}	7.86	28	0.280	59	R_{60}	13.14	28	0.469	38
R_{20}	3.80	28	0.135	80	R_{61}	13.54	28	0.447	41
R_{21}	6.74	28	0.240	65	R_{62}	16.24	28	0.580	27
R_{22}	6.57	28	0.234	66	R_{63}	4.58	28	0.163	75
R_{23}	7.42	28	0.265	62	R_{64}	5.42	28	0.193	71
R_{24}	4.52	28	0.161	77	R_{65}	5.34	28	0.190	72
R_{25}	4.56	28	0.162	76	R_{66}	5.61	28	0.200	70
R_{26}	25.02	28	0.893	3	R_{67}	4.58	28	0.163	75
R_{27}	19.54	28	0.697	15	R_{68}	4.24	28	0.151	78
R_{28}	14.01	28	0.500	33	R_{69}	5.24	28	0.187	73
R_{29}	20.72	28	0.740	11	R_{70}	5.98	28	0.213	68
R_{30}	22.66	28	0.809	9	R_{71}	7.00	28	0.253	63
R_{31}	18.21	28	0.647	20	R_{72}	7.02	28	0.250	64
R_{32}	18.72	28	0.668	19	R_{73}	21.84	28	0.780	10
R_{33}	17.36	28	0.620	22	R_{74}	8.52	28	0.304	54
R_{34}	19.44	28	0.694	16	R_{75}	22.82	28	0.815	8
R_{35}	20.72	28	0.694	16	R_{76}	5.80	28	0.207	69
R_{36}	13.72	28	0.490	34	R_{77}	12.80	28	0.431	45
R_{37}	11.04	28	0.390	47	R_{78}	15.36	28	0.548	28
R_{38}	7.47	28	0.266	61	R_{79}	12.34	28	0.440	44
R_{39}	14.10	28	0.503	32	R_{80}	8.32	28	0.297	56
R_{40}	19.08	28	0.681	17	R_{81}	9.47	28	0.338	51
R_{41}	12.62	28	0.450	40	R_{82}	14.20	28	0.507	31

5.3. Calculation and Results of Quantitative Risk Assessment

To determine the final priority of the identified CRFs, first the decision matrix should be formed based on the evaluation criteria and using the TOPSIS technique. Then, the normal matrix is extracted and the relative proximity of each option to the solution is determined.

The first step in the TOPSIS technique is the formation of a decision matrix, which is prepared by gathering the opinions of experts through a decision matrix questionnaire. This matrix is needed for evaluating the importance of risks based on criteria. Table A1 (Appendix A) shows the results of the questionnaires collected from 30 experts. Normalization is the second step in solving all MCDM techniques based on the decision matrix. In the present study, normalization is performed by the vector method, which results in normalization according to Table A2. In the TOPSIS method, to create a normal matrix, the weight of each criterion is multiplied by all the numbers below each of the same criteria. Accordingly, the weight of the proposed criteria is according to Table A3. After applying in the normal matrix, the normal matrix will be in accordance with Table A4.

In order to determine the risk rating, the relative proximity of each option to the ideal solution must be extracted. The Euclidean distance of each option from the positive and negative ideals was calculated, and the positive and negative ideals of each criterion were calculated according to Table A5. Formula (6) is also used to calculate the relative proximity of each option to the ideal solution. Finally, the rating of each risk is determined based on Confidence Interval (CL), which is a number between one and zero. The closer this value is to one, the higher the risk priority; conversely, the closer the value is to zero, the lower the risk significance. Table 7 shows the final results of the CRFs ranking of CRCBPs using the TOPSIS method.

$$d_i^+ = \sqrt{\sum_{j=1}^{n} \left(v_{ij} - v_j^+ \right)^2} \tag{4}$$

$$d_i^- = \sqrt{\sum_{j=1}^{n} \left(v_{ij} - v_j^- \right)^2} \tag{5}$$

$$CL_1^+ = \frac{d_1^-}{d_i^- + d_i^+} \tag{6}$$

Table 7. Prioritization of CRFs of CRCBPs using the TOPSIS method.

Code	d+	d−	CL	Final Rank	Code	d+	d−	CL	Final Rank
CRF_1	008/0	047/0	852/0	2	CRF_{14}	019/0	036/0	657/0	6
CRF_2	020/0	035/0	640/0	8	CRF_{15}	039/0	016/0	286/0	22
CRF_3	018/0	038/0	678/0	5	CRF_{16}	032/0	022/0	416/0	18
CRF_4	026/0	028/0	519/0	12	CRF_{17}	042/0	014/0	256/0	23
CRF_5	006/0	050/0	895/0	1	CRF_{18}	028/0	026/0	483/0	15
CRF_6	011/0	044/0	795/0	3	CRF_{19}	027/0	027/0	495/0	14
CRF_7	025/0	030/0	549/0	9	CRF_{20}	027/0	027/0	502/0	13
CRF_8	018/0	038/0	681/0	4	CRF_{21}	037/0	018/0	331/0	21
CRF_9	029/0	025/0	457/0	17	CRF_{22}	028/0	031/0	528/0	11
CRF_{10}	019/0	036/0	650/0	7	CRF_{23}	047/0	007/0	128/0	24
CRF_{11}	036/0	021/0	346/0	20	CRF_{24}	053/0	000/0	000/0	25
CRF_{12}	025/0	030/0	548/0	10	CRF_{25}	032/0	022/0	407/0	19
CRF_{13}	030/0	025/0	459/0	16					

Results shown in Table 7 highlight that the five CRFs with the highest importance for CRCBPs are (1) the risk of external threats due to international factors (with a relative distance of 0.895), (2) exchange rate fluctuations and changes (with a relative distance of 0.852), (3) bank interest rate fluctuations (with a relative distance of 0.795), (4) traffic licenses (with a relative distance of 0.681), and (5) access to skilled labor (with a relative distance of 0.678).

6. Discussions and Conclusions

What are the most severe risks in commercial and recreational complex building projects (CRCBPs)? This is the question at the center of the presented study, and results of the quantitative step (see Table 7) showed that the 10 most important CRFs of CRCBPs are (i) threats from international relations, (ii) exchange rate fluctuations and changes, (iii) bank interest rate fluctuations, (iv) traffic licenses, (v) access to skilled labor, (vi) changes in regional regulations, (vii) the condition of adjacent buildings, (viii) fluctuations and changes in inflation rates, (ix) failure to select a suitable and qualified advisor, and (x) previous experiences and records related to the employer.

The results of the current study are partly aligned with those of some previous scholars interested in identifying and ranking CRFs in construction projects, even if not specifically considering CRCBPs. For example, [10] reported that construction project risk indicators are mainly related to changes in domestic and international situations and the efficiency of a country's economics/workforce/construction characteristics/consultative and contractual services. Yet, Hatefi and Mohseni [73] also ranked the CRFs as high in relation to initial price fluctuations, rising inflation, exchange rate fluctuations, bank interest rate fluctuations, tax increases, and the uncertainty of fiscal policies. Results are slightly in contrast to Dey [74] who, by evaluating project risks using a MCDM method, identified that also risks connected with government bonds and equipment suppliers and technology selection have a high priority in construction projects (not ranked high in the proposed study).

With regard to CRCBPs, the produced results extend the contributions by Chen and Khumpaisal [35] and by Tamošaitienė et al. [12], who identified risks for CRCBPs without, however, providing their prioritization. Yet, when considering prior studies that identified and prioritized CRFs of CRCBPs, results of the proposed work are partly in accord; indeed, with regard to Comu et al. [17], this is despite 'exchange rate and inflation rate fluctuations' and 'political instability' not being included with other CRFs ranked as important in the proposed work (e.g., traffic licenses, access to skilled labor, changes in regional regulations, etc.). Differences in these findings can come from the different economic, social, and cultural contexts, as well as from the different features of the samples involved. Indeed, respondents for the proposed contribution are more experienced, they come from distinct stakeholders' categories (clients and consultants have been included), and they are more heterogeneous in terms of field of specialty. These individual differences, according to the literature [77], can lead to different perceptions and, as a consequence, to distinct prioritization of categories. Yet, project risks can also vary from time to time depending on the progress level of the project [78], and this is more important for financial risks, such as the exchange rate instabilities, which can occur suddenly due to unforeseen factors [79], most of which are often external ones.

If looking at risks identified and prioritized for megaprojects that have some parallelism with CRCBPs [80], results are slightly in contrast. Indeed, Cheng and Darsa [27] found that the most important risk factors in the Ethiopian context are 'change order', 'corruption/bribery', and 'delay in payment', while Chattapadhyay et al. [29] found that the most severe risks are delay in obtaining traffic regulation orders, inappropriate equipment, political and legal issues, political instability, government intervention, regulatory confirmation and regulation order delays, and wrong engineering designs. Obviously, these differences can be associated to the usual distinct nature of megaprojects and CRCBPS, mainly public and private, respectively [81].

Among the ten identified risks, some were external risks of CRCBPs while others were internal risks. However, all the identified risks have significant effects on project development in the setting of developed or developing countries. As expected, given the timing of the investigation into sanctions and severe economic problems in Iran, the most important CRFs were in the economic risk group. Sanctions imposed on Iran have reduced liquidity and increased inflation, which is reflected on the price of equipment and materials. Therefore, factors, such as exchange rate fluctuations and inflation, followed by changes in bank interest rates, certainly cause serious uncertainties in CRCBPs projects. The reason can be found in the fact that most of the mechanical, electrical, and even construction materials and equipment of these projects in Iran are supplied through imports from industrialized countries. Consequently, the cost of manufacturing CRCBPs in Iran is directly related to exchange rate fluctuations. Furthermore, the procurement of these projects is often faced with problems and therefore this issue causes negative effects during the implementation of the project and the construction of CRCBPs in Iran is always delayed. Given that such projects have a certain delicacy at the joining stage, the need for skilled and experienced labor is a condition for achieving the desired work. This is why, in Iran, the main experience of the staff is in uniform construction items and the need for training in this field is strongly felt. This is also evident in the installation of mechanical and electronic equipment. Yet, the increase in the exchange rate has made it practically impossible for public and private employers to employ non-Iranian specialized forces in the construction of CRCBPs, basically because the import of technical and engineering services in this situation will greatly affect the cost of manufacturing CRCBPs. Regarding the risk of the contractor's claim, acknowledging that economic problems will definitely lead to a reduction in the contractors' profit margins, various claims will therefore follow. The result of this issue will not only affect the executive affairs and the quality of the finished product but will also cause legal problems and difficulties for all parties.

The present study was conducted to identify and evaluate CRFs of CRCBPs, using a MCDM method, and identify the most severe ones. For this purpose, based on a careful study of the research literature and the implementation of the Delphi method, 82 risks of CRCBPs were identified and ranked; then, they were empirically analyzed through the TOPSIS method. Results showed that the most severe risks for CRCBPs are (1) the risk of external threats due to international factors, (2) exchange rate fluctuations and changes, (3) bank interest rate fluctuations, (4) traffic licenses, and (5) access to skilled labor.

In terms of theoretical implications, when considering similar works that prioritized CRFs of CRCBPs, the proposed contribution overcomes their main limitation in having considered just a small sample of risks to be assessed; see Comu et al. [17] who included only 21 risks, a small amount compared to the 82 risks included in this study. If considering results of this and prior studies on CRFs of CRCBPs, it can be stated that external risks, such as the exchange rate and inflation rate fluctuations and political instability, are the most severe. However, from the identified differences compared with prior literature, it can be put forward that the economic, social, and cultural contexts of the study and the socio-demographic/personality features of the sample involved are pivotal for the identification and prioritization of CRFs. This undermines the generalizability of results. Yet, this study also underlines the importance of considering the stage of the project life cycle [82] for which risks are assessed. The influence of different groups of identified risks cannot be separately studied in some phases of the CRCBPs. As an example, in the 'management' category of risks, individual identified risk factors can be important at different or for multiple phases of the CRCBPs' life cycle. This is very clear in the risk factor 'site unavailability and delay in delivery of land to the presenter', which has a significantly higher importance during the first phases of the CRCBPs compared to later phases. In contrast, a risk factor such as 'poor coordination and management' can be important in all CRCBPs' phases (e.g., design and engineering, and procurement and construction). This means that practitioners should aim to mitigate single risks that are more likely to occur in each phase of the CRCBPs' life cycle (but are not exclusive to that phase), while controlling the evolution of risks and their effects on project performance, even if the project is passed the phase during which these risks are expected to manifest. This can be done, for example, by the use of the real options method or a scenario-based approach [83]. In summary, the external and internal conditions of CRCBPs may vary significantly, resulting in the appearance of risks that were thought to be unlikely to occur. As a result, scholars interested in risk management should pay a great deal of attention to risks and changes in the internal and external environment of the project and be prepared for manifestation of such risk factors.

In terms of practical implications, it is worth noting that a number of external risks, such as fluctuations in currency exchange rates, changes in inflation rate, foreign threats due to international relations, and fluctuations of banking interest rates, are outside of the power of CRCBPs' managers while also having a large impact—especially on the project's conceptual planning and feasibility study's life cycle. Indeed, the presence of these economic and financial risks during the initial phases of the project can result in CRCBPs' decision makers abandoning the project before significant investments are made. Alternatively, decision makers can try implementing projects in countries and/or during periods of stable exchange rates that can facilitate the fulfillment of project objectives. Another practical strategy to reduce these risks is to insure CRCBPs' against possible economic and financial risks. The cost of such insurance is even more financially acceptable if investors in CRCBPs also have significant investments in other projects. The increase in the number of projects in the investors' portfolio allows them to control investments with different levels of risk manifestation and can reduce the risk of overall failure in practice. However, insurance protection is not possible for risks such as traffic permits and the condition of adjacent buildings, which are always outside a project management team's control. If such risks are verified, they can result in significant delays in CRCBPs (or undermine their fruition), and therefore decrease the overall value of the CRCBPs. In such a case, CRCBPs' decision makers must choose to either continue with the project while attempting to maintain economic and financial equilibrium or liquidate the project if the cost of these risks can reach or exceed the planned return on investments.

Finally, the risks of lack of access to skilled labor, lack of qualified consultants, and unrealistic preliminary estimation can also have significant effects, similar to the previously discussed risks, by delaying the execution of the CRCBPs while also undermining risk management and coordination efforts. However, since these risks are related to processes actively controlled by CRCBPs' management, their direct control is possible. For instance, the lack of access to resources or qualified consultants can often be resolved through human resource agencies, headhunters, or other qualified players capable of identifying and delivering suitable employees and consultants for participation in CRCBPs. Similarly, simple solutions are possible for unrealistic primary estimation, including conducting suitable feasibility studies on the CRCBPs. Furthermore, using skilled labor and qualified consultants can help minimize the forecasted mistakes and problems in the project's progress.

There are few limitations to this study. Despite the fact that the categories of identified risks overlap with the ones identified in the extensive and recent review on construction risks by Siraj et al. [70], all the risks identified by these scholars (i.e., 571) have not been included in our survey due to the: (i) Lack of a complete list of these risks (authors just propose a sample of 10 risks for each category), and (ii) methodological difficulty in proposing a related lengthy questionnaire for the ranking of risks. Always with regard to the method adopted, i.e., the Delphi technique, it has inner reliability and validity limits. In particular, considering the reliability problem of the Delphi study (i.e., two or more different groups of experts can lead to different results even if facing the same questions/phenomena); and the criteria for qualitative studies—i.e., truthfulness, applicability, consistency and confirmability—were followed to ensure that credible interpretations of the findings are produced [84]. These criteria are based on the following issues; however, as Keeney et al. [85] stated, following these criteria cannot totally limit the involvement of different panels that may lead to obtaining the same results. Despite that, results emerging from the Delphi study can be considered reliable, in as much as the best (in terms of knowledge and expertise) possible panelists are involved. With regard to the validity problem (i.e., whether the produced results are the right expression of the investigated phenomena), the involvement of a respondent with great knowledge in the field is the most used approach within the technique [85] and this solves also the problem of convergence of opinions that can occur over three rounds of the Delphi technique. However, it is true that this study involved a small number of experts, even though their expertise was in line with the study's aims and that this number is similar to works in the same field adopting the Delphi method [15]. Future studies should increase the validity of the results through interviewing a larger group of experts, at least around 50, or expanding their scope to that of other developing countries. Additionally, comparing the results of similar studies conducted in developed countries to that of developing countries could lead to interesting results. Furthermore, the socio-demographic characteristics of the experts participating in the initial phase of identifying risks of CRCBPs can play a role through their opinions regarding the existence and/or importance of certain risk factors. Therefore, an interesting future prospect will be to carry out future quantitative studies based on Upper Echelons Theory literature [77,86], regarding the effects of socio-demographic characteristics and/or other psychological variables on definition and evaluation of CRCBPs' risks at the individual and group levels. Another main limitation of this study is the adoption of the TOPSIS method. Indeed, as well accounted by Madi et al. [87], TOPSIS uses crisp information that is impractical in many real-world situations (e.g., human judgements are often vague and cannot estimate preferences in exact numerical form). Yet, TOPSIS suffers from the rank reversal problem that is related to the change in the ranking of alternatives when a criterion or an alternative is added or dropped; yet, since TOPSIS uses Euclidean distance (that does not consider correlations), results are affected due to information overlap [88]. To try to overcome this limitation, future research is encouraged to combine MCDM techniques [89]. Another solution is to adopt established developments of the TOPSIS method, such as fuzzy TOPSIS; the sets can be used to express preferences using linguistic variables [90]. The adoption of this more sophisticated technique can help also to overcome the limitation of this study having been based on a MCDM approach; indeed, the fuzzy set theory has led to a new decision theory, known today as fuzzy MCDM where decision-maker models are able to deal with incomplete and uncertain knowledge and information [91].

Author Contributions: Conceptualization, H.S. and J.T.; methodology, H.S. and M.K.; formal analysis, H.S. and M.K.; investigation, J.T., H.S., and D.W.M.C.; data curation, H.S. and D.W.M.C.; writing—original draft preparation, H.S. and J.T.; writing—review and editing, M.C. and H.S.; visualization, H.S. and M.C.; supervision, H.S.; project administration, J.T. All authors have read and agreed to the published version of the manuscript.

Funding: This research received no external funding.

Institutional Review Board Statement: Not applicable.

Informed Consent Statement: Not applicable.

Data Availability Statement: Not applicable.

Conflicts of Interest: The authors declare no conflict of interest.

Appendix A

Table A1. Decision matrix based on the results of the collected questionnaires.

Code	Risks	Vulnerability (−)	Probability of Occurrence of Risk (−)	Risk Detection (+)	Accepting Threat (−)	Impact on the Occurrence of Other Risks (−)	Risk Manageability (+)	Risk Response (+)
		C_1	C_2	C_3	C_4	C_5	C_6	C_7
CRF_1	Exchange rate fluctuation	1/333	1/625	6/375	2/25	2/5	4/625	4/25
CRF_2	Inflation fluctuation	2/667	2/375	6	4	3/625	4	3/875
CRF_3	Access to skilled worker	4/167	3/875	4/625	2/25	4/875	4/5	4
CRF_4	Contractor's claim	4/5	3/625	4/75	4/125	4/875	3/5	3/625
CRF_5	Foreign threats	2/667	2/25	5/375	1/375	2/625	5/625	5
CRF_6	Bank interest fluctuation	2/167	3/5	4/875	2/25	2/25	5	5/375
CRF_7	Lack of qualified consultant	1/833	3/375	5/125	4/125	5/25	3/375	3/25
CRF_8	Traffic permits	2/333	1/375	4/625	3/75	3	4/125	3/625
CRF_9	Unrealistic primary estimation	4/167	4/75	3/875	4/625	4/875	3/75	3/625
CRF_{10}	Adjacent building condition	3/333	3	5	3/375	5/375	4/625	4/375

Table A1. *Cont.*

Code	Risks	Vulnerability (−)	Probability of Occurrence of Risk (−)	Risk Detection (+)	Accepting Threat (−)	Impact on the Occurrence of Other Risks (−)	Risk Manage-ability (+)	Risk Response (+)
		C_1	C_2	C_3	C_4	C_5	C_6	C_7
CRF_{11}	Mismatch of job referrals to personnel with related specialized skills	5/833	5/625	4/25	5/5	5/75	4	4/125
CRF_{12}	Previous employer-related experience and background	4/667	4/625	5	4	4/25	4/25	4/5
CRF_{13}	Delays in construction	5	4/375	4/125	4/75	4/25	4/125	3/75
CRF_{14}	Regional standard changes (firefighting-master plans, etc.)	4/5	3/5	5/25	3/25	4	4/625	4/625
CRF_{15}	Changes in the legal obligations of contracts	5/633	5/5	3/375	5/75	6	3	3/625
CRF_{16}	Mismatch between demand and available resources	5/167	4/625	4	4/875	4/375	3/375	3
CRF_{17}	Lack of using appropriate methods in workshop management	6/167	5/865	3/375	6/125	6/75	3/25	3/625
CRF_{18}	Change in duties of imported equipment	4/5	4/125	4/625	4/5	4/625	3/5	3/875
CRF_{19}	Lack of on time finance	4/167	4	5/25	4/25	5	3/25	3/5
CRF_{20}	Inappropriate finance	4/167	4	4/625	4	5/625	3/375	3/5
CRF_{21}	Law changes and economic policies of materials	5/633	5/5	3/5	4/875	6/25	2/75	3/125
CRF_{22}	Bankruptcy	4/833	5/875	3	3/875	3/5	4/875	4
CRF_{23}	Incomplete plan	6/333	6/125	2	6/375	6/625	2/125	1/625
CRF_{24}	Poor technical specifications	6/5	6/375	1/75	7/25	7	1/375	1/125
CRF_{25}	Failure to complete work items in anticipated times	5	4/75	4/25	5	5	3/625	3

Table A2. Normalized decision matrix based on the results of the collected questionnaires.

Risks	−	−	+	−	−	+	+
	C_1	C_2	C_3	C_4	C_5	C_6	C_7
CRF_1	0/0590	0/074	0/284	0/101	0/101	0/238	0/225
CRF_2	0/118	0/108	0/267	0/179	0/147	0/205	0/205
CRF_3	0/184	0/176	0/206	0/101	0/197	0/231	0/211
CRF_4	0/199	0/165	0/212	0/185	0/197	0/180	0/192
CRF_5	0/118	0/102	0/239	0/062	0/106	0/289	0/264
CRF_6	0/096	0/159	0/217	0/101	0/091	0/257	0/284
CRF_7	0/081	0/153	0/228	0/185	0/212	0/173	0/172
CRF_8	0/103	0/063	0/206	0/168	0/121	0/212	0/192
CRF_9	0/184	0/216	0/173	0/207	0/197	0/193	0/192
CRF_{10}	0/148	0/136	0/223	0/151	0/217	0/228	0/231
CRF_{11}	0/258	0/256	0/189	0/246	0/233	0/205	0/218
CRF_{12}	0/207	0/210	0/223	0/179	0/172	0/218	0/238
CRF_{13}	0/221	0/199	0/184	0/213	0/212	0/212	0/198
CRF_{14}	0/199	0/159	0/234	0/145	0/162	0/238	0/244
CRF_{15}	0/249	0/250	0/150	0/257	0/243	0/154	0/192
CRF_{16}	0/229	0/210	0/178	0/218	0/177	0/173	0/158
CRF_{17}	0/273	0/267	0/150	0/274	0/273	0/167	0/192
CRF_{18}	0/199	0/188	0/206	0/201	0/187	0/180	0/205
CRF_{19}	0/184	0/182	0/234	0/190	0/202	0/167	0/185
CRF_{20}	0/184	0/182	0/206	0/179	0/228	0/173	0/185
CRF_{21}	0/249	0/25	0/156	0/218	0/253	0/141	0/165
CRF_{22}	0/214	0/267	0/134	0/173	0/142	1/250	0/211
CRF_{23}	0/280	0/279	0/089	0/285	0/268	0/109	0/086
CRF_{24}	0/288	0/290	0/078	0/325	0/283	0/071	0/059
CRF_{25}	0/221	0/216	0/189	0/224	0/202	0/186	0/158

Table A3. Weight of criteria.

Criteria	C_1	C_2	C_3	C_4	C_5	C_6	C_7
w_j	0/0633	0/089	0/0544	0/135	0/08	0/1	0/053

Table A4. Harmonic decision matrix.

Risks	(−)	(−)	(+)	(−)	(−)	(+)	(+)
	C_1	C_2	C_3	C_4	C_5	C_6	C_7
CRF_1	0/004	0/007	0/015	0/014	0/008	0/024	0/012
CRF_2	0/007	0/010	0/015	0/024	0/012	0/021	0/011
CRF_3	0/012	0/016	0/011	0/014	0/016	0/023	0/011
CRF_4	0/013	0/015	0/012	0/025	0/016	0/018	0/010
CRF_5	0/007	0/009	0/013	0/008	0/008	0/029	0/014
CRF_6	0/006	0/014	0/012	0/014	0/007	0/026	0/015
CRF_7	0/005	0/014	0/012	0/025	0/017	0/017	0/009
CRF_8	0/007	0/006	0/011	0/023	0/010	0/021	0/010
CRF_9	0/012	0/019	0/009	0/028	0/016	0/019	0/010
CRF_{10}	0/009	0/012	0/012	0/020	0/017	0/024	0/012
CRF_{11}	0/016	0/023	0/010	0/023	0/019	0/021	0/012
CRF_{12}	0/013	0/019	0/012	0/024	0/014	0/022	0/013
CRF_{13}	0/014	0/018	0/010	0/029	0/017	0/021	0/011
CRF_{14}	0/013	0/014	0/013	0/020	0/013	0/024	0/013
CRF_{15}	0/016	0/022	0/008	0/035	0/019	0/015	0/010
CRF_{16}	0/014	0/019	0/010	0/029	0/014	0/017	0/008
CRF_{17}	0/017	0/024	0/008	0/037	0/022	0/017	0/010
CRF_{18}	0/013	0/017	0/011	0/027	0/015	0/018	0/011
CRF_{19}	0/012	0/016	0/013	0/026	0/016	0/017	0/010
CRF_{20}	0/012	0/016	0/011	0/024	0/018	0/017	0/010
CRF_{21}	0/016	0/022	0/008	0/029	0/020	0/014	0/009
CRF_{22}	0/014	0/024	0/007	0/023	0/011	0/025	0/011
CRF_{23}	0/018	0/025	0/005	0/038	0/021	0/011	0/005
CRF_{24}	0/018	0/026	0/004	0/044	0/023	0/007	0/002
CRF_{25}	0/014	0/019	0/010	0/030	0/016	0/019	0/008

Table A5. Positive and negative ideals of each creation.

Positive and Negative Ideals	(−)	(−)	(+)	(−)	(−)	(+)	(+)
	C_1	C_2	C_3	C_4	C_5	C_6	C_7
A^+	0/004	0/006	0/015	0/008	0/007	0/029	0/015
A^-	0/018	0/026	0/004	0/044	0/023	0/007	0/003

References

1. Project Management Institute (PMI). *A Guide to the Project Management Body of Knowledge: PMBOK®Guide*, 6th ed.; Project Management Institute: Newton Square, PA, USA, 2017.
2. Yuan, J.; Chen, K.; Li, W.; Ji, C.; Wang, Z.; Skibniewski, M.J. Social network analysis for social risks of construction projects in high-density urban areas in China. *J. Clean. Prod.* **2018**, *198*, 940–961. [CrossRef]
3. Wuni, I.Y.; Shen, G.Q.P.; Mahmud, A.T. Critical risk factors in the application of modular integrated construction: A systematic review. *Int. J. Constr. Manag.* **2019**, 1–15. [CrossRef]
4. Ward, S.; Chapman, C. Transforming project risk management into project uncertainty management. *Int. J. Proj. Manag.* **2003**, *21*, 97–105. [CrossRef]
5. Khosravi, M.; Sarvari, H.; Chan, D.W.M.; Cristofaro, M.; Chen, Z. Determining and assessing the risks of commercial and recreational complex building projects in developing countries: A survey of experts in Iran. *J. Facil. Manag.* **2020**, *18*, 259–282. [CrossRef]
6. Hlaing, N.N.; Singh, D.; Tiong, R.L.K.; Ehrlich, M. Perceptions of Singapore construction contractors on construction risk identification. *J. Financ. Manag. Prop. Constr.* **2008**, *13*, 85–95. [CrossRef]

7. Chapman, R.J. The effectiveness of working group risk identification and assessment techniques. *Int. J. Proj. Manag.* **1998**, *16*, 333–343. [CrossRef]
8. Goh, C.S.; Abdul-Rahman, H.; Abdul Samad, Z. Applying risk management workshop for a public construction project: Case study. *J. Constr. Eng. Manag.* **2013**, *139*, 572–580. [CrossRef]
9. Marcelino-Sádaba, S.; Pérez-Ezcurdia, A.; Lazcano, A.M.E.; Villanueva, P. Project risk management methodology for small firms. *Int. J. Proj. Manag.* **2014**, *32*, 327–340. [CrossRef]
10. Zavadskas, E.K.; Turskis, Z.; Tamošaitienė, J. Risk assessment of construction projects. *J. Civ. Eng. Manag.* **2010**, *16*, 33–46. [CrossRef]
11. Valipour, A.; Yahaya, N.; Md Noor, N.; Kildienė, S.; Sarvari, H.; Mardani, A. A fuzzy analytic network process method for risk prioritization in freeway PPP projects: An Iranian case study. *J. Civ. Eng. Manag.* **2015**, *21*, 933–947. [CrossRef]
12. Tamošaitienė, J.; Zavadskas, E.K.; Turskis, Z. Multi-criteria risk assessment of a construction project. *Procedia Comput. Sci.* **2013**, *17*, 129–133. [CrossRef]
13. Tamošaitienė, J.; Sarvari, H.; Chan, D.W.M.; Cristofaro, M. Assessing the Barriers and Risks to Private Sector Participation in Infrastructure Construction Projects in Developing Countries of Middle East. *Sustainability* **2021**, *13*, 153. [CrossRef]
14. Tamošaitienė, J.; Yousefi, V.; Tabasi, H. Project Portfolio Construction Using Extreme Value Theory. *Sustainability* **2021**, *13*, 855. [CrossRef]
15. Tamošaitienė, J.; Sarvari, H.; Cristofaro, M.; Chan, D.W.M. Identifying and prioritizing the selection criteria of appropriate repair and maintenance methods for commercial buildings. *Int. J. Strateg. Prop. Manag.* **2021**, *25*, 413–431. [CrossRef]
16. Sarvari, H.; Valipour, A.; Yahya, N.; Noor, N.; Beer, M.; Banaitiene, N. Approaches to Risk Identification in Public–Private Partnership Projects: Malaysian Private Partners' Overview. *Adm. Sci.* **2019**, *9*, 17. [CrossRef]
17. Comu, S.; Elibol, A.Y.; Yucel, B. A risk assessment model of commercial real estate development projects in developing countries. *J. Constr. Eng.* **2021**, *4*, 52–67. [CrossRef]
18. Sarvari, H.; Chan, D.W.M.; Alaeos, A.K.F.; Olawumi, T.O.; Abdalridah Aldaud, A.A. Critical success factors for managing construction small and medium-sized enterprises in developing countries of Middle East: Evidence from Iranian construction enterprises. *J. Build. Eng.* **2021**, *43*, 103152. [CrossRef]
19. Mirkatouli, J.; Samadi, R.; Hosseini, A. Evaluating and analysis of socio-economic variables on land and housing prices in Mashhad, Iran. *Sustain. Cities Soc.* **2018**, *41*, 695–705. [CrossRef]
20. United Nations. *Central Product Classification*; United Nations Statistics Division: New York, NY, USA, 2015.
21. Sarvari, H.; Mehrabi, A.; Chan, D.W.M.; Cristofaro, M. Evaluating urban housing development patterns in developing countries: Case study of Worn-out Urban Fabrics in Iran. *Sustain. Cities Soc.* **2021**, *70*, 102941. [CrossRef]
22. Kumaraswamy, M.M.; Zhang, X.Q. Governmental role in BOT-led infrastructure development. *Int. J. Proj. Manag.* **2001**, *19*, 195–205. [CrossRef]
23. Ghaed Rahmati, S.; Daneshmandi, N. Analysis of urban tourism spatial pattern (case study: Urban tourism space of Isfahan city). *Hum. Geogr. Res.* **2018**, *50*, 945–961.
24. Sarvari, H.; Rakhshanifar, M.; Tamošaitienė, J.; Chan, D.W.M.; Beer, M. A risk based approach to evaluating the impacts of Zayanderood drought on sustainable development indicators of riverside urban in Isfahan-Iran. *Sustainability* **2019**, *11*, 6797. [CrossRef]
25. Marle, F.; Gidel, T. A multi-criteria decision-making process for project risk management method selection. *Int. J. Multicriteria Decis. Mak.* **2012**, *2*, 189–223. [CrossRef]
26. Yatsalo, B.; Gritsyuk, S.; Sullivan, T.; Trump, B.; Linkov, I. Multi-criteria risk management with the use of DecernsMCDA: Methods and case studies. *Environ. Syst. Decis.* **2016**, *36*, 266–276. [CrossRef]
27. Cheng, M.Y.; Darsa, M.H. Construction Schedule Risk Assessment and Management Strategy for Foreign General Contractors Working in the Ethiopian Construction Industry. *Sustainability* **2021**, *13*, 7830. [CrossRef]
28. Czajkowska, A.; Ingaldi, M. Structural Failures Risk Analysis as a Tool Supporting Corporate Responsibility. *J. Risk Financ. Manag.* **2021**, *14*, 187. [CrossRef]
29. Chattapadhyay, D.B.; Putta, J.; Paneem, R.M. Risk Identification, Assessments, and Prediction for Mega Construction Projects: A Risk Prediction Paradigm Based on Cross Analytical-Machine Learning Model. *Buildings* **2021**, *11*, 172. [CrossRef]
30. Tserng, H.P.; Cho, I.; Chen, C.H.; Liu, Y.F. Developing a Risk Management Process for Infrastructure Projects Using IDEF0. *Sustainability* **2021**, *13*, 6958. [CrossRef]
31. Chan, D.W.M.; Chan, J.H.L.; Ma, T. Developing a fuzzy risk assessment model for guaranteed maximum price and target cost contracts in South Australia. *Facilities* **2014**, *32*, 624–646. [CrossRef]
32. Siu, F.M.F.; Leung, J.W.Y.; Chan, D.W.M. A Data-driven approach to identify-quantify-analyse construction risk for Hong Kong NEC projects. *J. Civ. Eng. Manag.* **2018**, *24*, 592–606. [CrossRef]
33. Chatterjee, K.; Zavadskas, E.K.; Tamošaitienė, J.; Adhikary, K.; Kar, S. A hybrid MCDM technique for risk management in construction projects. *Symmetry* **2018**, *10*, 46. [CrossRef]
34. Zhou, H.; Zhao, Y.; Shen, Q.; Yang, L.; Cai, H. Risk assessment and management via multi-source information fusion for undersea tunnel construction. *Autom. Constr.* **2020**, *111*, 103050. [CrossRef]
35. Chen, Z.; Khumpaisal, S. An analytic network process for risks assessment in commercial real estate development. *J. Prop. Invest. Financ.* **2009**, *27*, 238–258. [CrossRef]

36. Irimia Diéguez, A.I.; Sánchez Cazorla, Á.; Alfalla Luque, R. Risk management in megaprojects. *Procedia Soc. Behav. Sci.* **2014**, *119*, 407–416. [CrossRef]
37. Krane, H.P.; Rolstadås, A.; Olsson, N.O. Categorizing risks in seven large projects—Which risks do the projects focus on? *Proj. Manag. J.* **2010**, *41*, 81–86. [CrossRef]
38. Wang, H.; Tong, Y. Algorithm Study on Models of Multiple Objective Risk Decision under Principal and Subordinate Hierarch Decision-making. *Oper. Res. Manag. Sci.* **2007**, *16*, 1–8.
39. Wideman, R.M. *A Guide to Managing Project Risks and Opportunities*; Project Management Institute: Newtown Square, PA, USA, 1992.
40. Sigmund, Z.; Radujković, M. Risk breakdown structure for construction projects on existing buildings. *Procedia-Soc. Behav. Sci.* **2014**, *119*, 894–901. [CrossRef]
41. Hillson, D.; Grimaldi, S.; Rafele, C. Managing project risks using a cross risk breakdown matrix. *Risk Manag.* **2006**, *8*, 61–76. [CrossRef]
42. Dabiri, M.; Oghabi, M.; Sarvari, H.; Sabeti, M.; Kashefi, H. A combination risk-based approach to post-earthquake temporary accommodation site selection: A case study in Iran. *Iran. J. Fuzzy Syst.* **2020**, *17*, 54–74. [CrossRef]
43. Lyu, H.M.; Sun, W.J.; Shen, S.L.; Zhou, A.N. Risk assessment using a new consulting process in fuzzy AHP. *J. Constr. Eng. Manag.* **2020**, *146*, 04019112. [CrossRef]
44. Milion, R.N.; Alves, T.D.C.; Paliari, J.C.; Liboni, L.H. CBA-Based Evaluation Method of the Impact of Defects in Residential Buildings: Assessing Risks towards Making Sustainable Decisions on Continuous Improvement Activities. *Sustainability* **2021**, *13*, 6597. [CrossRef]
45. Taylan, O.; Bafail, A.O.; Abdulaal, R.M.; Kabli, M.R. Construction projects selection and risk assessment by fuzzy AHP and fuzzy TOPSIS methodologies. *Appl. Soft Comput.* **2014**, *17*, 105–116. [CrossRef]
46. Mata, P.; Silva, P.F.; Pinho, F.F. Risk Management of Bored Piling Construction on Sandy Soils with Real-Time Cost Control. *Infrastructures* **2021**, *6*, 77. [CrossRef]
47. Liu, Z.; Meng, X.; Xing, Z.; Jiang, A. Digital Twin-Based Safety Risk Coupling of Prefabricated Building Hoisting. *Sensors* **2021**, *21*, 3583. [CrossRef] [PubMed]
48. Draji Jahromi, A.; Valipour, A.; Pakdel, A. Ranking of risk assessment criteria in construction projects using network analysis. In Proceedings of the 4th International Congress on Civil Engineering, Architecture & Urban Development, Tehran, Iran, 27–29 December 2016.
49. Mohammadi Talvar, Z.; Panahi, J. Identifying effective criteria in assessing and ranking the risk management of construction projects. In Proceedings of the Conference on Civil Engineering, Architecture and Urbanism of the Islamic Countries, Tabriz, Iran, 10 May 2018.
50. Muriana, C.; Vizzini, G. Project risk management: A deterministic quantitative technique for assessment and mitigation. *Int. J. Proj. Manag.* **2017**, *35*, 320–340. [CrossRef]
51. Sarvari, H.; Cristofaro, M.; Chan, D.W.M.; Noor, N.M.; Amini, M. Completing abandoned public facility projects by the private sector: Results of a Delphi survey in the Iranian Water and Wastewater Company. *J. Facil. Manag.* **2020**, *18*, 547–566. [CrossRef]
52. Cox, L.A., Jr.; Babayev, D.; Huber, W. Some limitations of qualitative risk rating systems. *Risk Anal. Int. J.* **2005**, *25*, 651–662. [CrossRef]
53. Zou, P.X.; Zhang, G.; Wang, J. Understanding the key risks in construction projects in China. *Int. J. Proj. Manag.* **2007**, *25*, 601–614. [CrossRef]
54. Behzadian, M.; Otaghsara, S.K.; Yazdani, M.; Ignatius, J. A state-of the-art survey of TOPSIS applications. *Expert Syst. Appl.* **2012**, *39*, 13051–13069. [CrossRef]
55. Jozaghi, A.; Alizadeh, B.; Hatami, M.; Flood, I.; Khorrami, M.; Khodaei, N.; Ghasemi Tousi, E.A. Comparative Study of the AHP and TOPSIS Techniques for Dam Site Selection Using GIS: A Case Study of Sistan and Baluchestan Province, Iran. *Geosciences* **2018**, *8*, 494. [CrossRef]
56. Gebrehiwet, T.; Luo, H. Risk level evaluation on construction project lifecycle using fuzzy comprehensive evaluation and TOPSIS. *Symmetry* **2019**, *11*, 12. [CrossRef]
57. Dandage, R.; Mantha, S.S.; Rane, S.B. Ranking the risk categories in international projects using the TOPSIS method. *Int. J. Manag. Proj. Bus.* **2018**, *11*, 317–331. [CrossRef]
58. Jalal, M.P.; Shoar, S. A hybrid framework to model factors affecting construction labour productivity: Case study of Iran. *J. Financ. Manag. Prop. Constr.* **2019**, *24*, 630–654. [CrossRef]
59. Ghasseminejad, S.; Jahan-Parvar, M.R. The impact of financial sanctions: *Case Iran. J. Policy Modeling* **2021**, *43*, 601–621. [CrossRef]
60. Arnott, R. Housing policy in developing countries: The importance of the informal economy. In *Urbanization and Growth*; Spence, M., Annez, P.C., Buckley, R.M., Eds.; The World Bank: Washington, DC, USA, 2008; pp. 167–196.
61. Yeung, J.F.Y.; Chan, A.P.C.; Chan, D.W.M.; Li, L.K. Development of a partnering performance index (PPI) for construction projects in Hong Kong: A Delphi study. *Constr. Manag. Econ.* **2007**, *25*, 1219–1237. [CrossRef]
62. Chan, D.W.M.; Chan, J.H.L. Developing a Performance Measurement Index (PMI) for target cost contracts in construction: A Delphi study. *Constr. Law J.* **2012**, *28*, 590–613.
63. Olawumi, T.O.; Chan, D.W.M. Critical success factors for implementing building information modeling and sustainability practices in construction projects: A Delphi survey. *Sustain. Dev.* **2019**, *27*, 587–602. [CrossRef]

64. Ezeldin, S.; Ibrahim, H.H. Risk analysis for mega shopping mall projects in Egypt. *J. Civ. Eng. Archit.* **2015**, *9*, 444–651.
65. Schmidt, R.C. Managing Delphi surveys using nonparametric statistical techniques. *Decis. Sci.* **1997**, *28*, 763–774. [CrossRef]
66. Chan, D.W.M.; Chan, A.P.C.; Lam, P.T.I.; Wong, J.M.W. An empirical survey of the motives and benefits of adopting guaranteed maximum price and target cost contracts in construction. *Int. J. Proj. Manag.* **2011**, *29*, 577–590. [CrossRef]
67. Yeung, J.F.Y.; Chan, A.P.C.; Chan, D.W.M. Developing a performance index for relationship-based construction projects in Australia: Delphi study. *J. Manag. Eng.* **2009**, *25*, 59–68. [CrossRef]
68. Chan, A.P.C.; Lam, P.T.I.; Chan, D.W.M.; Cheung, E.; Ke, Y. Potential obstacles to successful implementation of public-private partnerships in Beijing and the Hong Kong special administrative region. *J. Manag. Eng.* **2010**, *26*, 30–40. [CrossRef]
69. Durdyev, S.; Mbachu, J.; Thurnell, D.; Zhao, L.; Hosseini, M.R. BIM Adoption in the Cambodian Construction Industry: Key Drivers and Barriers. *ISPRS Int. J. Geo-Inf.* **2021**, *10*, 215. [CrossRef]
70. Siraj, N.B.; Fayek, A.R. Risk identification and common risks in construction: Literature review and content analysis. *J. Constr. Eng. Manag.* **2019**, *145*, 03119004. [CrossRef]
71. Shin, D.W.; Shin, Y.; Kim, G.H. Comparison of risk assessment for a nuclear power plant construction project based on analytic hierarchy process and fuzzy analytic hierarchy process. *J. Build. Constr. Plan. Res.* **2016**, *4*, 157–171. [CrossRef]
72. Tian, J.; Yan, Z.F. Fuzzy analytic hierarchy process for risk assessment to general-assembling of satellite. *J. Appl. Res. Technol.* **2013**, *11*, 568–577. [CrossRef]
73. Hatefi, S.; Mohseni, H. Evaluating and prioritizing the risks of BOT projects using structural equations and integrated model of fuzzy AHP and fuzzy TOPSIS. *J. Struct. Constr. Eng.* **2019**, *6*, 111–130. [CrossRef]
74. Dey, P.K. Project risk management using multiple criteria decision-making technique and decision tree analysis: A case study of Indian oil refinery. *Prod. Plan. Control* **2012**, *23*, 903–921. [CrossRef]
75. Hwang, C.L.; Yoon, K. Methods for multiple attribute decision making. In *Multiple Attribute Decision Making*; Springer: Berlin/Heidelberg, Germany, 1981; pp. 58–191. [CrossRef]
76. Taroun, A. Towards a better modelling and assessment of construction risk: Insights from a literature review. *Int. J. Proj. Manag.* **2014**, *32*, 101–115. [CrossRef]
77. Abatecola, G.; Cristofaro, M. Upper echelons and executive profiles in the construction value chain: Evidence from Italy. *Proj. Manag. J.* **2016**, *47*, 13–26. [CrossRef]
78. Perroni, M.; Dalazen, L.L.; Da Silva, W.V.; Gouv, S.; Da Veiga, C.P. Evolution of risks for energy companies from the energy efficiency perspective: The Brazilian case. *Int. J. Energy Econ. Policy* **2015**, *5*, 612–623.
79. Froot, K.A. The intermediation of financial risks: Evolution in the catastrophe reinsurance market. *Risk Manag. Insur. Rev.* **2008**, *11*, 281–294. [CrossRef]
80. Shatkin, G. The city and the bottom line: Urban megaprojects and the privatization of planning in Southeast Asia. *Environ. Plan. A* **2008**, *40*, 383–401. [CrossRef]
81. Brunet, M. Making sense of a governance framework for megaprojects: The challenge of finding equilibrium. *Int. J. Proj. Manag.* **2021**, *39*, 406–416. [CrossRef]
82. Wetzel, E.M.; Thabet, W.Y. The use of a BIM-based framework to support safe facility management processes. *Autom. Constr.* **2015**, *60*, 12–24. [CrossRef]
83. Chen, T.; Zhang, J.; Lai, K.K. An integrated real options evaluating model for information technology projects under multiple risks. *Int. J. Proj. Manag.* **2009**, *27*, 776–786. [CrossRef]
84. Lincoln, Y.S.; Guba, E.G. *Naturalistic Inquiry*; Sage: London, UK, 1985.
85. Keeney, S.; McKenna, H.; Hasson, F. *The Delphi Technique in Nursing and Health Research*; John Wiley and Sons: London, UK, 2011.
86. Abatecola, G.; Cristofaro, M. Hambrick and Mason's "Upper Echelons Theory": Evolution and open avenues. *J. Manag. Hist.* **2020**, *26*, 116–136. [CrossRef]
87. Madi, E.N.; Garibaldi, J.M.; Wagner, C. An exploration of issues and limitations in current methods of TOPSIS and fuzzy TOPSIS. In Proceedings of the 2016 IEEE International Conference on Fuzzy Systems (FUZZ-IEEE), Vancouver, BC, Canada, 24–29 July 2016; pp. 2098–2105. [CrossRef]
88. Çelikbilek, Y.; Tüysüz, F. An in-depth review of theory of the TOPSIS method: An experimental analysis. *J. Manag. Anal.* **2020**, *7*, 281–300. [CrossRef]
89. Yadav, S.K.; Joseph, D.; Jigeesh, N. A review on industrial applications of TOPSIS approach. *Int. J. Serv. Oper. Manag.* **2018**, *30*, 23–28. [CrossRef]
90. Nădăban, S.; Dzitac, S.; Dzitac, I. Fuzzy TOPSIS: A general view. *Procedia Comput. Sci.* **2016**, *91*, 823–831. [CrossRef]
91. Sun, C.C. A performance evaluation model by integrating fuzzy AHP and fuzzy TOPSIS methods. *Expert Syst. Appl.* **2010**, *37*, 7745–7754. [CrossRef]

Article

Substitution of Material Solutions in the Operating Phase of a Building

Anna Sobotka, Kazimierz Linczowski and Aleksandra Radziejowska *

Department of Geomechanics, Civil Engineering and Geotechnics, Faculty of Mining and Geoengineering, AGH University of Science and Technology in Cracow, Al. Mickiewicza 30, 30-059 Cracow, Poland; sobotka@agh.edu.pl (A.S.); klinczowski@gmail.com (K.L.)
* Correspondence: aradziej@agh.edu.pl

Abstract: During the operation of buildings, repairs, modernizations, adaptations, renovations, and reconstructions of parts of historic objects are performed. There is often the problem of using a different material or construction technology than was originally used, for a variety of reasons. For example, these are materials not currently manufactured, with necessary higher performance values (insulation, strength). The aim of the article was to analyze and evaluate the possibility of material substitution in repair works and to analyze the cause and effect analysis of its application in the context of different conditions. The article analyzes the causes and conditions of the substitution of materials in various stages of the exploitation phase of buildings, including historic buildings. A SWOT (Strengths, Weaknesses, Opportunities, Threats) matrix was developed for the phenomenon of material substitution during the operational phase. With aid from the DEMATEL (Decision Making Trial and Evaluation Laboratory) method, identification of cause–effect relationships regarding the issue of the possibility of applying the substitution of material solutions in building objects was carried out. The analysis carried out by the authors allows us to conclude that the use of substitution in the construction sector is justified and shows great opportunities in its implementation and development.

Keywords: substitution; operation and maintenance phase; cause–effect relationships; historical buildings

Citation: Sobotka, A.; Linczowski, K.; Radziejowska, A. Substitution of Material Solutions in the Operating Phase of a Building. *Appl. Sci.* **2021**, *11*, 2812. https://doi.org/10.3390/app11062812

Academic Editor: Francesco Colangelo

Received: 8 January 2021
Accepted: 17 March 2021
Published: 22 March 2021

1. Introduction

The phenomenon of substitution is common in various fields of social and economic activity [1–4]. In the case of material economic activity, it is the mutual substitutability of goods with similar properties. The subject of the article is the substitution of constructional and material solutions in the implementation of construction projects, understood as a phenomenon consisting of replacing the designed object structure (element) with another one that meets the same or similar technical and functional requirements, as well as aesthetic requirements [5].

In construction, the application of substitution occurs throughout the life cycle of an object and addresses various issues. Both in the preparation phase, e.g., during choosing the location of a construction investment, variants of functions and/or construction, technology, as well as during the implementation of facilities and construction works, especially when the contractor is left with the choice of construction products. The selection and supply of construction sites with resources is related to the phenomenon of the substitution of suppliers and entire supply chains.

The exploitation phase of a building object is the longest period of its life cycle. However, the scope of construction works, at this stage, is not too large compared to the construction of the facility. Decisions related to undertaking repairs, including reconstruction, changes in the functions of rooms and facilities, and the choice of material solutions

Appl. Sci. **2021**, *11*, 2812. https://doi.org/10.3390/app11062812

of the structure, their repair, replacement, or renovation of finishing elements, etc., are difficult and require many aspects to be taken into account.

Despite the phenomenon of substitution that has been present in construction projects for years, there is a need to develop theoretical foundations and methods and tools to support decision-making in construction practice. Analysis and selection of substitute materials should consider the full life cycle of the object. They should also refer to current socio-economic concepts such as sustainable development and the circular economy.

Substitution can significantly affect the quality, cost, and time of individual construction projects. It also has a broader multi-faceted impact on the delivery of construction in environmental, economic, and social contexts. For example, the use of material substitution may make it possible to meet a construction completion date in the event of a market collapse or to purchase equally suitable but less expensive products. This may result in improved user comfort or use of products whose manufacture and use do not result in harmful emissions. This last example has a very large contribution to environmental protection—the implementation of sustainable development principles.

In the presented article, the authors focus on the application of the possibility of substituting construction products. It may be caused by the desire to use materials that raise the standard of the facility and cost conditions, as well as limitations due to the unavailability of original materials used during construction. The last aspect concerns, in particular, the refurbishment of buildings entered in the register of monuments. The use of replacement construction products in these types of buildings is a challenge, not only because of the difficulty in selecting an appropriate substitute, but also because of meeting the procedural requirements approved by the restorer. Thus, many factors and conditions of different natures influence the selection of the best substitute under given conditions, taking into account the consequences in terms of durability, strength, etc. during their further use [6], and, therefore, on the life cycle costs of the facility.

The purpose of this paper is to analyze and evaluate the possibility of material substitution in repair works and to analyze the cause and effect analysis of its application in the context of various conditions. A division of the service life of a building was made in the context of the execution of construction works, their contractors, and investors. Conditions and factors occurring in the decision-making process of maintenance of the object in the deteriorated condition, selection of works, and building materials were analyzed. Attention is drawn to the possibility and necessity of material substitution in relation to historical buildings. The developed SWOT matrix and its analysis allowed us to systematize factors (conditions and limitations) of substitution in the exploitation phase and its influence on the life cycle of buildings. The factors covering various substitution determinants, included in the SWOT matrix, were used for identification of cause–effect relationships in the issue of possibility to apply the substitution of material solutions in building objects. For this purpose, the DEMATEL method was used.

2. Substitution of Construction Products in the Exploitation Phase of a Building Object

One of the activities aimed at caring for the environment is striving to extend the life cycle. The products of the construction industry are one of the elements that allow us to take care of this trend. Existing buildings are designed for many years, and thanks to appropriate maintenance and refurbishment measures, they can survive many times longer. One of the ways of extending the life cycle of building objects is to carry out a refurbishment policy, during which it is necessary to take care of the proper selection of material solutions.

Depending on the stage of exploitation under consideration, the participant of the investment process, which may be the user, owner/investor, or property manager, will make decisions in which sooner or later will meet the need, or even the necessity, to use the substitution of construction products. Considering the wide market offer of construction products, the decision-maker will have to consider many criteria before deciding to use a product other than the originally built-in product.

Due to many different conditions, it is proposed in the research that the substitution of construction products in the operation phase should be considered by distinguishing its three stages/periods:

Substitution of construction products during the life of a building object is strictly connected with the division presented in Figure 1.

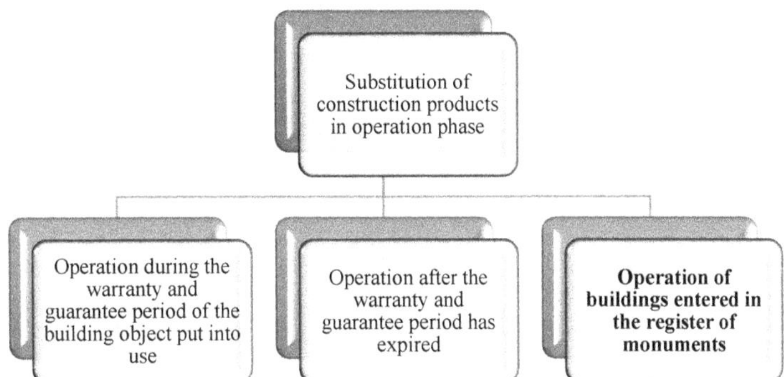

Figure 1. Division of substitution during the lifetime of a building structure.

During the warranty and guarantee period in a newly commissioned building, all necessary repairs should be carried out by the contractor who carried out this investment. Consequently, all costs associated with the construction work under consideration are not financially chargeable to the property owner. In the situation described above, due to the short period of time from putting the facility into use, construction products used for repairs and troubleshooting should still be available on the market.

The substitution of construction products during the warranty and guarantee period should result from a possible lack of availability of the original product at the moment of repairing the defect resulting from e.g., the necessity to wait too long for the construction product originally used in the facility, change of the manufacturer's brand, completion of production of a specific construction product, a clear wish of the facility owner, or a change of e.g., fire safety regulations.

However, during the warranty and guarantee period, construction work may already occur that does not merely involve the removal and repair of faults. The owner of the property may decide to reconstruct, expand, or even change the use of a building that has just been put into use. In such a situation, the guarantee and warranty for the current scope of construction works is lost, and as a result, substitutes for the construction products originally used may be introduced.

The next stage of the operation of a building object, after the warranty period, which will usually last for several or even several dozen years, is a natural period during which substitution of construction products is a common phenomenon. It results from the natural wear and tear of a given element and the desire to replace it with other products that raise the standard of use, e.g., safety, convenience, aesthetics, comfort, and even fashion. After the expiry of the warranty and guarantee period, the construction products used for repairs are the responsibility of the property owners and to a large extent their choice is also dependent on the purchase price. In this phase of building operation, all factors that affect the price of the construction service (refurbishment, reconstruction, etc.) are crucial. It can be stated that the investor, when determining the scope of planned works, in most cases initiates a tender procedure, which differs from the one used during the construction of a new facility only in the scope of planned works. The very stage of collecting offers, their consideration, and selection of a potential contractor is analogous to that of any new construction project under construction.

It is important to note that the selection of construction products during this phase is critical in terms of the life cycle of the facility [7]. The proper selection of these for refurbishment and/or modernization works will have a significant impact on the extension or shortening of this phase of the life cycle as well as on costs [8–10]. Saving at the refurbishment stage may result in the necessity to perform another refurbishment quickly.

The use of substitutes for construction products of better quality and technical values may postpone the need for further refurbishment as well as reduce maintenance costs and also raise the standard of the facility.

Figure 2 presents the change of utility values of a building object in its life cycle, which is connected with two main processes, i.e., constant decrease of utility properties—from the moment of putting the object into use (curve b) and simultaneous increase of the object users' requirements while taking into account changes in regulations and standards (curve a*). The drop in the value of curve b is caused by the wear and tear of individual building elements during the operation phase. The continuous line Z shows the performance assessment at the moment the building is put into operation. It was assumed that the building was designed and constructed in accordance with the relevant standards (Eurocodes) with the application of the required supervision procedures and control throughout the construction process. The dashed straight line Z' specifies the minimum level of utility requirements that a building should meet. If the assessment of performance is below the Z' level, further use of the object is unacceptable.

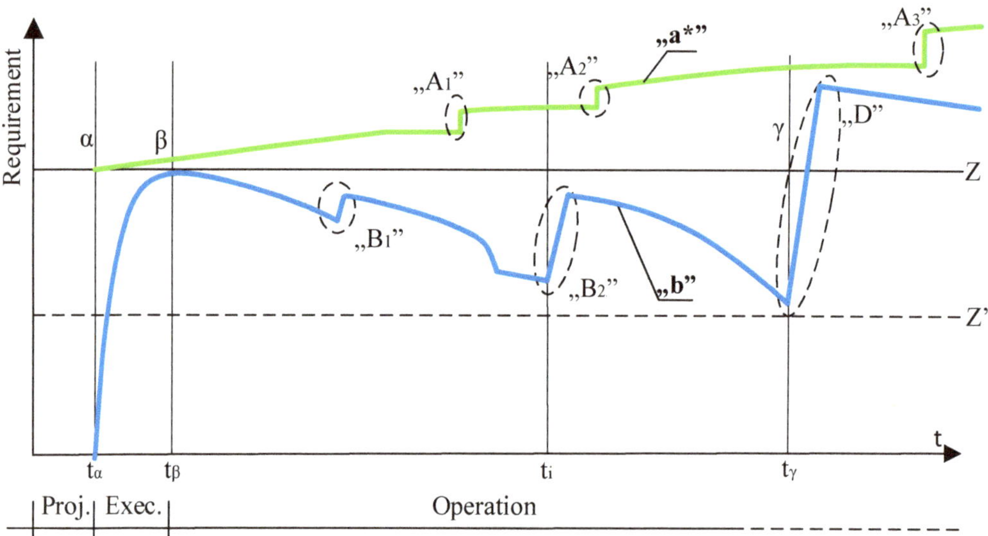

Figure 2. Schematic diagram of the increase in building performance requirements (a-curve) and changes in technical condition due to aging and renovation during the building's service life (b-curve) [11,12].

The decrease in the value of curve b is caused by the wear and tear of individual building elements during the exploitation phase. We can observe "jumps" on it, i.e., an increase in the usable value of the object as a result of repairs and renovations—points B1 and B2—and modernization—point D [12]. Modernization is caused not only by the increase of users' requirements but also by the increase of requirements regarding the object's features as a result of stricter legal regulations (e.g., regarding fire protection).

Construction objects are characterized by a long service life when properly operated. Very often they perform a completely different function than those for which they were designed. The durability of their construction exceeds the often assumed periods [13,14]. We have many examples in the world of such age-old buildings and structures. In Europe, in particular, for many years there has been a desire to take care of the historical substance, objects of historical, cultural, and religious significance that bear witness to past eras. Many of the objects among those existing in the building stock, that due to their exceptional value, are entered in the register of monuments kept by the relevant governmental administration bodies. There is no specific time after which the building is considered a monument. The Act on the Protection and Care of Historical Monuments states that any building which is important for history and science can become a monument, and thus should be preserved [15]. It can also be a building built in the 1950s or 1960s if it presents features characteristic for the architecture of a given period and can be important for its history. Buildings entered in the register of monuments are subject to the Act on the Protection and Care of Historical Monuments [16] and all activities related to the use and in particular their maintenance in a proper technical condition and standard are subject to the supervision of the conservator. Thus, in the phase of exploitation of buildings, the period of their functioning as a monument should be distinguished for a group of exceptional objects.

Substitution starts to appear much more often in the case of buildings already in use for a longer period of time [17–21]. Among the exploited properties, we can observe a certain phenomenon, in which the trend is manifested by the growing deficiencies in the documentation of the exploitation of the building with its age. For example, the documentation of mass-produced buildings built in the 1970s and 1980s in large panel technology is often incomplete and inconsistent. Therefore, owners often look for construction products similar to the original ones while carrying out renovation works, usually guided by the criterion of aesthetics and price. In this case, someone else is also responsible for financing the work on the facility. In cooperative buildings or those owned by housing communities, the costs of all repairs and renovations is borne by the property owners. Most often this is done through the so-called "renovation fund". Such works are very often performed in the order of "from the most urgent", unfortunately in many cases without taking into account the durability of the construction products used for this purpose.

Moreover, one of the major problems of substitution is the choice of substitute material. During the design phase, the architect is almost free to choose a replacement material. In contrast, there are many more factors to consider during the renovation phase. Thanks to advances in material engineering, manufacturers offer a large selection of substitution products with different properties. There is a need to select criteria to evaluate possible alternate materials and make a decision. This is done by multi-criteria analysis using different methods [22]. Among the adopted criteria, the important ones are those that take into account the principles of sustainable development. Therefore, ecological materials, modern technologies or modernization of traditional ones with addition of raw materials from different branches of economy (tea to brick) are being sought.

Two types of approach to substitution can be observed. The first, in a more general sense, is an attempt to:

- produce new materials and building elements capable of performing the appropriate function in the construction of a building. Improve their physical, chemical, etc. properties and usability (durability, aesthetics, usability, operation, etc.), thanks to the development of materials engineering, using the achievements of science, nanotechnology, etc. [23–26]. They can be used interchangeably with traditional materials (instead of clay bricks, e.g., cellular concrete).

The second, however, related to the idea of sustainable development through:

- development and use of materials and elements of the structure of a building and its equipment in building installations which minimize energy consumption (energy efficient) [27–29];
- production of raw and building materials using wastes (as additives e.g., to cement and aggregates or entirely made from waste) [30–33].

The second approach to substitution is a partial restriction on the choice of a substitute by, for example, an architect, developer, or user by placing a condition (of an aesthetic, logistical, etc. nature). This situation relates to a specific building or material solution [34,35]. Here also the selection can be made in terms of one or more optimization criteria [36]. The criteria are based on the individual requirements of the user, the investor or on current social and economic concepts: sustainable development, circular economy [37]. Applying material substitution, it is useful to have knowledge about the determinants of its use, the cause–effect relationships of the factors that have an impact on its use. Such research and results are presented in Section 4 of the article.

3. Substitution in Historical Buildings

It should be noted that in most European countries it is obligatory to replace the materials used in historic buildings with the same ones that were used originally. In Poland, however, the law allows the use of substitutes [16] depending on various conditions.

Factors that affect the possibility of using construction product substitutes are defined in the so-called conservation program, which is developed for each renovation of an object entered in the register of monuments. Each proposed substitute for a construction product must be prepared in the form of a sample and accepted by the Conservator.

Positive aspects of the application of construction product substitution in the renovation of buildings entered in the register of monuments are the factors that primarily enable the refurbishment. Historic sites were built in different construction realities, at a time when available building products were based on natural resources (e.g., stone, rock, clay) and the technology to produce them was simpler. It was common practice to import construction products from other areas of Europe. Even today it is costly and environmentally unfriendly and, due to the environmental protection of certain areas, exploitation is prohibited. However, it is worth considering the use of a substitute material and conducting an analysis of the impact of using such a solution on social, environmental, and economic factors [38–41].

Ownership of the most valuable objects entered in the register of monuments is mostly in the hands of the State or various institutions such as the churches. It should be remembered that the number of facilities under consideration is large and the possibilities of financing renovations are limited, hence the price will always be an important component of planning a refurbishment. The use of original construction products in one object may lead to abandonment or postponement of the renovation in other objects. Such a situation may lead to degradation of the remaining buildings and, consequently, increase the costs of renovations that are planned in them. Therefore, the introduction of substitutes for construction products in historic buildings, which give positive aesthetic and visual values and are less of a financial burden, gives the opportunity to conduct a more effective and larger-scale renovation policy.

In the case of the described refurbishments, a significant price-creating factor is also the time of completion. It is obvious that a longer period of renovation of one object can postpone the start of renovation in another object, which also requires this renovation. Substitution of construction products may increase the pace of renovation works in connection with, among others, less complicated technology of conducting works, faster pace of assembly of built-in elements, and the possibility of conducting works in less favorable weather conditions. Reducing the duration of the renovation gives further savings, thanks to which it is possible to predict that the renovation of a monumental object (the process of renovation of an object entered in the register of monuments takes a very long time because not all the necessary construction works can be predicted at the stage of designing the renovation) will be completed within the assumed time.

Construction products used in historic buildings have often survived years or even centuries. Thus, these are durable products that have been subject to gradual degradation over the years due to lack of refurbishment or minor damage, which has increased the impact zone from year to year [42,43]. The weakness of the construction product substitutes may be their durability and resistance to weather conditions in comparison with primary products and other influences e.g., related to the intensity of car exhaust or air pollution [44]. Renovations of buildings included in the register of monuments should be carried out by companies specializing in this type of construction works. Due to the specific nature of renovation work in historic buildings, the contractor may encounter problems at each stage of the work that are unusual for newly erected buildings.

Substitutes of construction products used in the renovation of historic buildings give a wide range of possibilities. Substitutes can be manufactured from recycled, environmentally-friendly materials and produced by local entrepreneurs [7,45]. The current technology of conducting construction works and the variety of construction products makes it possible to carry out a renovation of basically any building, including historic buildings.

In the current market situation, the cost and time of implementation are critical in any type of construction project. In the case of renovations of objects entered in the register of monuments, the specificity of the conducted construction works and a certain unpredictability of additional construction works, which may appear at each stage of the renovation, are still imposed.

The authors met with an opinion that a historic object that has undergone renovation with the use of construction product substitutes loses its historical value and should no longer be treated as a monument. The basic issue to consider in such a situation is the possibility of renovation.

In old, historic buildings, especially those protected by law (in Poland the register of historic monuments), the use of substitution of materials has a long history. There is an extensive literature in this area, including concepts of substitution principles, developed and proven methodologies for design, testing, analysis, and selection of substitutes [46–48].

4. Evaluation of the Possibility of Using Material Substitution in the Maintenance of Buildings

Preceding the decision to use substitution, the authors suggest performing an assessment and identifying key factors that influence the effectiveness of its use. On the basis of the presented conditions and factors influencing the application of substitution in construction objects in the exploitation phase, a SWOT (Strengths, Weaknesses, Opportunities, Threats) matrix was developed (see Table 1). It contains factors that constitute strengths and weaknesses of the substitution phenomenon and opportunities and threats in its application.

Table 1. SWOT (Strengths, Weaknesses, Opportunities, Threats) analysis of the application of material substitution in renovation works during the operation of buildings.

	Positive	Negative
	Strenghts	**Weaknesses**
Inside	1.1. possibility of refurbishment, 1.2. lower price of the renovation, 1.3. faster pace of construction works (less complicated technology), 1.4. to achieve the desired visual effect, 1.5. replacement of products that are no longer found in the market, 1.6. shortening of supply chains	2.1. faster degradation of the object's substance by inappropriate selection of built products, 2.2. involvement in the work of specialized companies, 2.3. conducting works by experienced supervision, 2.4. investment revaluation, 2.5. possible lack of a proposed substitute in a previously approved solution
	Opportunities	**Threats**
Outside	3.1. market (access and development) of modern construction products 3.2. recycled products, 3.3. products more environmentally friendly, 3.4. establishing cooperation with local entrepreneurs, 3.5. revitalization of degraded areas under conservation care	4.1. loss of authenticity and historical value and object 4.2. specific requirements for carrying out works, especially renovations during adverse weather conditions, 4.3. lack of competent professionals, inspectors, conservators, 4.4. lack of legal regulations concerning the applied solutions

The analysis of the matrix, in particular the comparison of factors from different fields of the matrix gives an opportunity to determine the type of a possible general strategy in substitution activity, as well as detailed strategies in organizations dealing with the management of building real estate, including historic buildings.

If a strategy is established, reference should be made to a specific object.

On the basis of the analysis of information from the presented SWOT matrix, conclusions can be drawn with regard to the possibilities for developing material substitution in the construction industry. Undoubtedly, in the situation of emerging new and modern technologies, more and more diversified offers of the manufacturers' market allows for flexible and quick adaptation of investors to the dynamics of social and economic changes, especially for such long-lasting products as building structures. Undoubtedly, there is an advantage to the benefits of substitution in various aspects of the investment and construction process, both in terms of execution and ancillary activities, including logistics. Out of the four presented threats, two factors concern historic buildings, and one needs to be supplemented in legal regulations. The fourth one related to the requirements of relevant competences requires the support of the educational system.

The information contained in the presented SWOT matrix can be used in two ways. It can be used to analyze and generally evaluate the development of a certain phenomenon. It can also be used in the strategic analysis of an individual specific enterprise, company, or system.

This paper will use the data from the SWOT matrix to assess the overall feasibility of using substitution in building repair work (Section 5). The factors collected in the SWOT matrix can be used to establish cause–effect relationships between them. Identifying the causal chain will allow us to identify those factors that have the greatest impact on the process of substitution.

For an individual facility, on the other hand, this analysis will determine whether the planned substitution will have a more positive or negative impact on the renovated facility. It will also allow the investor to look at all the pros and cons of using substitution and assist him in making a final decision on the renovation policy on the chosen facility.

5. Cause-and-Effect Analysis of the Use of Substitution

5.1. Research Methodology

To identify cause–effect relationships in the issue of possible substitution of material solutions of buildings the authors propose to use the DEMATEL method [49–52]. When analyzing a multi-factor problem, a multi-criteria analysis is used to evaluate the problem using different methods that allow ranking of solutions. On the other hand, the DEMATEL method chosen by the authors also enables a cause-and-effect analysis of the phenomenon under study.

The computational flow is as follows:

1. Determining a set of influence factors, in the proposed study based on the SWOT matrix (Figure 3);

2. Development of a direct influence graph, according to the DEMATEL method, which allows us to express the targeted influence of the considered factors on each other, in a cause-and-effect context. A scale with a parameter value of N = 3 (where: 0—no influence, 1—weak influence, 2—influence, 3—strong influence) was used to assess the "strength" of the influence of each factor. The values of the direct influence relationships within each pair of factors were determined based on the evaluations of the expert group and they were calculated using fuzzy logic;

3. Based on the relationships determined with the graph, a matrix of direct mutual influence of factors on each other A_D was created (Figure 4);

4. Determination of the normalized direct influence matrix A'_D, which contains all parameters that take values that are in the range [0, 1] (Table 2). The normalizing number (n) is taken as the largest of the sum of the rows or columns of the matrix A_D:

$$A'_D = \frac{A_D}{n},\tag{1}$$

$$n - \max\left\{\sum_{i=1}^{n} a_{ij}; \sum_{j=1}^{n} a_{ij};\right\},\tag{2}$$

5. It is also possible to develop an indirect impact matrix ΔT:

$$\Delta T = A'^2_D \cdot (I - A'_D),\tag{3}$$

6. Determination of the total influence matrix T (Table 3):

$$T = A'_D \cdot (I - A'_D),\tag{4}$$

7. On the basis of the above matrices, the determination of the indices of position and relationship, respectively, which express in turn: s^+—tells about the role of a given factor in the process of determining the structure of links between objects, while s^-—expresses the total influence of a given factor on the others. These values are determined according to the formulas (Table 4):

$$s^+ = \sum_{j=1}^{n} t_{ij} + \sum_{j=1}^{n} t_{ji} = R_{T_i} + C_{T_i},\tag{5}$$

$$s^- = \sum_{j=1}^{n} t_{ij} - \sum_{j=1}^{n} t_{ji} = R_{T_i} - C_{T_i},\tag{6}$$

When these values are plotted on a graphical representation, it is easy to see which factors have the greatest influence on the others and to determine which are the causes and which are the effects of the actions taken (Figure 5).

8. Finally, the net impact value is also determined, which tells the factor that has the greatest impact on the others considering both the causal and effect nature (Table 4):

$$netto = s^+ + s^-\tag{7}$$

Table 2. The fragment of normalized direct influence matrix A'_D.

	1.1	1.2	1.3	1.4	1.5	1.6	2.1	
1.1	−0.0156	−0.0017	0.0000	0.0313	0.0000	0.0000	−0.0035	… …
1.2	0.1250	0.0000	0.0000	−0.0052	0.0000	0.0000	0.0833	… …
1.3	0.0729	0.0833	0.0000	−0.0035	0.0000	0.0000	−0.0069	… …
1.4	0.0000	0.0000	0.0000	0.0000	0.0000	0.0000	0.0000	… …
1.5	0.1128	0.0330	0.0816	0.0781	−0.0017	−0.0035	−0.0052	… …
… …	… …	… …	… …	… …	… …	…	… …	… …

Table 3. Total influence matrix T (fragment).

	1.1	1.2	1.3	1.4	1.5	1.6	2.1	… …
1.1	0.0000	0.0000	0.0000	0.0417	0.0000	0.0000	0.0000	… …
1.2	0.1250	0.0000	0.0000	0.0000	0.0000	0.0000	0.0833	… …
1.3	0.0833	0.0833	0.0000	0.0000	0.0000	0.0000	0.0000	… …
1.4	0.0000	0.0000	0.0000	0.0000	0.0000	0.0000	0.0000	… …
1.5	0.1250	0.0417	0.0833	0.0833	0.0000	0.0000	0.0000	… …
… …	… …	… …	… …	… …	… …	… …	… …	… …

Table 4. Summary of DEMATEL analysis results.

Criterion i	R_{T_i}	C_{T_i}	s^+	s^-	Netto
1.1	0.2830	0.6042	**0.8872**	-0.3212	0.5660
1.2	0.2396	0.4288	0.6684	-0.1892	0.4792
1.3	0.1146	0.1198	0.2344	-0.0052	0.2292
1.4	0.1198	0.2899	0.4097	-0.1701	0.2396
1.5	0.5243	0.0399	0.5642	**0.4844**	**1.0486**
1.6	0.0590	0.1493	0.2083	-0.0903	0.1181
2.1	0.0799	0.2708	0.3507	-0.1910	0.1597
2.2	0.1181	0.1632	0.2813	-0.0451	0.2361
2.3	0.1198	0.0330	0.1528	0.0868	0.2396
2.4	0.0000	**0.0660**	0.0660	-0.0660	**0.0000**
2.5	0.0000	0.1163	0.1163	-0.1163	0.0000
3.1	0.4757	0.0000	0.4757	0.4757	0.9514
3.2	0.2865	0.0122	0.2986	0.2743	0.5729
3.3	0.3038	0.0399	0.3438	0.2639	0.6076
3.4	0.0347	0.0399	0.0747	-0.0052	0.0694
3.5	0.1563	0.0365	0.1927	0.1198	0.3125
4.1	0.0365	0.7743	0.8108	**-0.7378**	0.0729
4.2	0.1927	0.1632	0.3559	0.0295	0.3854
4.3	0.1094	0.0000	0.1094	0.1094	0.2188
4.4	0.2153	0.1215	0.3368	0.0938	0.4306

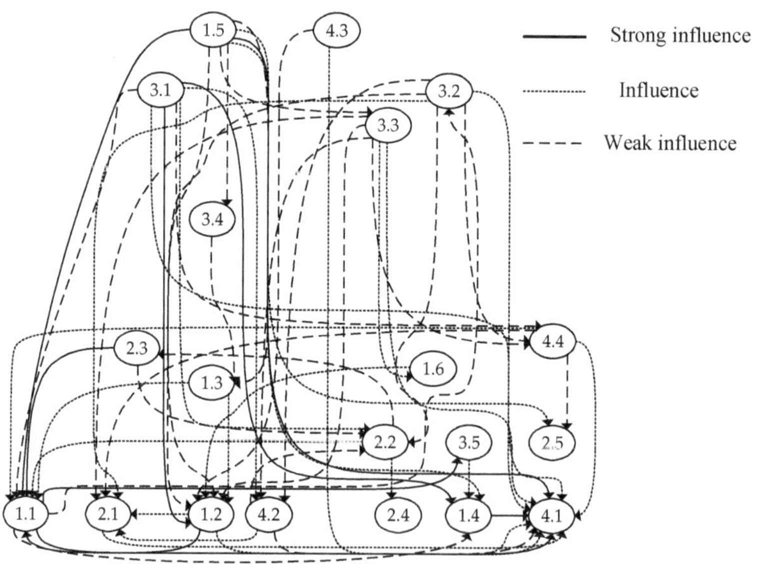

Figure 3. Direct influence graph—expert evaluation results.

Figure 4. The matrix of direct effects of factors on each other.

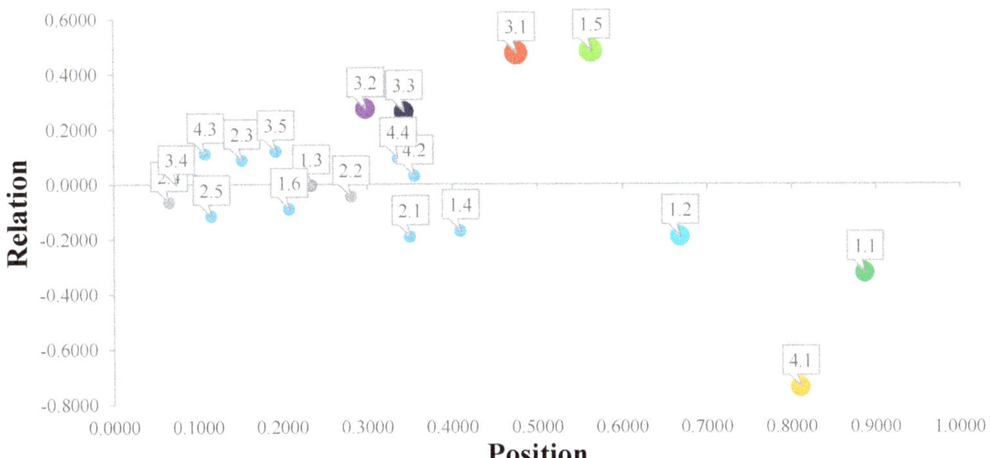

Figure 5. Graphical interpretation of DEMATEL results.

5.2. Study Results and Its Analysis

In supporting the decision to use substitution to examine the cause and effect relationships, all the factors summarized in the SWOT matrix were considered. These factors, as in the case of the SWOT matrix, were divided into the same four groups. To simplify the recording of the factors in further analysis with the help of the DEMATEL method, only the number from the SWOT table is marked (see Table 1).

The factors identified during the SWOT analysis are subjected to an assessment of the strength of their impact on each other.

For the analyzed issue—application of substitution, e.g., in repair works, the form of direct influence graph is presented in Figure 3. The intensity of relationships was coded using different hatchings of arc lines.

Based on the relationships illustrated above, a direct influence matrix A_D was created (step 3).

Table 2 shows fragment of element values of the normalized matrix (step 4):

Next, based on Equation (3), the matrix of total relations T was determined:

A summary of the values to build an illustration of the causal nature is shown in Table 4 (step 7).

The analysis was performed by using summative, linear aggregation of the values of the position and relationship indicators (s^+ and s^-). The calculations in general are expressed in the graph shown in Figure 5, which shows the values of the position and relationship indicators. Based on the aggregated values of the item index, it was found that the greatest role in determining the nature of the factors is played by: 3.1 (market-access and development of modern construction products) and 1.5 (replacement of products no longer manufactured).

Factors 3.2 (recycled products) and 3.3 (more environmentally friendly products) have slightly less influence. The clearly positive values of the relationship index for these factors indicate their causal character.

Almost half of the analyzed factors show a negative value of the relationship index, hence they should be treated as possible effects of the causes.

Of the factors with a negative relationship index value, a significantly outstanding negative value was obtained by 4.1 (loss of authenticity and historical value and object), which represents the largest negative possible effect of using substitution.

Factors with a positive sign but close to the zero value can be treated as elements of a mixed nature, partly causal, partly effectual, but both as causes and effects of far less importance.

The situation is different if we look at the values of the factors they obtain in the position axis (Figure 5). Factors with above average values of the item index testify to their leading role in determining the nature of individual factors. Among the prominent factors of the position indicator are 1.1 (possibility of renovation), 4.1 (loss of authenticity and historical value of the object), 1.2 (lower price of renovation), 1.5 (replacement of products no longer manufactured), and 3.1 (market-access and development of modern construction products). Again, as far as the others are concerned, they have far less active participation in the process of identifying the role of factors.

The aggregated values of the relation index allowed for distinguishing three groups of factors: key, average and insignificant for shaping the renovation policy. In particular, the key factors as reasons for decision-making turned out to be: 3.1, 1.5. Key factors as reasons for taking the group of average significant factors form: 3.2 i 3.3. The other factors can be considered by far the least important.

The possible impacts of the decision are definitely influenced by factor 4.1, which reflects the fear of losing the authenticity of the historic substance, as well as 1.1, which represents the opportunity for renovation. The fear of loss of authenticity should be the starting point in selecting the right, in this case the closest substitution to the original. The effect of being able to renovate is a decisive advantage of substitution and can often be the only solution to improve the technical condition of an object and extend its life cycle.

6. Summary

Substitution of construction products is a common phenomenon in the construction industry at every stage of a building's life cycle. Moreover, sometimes the use of substitution may be the only feasible solution to save a facility. In a wider context, it can have a great impact on the implementation of the principles of sustainability and circular economy in the maintenance of building stock.

The conducted observations show that during the warranty and guarantee period in newly constructed buildings, the substitution of construction products is much lower than in the next period of the facility's operation. The phenomenon of substitution, however, is often encountered in the long-term perspective of facility operation, especially during all repair and overhaul works.

One of the ways of extending the life cycle of buildings is a proper renovation policy, which through proper selection of material solutions will ensure longer durability of components and the entire facility.

A special case is the substitution in the renovation works of objects entered in the register of monuments, which gives the possibility to protect the historic substance while maintaining the structural and aesthetic values.

Product substitution, which is often cheaper than the original, may also allow for a wider range of renovations, which directly contributes to improving the technical and functional condition of the object and thus allows for extending its life cycle.

The SWOT analysis conducted by the authors allows us to conclude that substitution in the construction industry is justified and that there are great opportunities for its implementation and development. A detailed analysis of the SWOT matrix factors, using the DEMATEL method, allowed for an overall assessment of substitution possibilities with the determination of the cause–effect relationship of factors from particular groups characterizing the strengths and weaknesses of substitution as well as the opportunities and threats. Undoubtedly, the use of substitute materials, especially in historic buildings, will result in a decrease in their authenticity, but will ultimately restore them to safe operation and in other buildings allow for an extended life cycle.

Despite the possibility of product substitution thanks to materials engineering and technology development, the use of substitution should not be approached uncritically. The authors recommend a case-by-case approach, conducting a comprehensive analysis and making decisions based on, among other things, the tools proposed in the article and evaluating the cause-and-effect relationships that will occur when substitution is applied.

The above comment also applies to using a different approach to material substitution in historic buildings. The decision whether or not to use substitution and the freedom to choose substitution solutions are influenced by the conservation concepts and legal regulations of the respective country or type of object.

Author Contributions: Conceptualization, A.S., K.L. and A.R.; methodology, A.S., K.L. and A.R.; validation, A.S., K.L. and A.R.; formal analysis, A.S., K.L. and A.R.; investigation, A.S., K.L. and A.R.; resources, K.L. and A.R.; writing—original draft preparation, A.S., K.L. and A.R.; writing—review and editing, A.S. and A.R.; visualization, K.L. and A.R.; supervision, A.S. All authors have read and agreed to the published version of the manuscript.

Funding: This research received no external funding.

Institutional Review Board Statement: Not applicable.

Informed Consent Statement: Not applicable.

Data Availability Statement: Data sharing not applicable.

Acknowledgments: The work was carried out as part of statutory research no. 11.11.100.197 in the Department of Geomechanics, Civil Engineering and Geotechnics of Faculty of Mining and Geoengineering, AGH University of Science and Technology in Cracow.

Conflicts of Interest: The authors declare no conflict of interest.

References

1. Linczowski, K.; Sobotka, A. Relations between substitution in technological processes and construction supply logistics. *Logistics* **2014**, *6*, 9816–9825.
2. Gustavsson, L.; Pingoud, K.; Sathre, R. Carbon dioxide balance of wood substitution: Comparing concreto- and wood-framed buildings. *Mitig. Adapt. Strateg. Glob. Chang.* **2006**, *11*, 667–691. [CrossRef]
3. Wadr, H.J. The Use of Substitute Materials on Historic Building Exteriors. 1988. Available online: http://www.nps.gov/tps/how-to-preserve/briefs/16-substitute-materials.htm#top (accessed on 14 February 2021).
4. New Zealand Government, Ministry of Business Innovation & Employment. *Quick Guide to Product Substitution*; New Zealand Government: Wellington, New Zealand, 2004.

5. Linczowski, K.; Sobotka, A. Analysis of logistics for construction site supply with reinforcing steel. *Tech. Trans. Civ. Eng.* **2014**, *6*, 225–232.
6. Arendalski, J. *Durability and Reliability of Residential Buildings*; Arkady: Warsaw, Poland, 1978.
7. Petersen, A.K.; Solberg, B. Environmental and economic impacts of substitution between wood products and alternative materials: A review of micro-level analyses from Norway and Sweden. *For. Policy Econ.* **2005**, *3*, 249–259. [CrossRef]
8. Fregonara, E.; Patono, S. A sustainability indicator for building projects in presence of risk/uncertainty over time: A research experience. *AESTIMUM* **2018**, *19*, 173–205.
9. Grzyl, B.; Apollo, M.; Miszewska-Urbańska, E.; Kristowskim, A. Object operation management in terms of life cycle costs. *Acta Sci. Pol.* **2017**, *16*, 85–89.
10. Plebankiewicz, E.; Meszek, W.; Zima, K.; Wieczorek, D. Probabilistic and fuzzy approaches for estimating the life cycle costs of buildings under conditions of exposure to risk. *Sustainability* **2020**, *12*, 2504. [CrossRef]
11. Orłowski, Z.; Szklennik, N. Analysis of the building value in use as a function of time. *Tech. Trans.* **2010**, *6*, 303–312.
12. Radziejowska, A. The Method of Assessing the Social Performance of Residential Buildings in the Aspect of Sustainable Construction. Ph.D. Thesis, AGH, Cracow, Poland, 2018.
13. Economakis, R. *Durability in Construction: Traditions and Sustainability in 21st Century Architecture*; Papadakis Publisher: Winterbourne, Berkshire, UK, 2015; ISBN-13: 978-1906506551.
14. Bai, J. Durability of sustainable construction materials. *Sustain. Constr. Mater.* **2016**, *38*, 397–414.
15. Gmiter, M. *Guide to the Investment Process in Historic Buildings*; Presscom Sp. Z.o.o.: Poland, Wrocław, 2018.
16. Act on the Protection and Care of Historical Monuments, Old Monuments Law (Ustawa, o Ochronie Zabytków; Opiece Nad Zabytkami) of 23 July 2003, with Later Changes; Journal of Laws: 2014; pos. 1446. Available online: https://isap.sejm.gov.pl/isap.nsf/download.xsp/WDU20031621568/U/D20031568Lj.pdf (accessed on 16 February 2021).
17. Technical Preservation Services. *Evaluating Substitute Materials in Historic Buildings*; Historic Preservation Tax Incentives Programme: Washington, DC, USA, 2007.
18. Cluver, J.H. *No Substitute: Inexpensive and Maintenance Free or Short Sighted and Maintenance Proof: How Do Substitute Materials Stack Up in the Long Run?* Labine's Period Homes: Boulder, CO, USA, 2005; pp. 12–16.
19. Dierickx, M.B. Substitute materials for wooden buildings: The system or the artifact? *APT Bull.* **1986**, *18*, 4–5. [CrossRef]
20. Capen, J. Using substitute materials in a historic district. *Hill Rag* **2008**, *5*, 134–135.
21. S & D Design Guidelines for Landmark. *Guidelines for Preserving Historic Buildings—Alterations to Landmark Structures and Contributing Structures in Historic Districts*; S & D Design Guidelines for Landmark: Denver, CO, USA, 2016.
22. Zavadskas, E.K.; Turkis, Z.; Tamosaitiene, J. Selection of construction enterprises management strategy based on the SWOT and multi-criteria analysis. *Arch. Civ. Mech. Eng.* **2011**, *11*, 1063–1082. [CrossRef]
23. Nawar, A.H. Nano-technologies and nano-materials for civil engineering construction works applications. *Mater. Today Proc.* **2020**. [CrossRef]
24. Razakhanova, F.M. Issues concerning import substitution of building materials in the construction market. *Vestn. Dagest. Gos. Teh. Univ. Teh. Nauk.* **2018**, *44*, 223–233. [CrossRef]
25. Li, H.; Xu, B. *Advanced Building Materials and Structural Engineering*; Advanced Materials Research; Trans Tech Publications Ltd.: Zurich, Switzerland, 2012; Volume 461.
26. Enríquez, E.; Torres-Carrasco, M.; Cabrera, M.; Muñoz, D.; Fernández, J. Towards more sustainable building based on modified Portland cements through partial substitution by engineered feldspars. *Constr. Build. Mater.* **2020**. [CrossRef]
27. Piccardo, C.; Dodoo, A.; Gustavsson, L.; Tettey, U. Retrofitting with different building materials: Life-cycle primary energy implications. *Energy* **2020**, *19*, 5–10. [CrossRef]
28. Blomqvist, S.; La Fleur, L.; Amiri, S.; Rohdin, P.; Ödlund, L. The impact on system performance when renovating a multifamily building stock in a district heated region. *Sustainability* **2019**, *11*, 2199. [CrossRef]
29. Rubio-Cintas, M.D.; Parron-Rubio, M.E.; Perez-Garcia, F.; Ribeiro, A.B.; Oliveira, M.J. Influence of steel slag type on concrete shrinkage. *Sustainability* **2021**, *13*, 214.
30. Bouasria, M.; Khadraoui, F.; Benzaama, M.-H.; Touati, K.; Chateigner, D.; Gascoin, S.; Pralong, V.; Orberger, B.; Babouri, L.; El Mendili, Y. Partial substitution of cement by the association of Ferronickel slags and *Crepidula fornicata* shells. *J. Build. Eng.* **2021**, *4*, 103. [CrossRef]
31. Chen, W.; Li, Y.; Chen, S.; Zheng, C. Properties and economics evaluation of utilization of oil shale waste as an alternative environmentally-friendly building materials in pavement engineering. *Constr. Build. Mater.* **2020**. [CrossRef]
32. Patel, D.; Shrivastava, R.; Tiwari, R.; Yadav, R. Properties of cement mortar in substitution with waste fine glass powder and environmental impact study. *J. Build. Eng.* **2020**. [CrossRef]
33. Raheem, A.A.; Ikotun, B.D. Incorporation of agricultural residues as partial substitution for cement in concrete and mortar. *J. Build. Eng.* **2020**. [CrossRef]
34. Hafner, A.; Schäfer, S. Comparative LCA study of different timber and mineral buildings and calculation method for substitution factors on building level. *J. Clean. Prod.* **2017**, *167*, 630–642. [CrossRef]
35. D'Amico, B.; Pomponi, F.; Hart, J. Global potential for material substitution in building construction: The case of cross laminated timber. *J. Clean. Prod.* **2021**, *7*, 279. [CrossRef]

36. Vachris, J. Asbestos substitution serviceability of new materials. In Proceedings of the Electrical Insulation Conference and Electrical Manufacturing & Coil Winding Conference, Cincinnati, OH, USA, 26–28 October 1999.
37. Omer, M.A.; Noguchi, T. A conceptual framework for understanding the contribution of building materials in the achievement of Sustainable Development Goals (SDGs). *Elsevier Sustain. Cities Soc.* **2020**, *52*, 244–302. [CrossRef]
38. Van Domelen, S.K. The Choice Is Yours: Considerations & Methods for the Evaluation & Selection of Substitute Materials for Historic Preservation. Ph.D. Thesis, University of Pennsylvania, Philadelphia, PA, USA, 2009. Available online: http://repository.upenn.edu/hp_theses (accessed on 14 February 2021).
39. Liu, L.; Li, H.; Lazzaretto, A.; Manente, G.; Tong, C.; Liu, Q.; Li, N. The development history and prospects of biomass-based insulation materials for buildings. *Renew. Sustain. Energy Rev.* **2017**, *69*, 912–932. [CrossRef]
40. Iucolano, F.; Boccarusso, L.; Langella, A. Hemp as eco-friendly substitute of glass fibres for gypsum reinforcement: Impact and flexural behaviour. *Compos. Part B Eng.* **2019**, *175*, 25. [CrossRef]
41. Mathur, V. Composite materials from local resources. *Constr. Build. Mater.* **2006**, *20*, 470–477. [CrossRef]
42. Myers, J.; Aluminum and Vinyl Siding on Historic Buildings. The Appropriateness of Substitute Materials for Resurfacing Historic Wood Frame Buildings. 1984. Available online: https://www.nps.gov/TPS/how-to-preserve/briefs/8-aluminum-vinyl-siding.htm (accessed on 14 February 2021).
43. Nowogońska, B. Value of technical wear and costs of restoring performance characteristics to residential buildings. *Buildings* **2020**, *10*, 9. [CrossRef]
44. Demir, I. An investigation on the production of construction brick with processed waste tea. *Build. Environ.* **2006**, *9*, 1274–1278. [CrossRef]
45. Teha, S.H.; Wiedmann, T.; Schinabeck, J.; Moorea, S. Replacement scenarios for construction materials based on economy-wide hybrid LCA. *Procedia Eng.* **2017**, *180*, 179–189. [CrossRef]
46. Friedman, D. *Historical Building Construction: Design, Materials, and Technology*; W.W. Norton & Company: New York, NY, USA, 1995.
47. Fisher, T. The sincerest form of flattery. *Progress. Archit.* **1985**, *85*, 118–123.
48. Brownell, B. *Transmaterial: A Catalog of Materials that Redefine our Physical Environment*; Princeton Architectural: New York, NY, USA, 2006; Volume 6.
49. Dytczak, M.; Ginda, G.; Gotowała, B.; Szklennik, N. Application potential of Dematel method and its extensions for analysis of decision problems in civil engineering. *Bud. Inżynieria Sr.* **2011**, *2*, 235–240.
50. Sheng-Li, S.; Xiao-Yue, Y.; Hu-Chen, L.; Ping, Z. Dematel technique: A systematic review of the state-of-the-art literature on methodologies and applications. *Math. Probl. Eng.* **2018**, *23*, 1–33.
51. Jiunn, I.S.; Hsin-Hung, W.; Kuan-Kai, H. A Dematel method in identifying key success factors of hospital service quality. *Knowl. Based Syst.* **2010**, *23*, 277–282.
52. Kijewska, K.; Torbacki, W.; Iwan, S. Application of AHP and Dematel methods in choosing and analysing the measures for the distribution of goods in Szczecin region. *Sustainability* **2018**, *10*, 2365. [CrossRef]

Article

Bayes Conditional Probability of Fuzzy Damage and Technical Wear of Residential Buildings

Jarosław Konior * and Tomasz Stachoń

Department of Building Engineering, Faculty of Civil Engineering, Wroclaw University of Science and Technology, 50-370 Wrocław, Poland; tomasz.stachon@pwr.edu.pl
* Correspondence: jaroslaw.konior@pwr.edu.pl; Tel.: +48-71-320-23-69

Abstract: The purpose of the research presented in the article is to identify the impact of the processes associated with the broadly understood maintenance of old residential buildings with a traditional construction on the size and intensity of the wear of their elements. The goal was achieved by analyzing the symptoms of the technical wear process, which involved the understanding of the mechanism of the occurrence of the phenomenon of damage, and the identification of the size and intensity of the damage to the elements of the evaluated buildings. The consequence of systematizing the most important processes that influence the loss of functional properties of residential buildings was the creation of the authors' own qualitative model and its transformation into a quantitative model. This, in turn, enabled a multi-criteria quantitative analysis of the cause and effect phenomena—"damage-technical wear"—of the most important elements of downtown tenement buildings to be carried out in fuzzy conditions, i.e., uncertainty concerning the occurrence of damage and the wear process. The following key question was answered in the subjective expert assessment of the technical condition of an evaluated residential building: what is the probability of the wear of an element, which may be more or less correlated with its average maintenance conditions, or more simply, what is the probability that the element is more or less (approximately) worn? It has been proven that the conditional probability of the technical wear of an element in relation to its damage increases with the deterioration of the maintenance conditions of the building, and this increase is very regular, even in the case of different building elements. This probability is characterized by a low standard deviation and a narrow range of the dispersion of results in the case of various elements with regards to each of the considered building maintenance conditions.

Keywords: residential buildings; technical wear; damage; Bayes conditional probability; fuzzy sets

Citation: Konior, J.; Stachoń, T. Bayes Conditional Probability of Fuzzy Damage and Technical Wear of Residential Buildings. *Appl. Sci.* **2021**, *11*, 2518. https://doi.org/10.3390/app11062518

Academic Editor: Marco Vona

Received: 17 February 2021
Accepted: 8 March 2021
Published: 11 March 2021

1. Introduction

1.1. Literature Survey

The literature survey was based on the theory of decision-making in the conditions of uncertainty and fuzziness, which is given by Kacprzyk [1], and which defines the following decision situations [2]:

- Certainty: all the information that describes the issue of decision-making is deterministic, i.e., options for choosing a decision, and what each choice gives in terms of certain usefulness (e.g., value analysis) etc., is known. In this case, making decisions comes down to the direct maximization of the utility function;
- Risk: information that describes the decision-making issue is probabilistic, i.e., appropriate probability distributions are provided. In this case, making decisions comes down to maximizing the expected value of the utility function;
- Indeterminacy: even probabilities are not known. Decision-making usually comes down to applying a minimax strategy in order to ensure the highest utility value under the most unfavorable conditions;

Appl. Sci. **2021**, *11*, 2518. https://doi.org/10.3390/app11062518 — https://www.mdpi.com/journal/applsci

- Fuzziness: indeterminacy not only relates to the occurrence of an event, but also to its meaning in general, which can no longer be described using probabilistic methods. Of course, further extensions are possible here, such as adding risk to fuzziness.

When assessing the degree of the technical wear of building elements, apart from measurable (quantitative) criteria, non-measurable (qualitative) criteria are also used. They are expressed in the analysis of symptoms by, i.e., damage that reduces the technical condition and utility value of building elements. Only some of these criteria can be roughly quantified. However, most of these criteria are qualitative. Their value is determined verbally by using terms such as "significant", "poor", "strong", "almost not at all", "partial", or "complete", and always appears in the description of damage phenomena. The interpretation of the effects of these phenomena, which is performed according to subjective and qualitative premises, leads to the indiscriminate categorization of the technical maintenance conditions for buildings and their elements, i.e., good, satisfactory, average, poor, or bad. Striving for a quantification of criteria that are inherently qualitative and immeasurable, and trying to determine the relations between them, led to the use of the category of fuzzy sets (the basis of which were formulated by Zadeh [3,4] and Yager [5,6]) with regards to this issue. Their properties enable damage to building elements, as well as the conditions of their technical maintenance, to be described within an unambiguous measurable quantitative aspect.

In the methodical approach to the technical assessment of residential buildings, research by Nowogonska [7–11] was used, which provides methods and models for the estimation of the degree of the technical wear of buildings. However, it should be remembered that the presented methodical approach of Nowogonska is exclusively deterministic, and therefore simplified and also practical. This approach is confirmed by the research of Lee and Kim [12], who indicated the degree of risk that is associated with damage to a building element. The assessment of the entire service life of a building structure includes a fuzzy calculation, which was presented in the publications of Plebankiewicz, Wieczorek, and Zima [13–16] in order to determine the impact and significance of the risk of the emergency operation of a building. The works by Ibadov [17–20] concerning the building investment process with a fuzzy phase allowed for the practical application of uncertain and subjective events when determining the degree of damage to the tested tenement houses. The assessment of the risk and costs of maintaining construction facilities, and also the conducting of the construction process in fuzzy conditions, were also presented by Kamal and Jain [21], Andrić, Wang, Zou and Zhang [22], J. Marzouk and Amin [23], Knight, Robinson, and Fayek [24], Sharma and Goyal [25], Al-Humaidi and Hadipriono [26], Ammar, Zayed and Moselhi [27], Chan, Kwong, Dillon and Fung [28], and Naszrzadeh, Afshar, Khanzadi, and Howick [29].

Methods, models, and methodological tools for the assessment of the technical condition of buildings, which are considered in the article with regards to the research sample, were described and summarized by Konior in papers [30–34] with co-authors [35–37] and in a collective study under the supervision of Kapliński, which is entitled "Methods and Models in the Engineering of Building Processes" [38].

1.2. Research Sample

The research sample, which included 102 technically assessed residential buildings from the "Srodmiescie" district of Wroclaw, was selected from a group of 160 examined objects [39]. The overriding criterion for sampling involved the obtaining of a comparable group of objects. Mutual comparability of the downtown tenement houses meant:

- age coherence, i.e., a similar period of erection, maintenance, and use with regards to historical and social aspects;
- compactness of the building development in the urban layout that remained unchanged for years;
- similar location along downtown street routes with an urban, but not representative, character;
- construction and material homogeneity, especially with regards to the load-bearing structure of buildings;

- identical functional solutions, understood as the standard of apartment amenities and furnishings (for that time), and a defined standard of living for residents.

The method of selecting the research sample at the level of greater detail was based on the mutual similarity of all the technical solutions of the downtown tenement houses. The selected research sample, according to the criteria presented above, is a representative sample with regards to the concept of representativeness that is specific to the adopted purpose of the study [40,41]. It contains all the values of the variables that could be recreated from the research carried out earlier using a different objective function than the one adopted in the study.

These values and variables were then compiled and processed in such a way that it was possible to make conclusions about the cause–effect relationships between them in the general population.

Therefore, it can be considered as a typologically representative sample that includes the desired types of homogeneous variables. Due to the fact that the structure of the population and its properties were previously well recognized, such a selection of the research sample can also be seen as a deliberate selection. It should be noted that the sample may not be representative in terms of the distributions of the studied variables, which may—for the adopted level of significance—not correspond to analogous distributions in the general population. It is also not known—at this stage of the research—whether the selected sample is representative due to the relationship between its variables and the identically defined variables in the entire set of downtown residential buildings. Therefore, at the very beginning of the research, it was assumed that a specific research sample occurs in the existing population with the fuzzy phase.

Tested buildings have been classified into classes, determined by the degree of the technical wear. The technical wear 0–15% has been classified to the class I, 16–30% to the class II, 31–50% to the class III, 51–70% to the class IV, 71–100% to the class V. Owing to the fact that all considering apartment houses belong to the same group of their age it is possible to assume that the class of the technical wear corresponds to the conditions of building maintenance. Therefore, the equivalence has been defined: a poor maintenance—the class IV, V, an average maintenance—the class III, an above than an average maintenance—the class II, a very well cared maintenance—the class I.

2. Research Method

2.1. Problem Identification

The research methodology at a level of greater detail was prepared in such a way that allowed the previously prepared qualitative model to be transformed into a quantitative model. Therefore, the diagnosis of the impact of the maintenance of the residential buildings on the amount of their technical wear was carried out using quantitative methods in fuzzy set categories, and also by using the authors' own model that was created in the conditions of fuzziness. The model allowed for the determination of the conditional probabilities of the process of technical wear, and also the set of damage according to both Bayes formulas [40–42] and the combined approach of Zadeh [3,4] and Yager [5,6];

As mentioned in the introduction, when visually assessing the technical wear of building elements, the symptoms of their destruction are taken into account, i.e., individual damage that can be categorized into the following groups (groups) of damage:

- UM—mechanical damage to the structure and texture of building elements;
- UW—damage to building elements caused by water penetration and moisture penetration;
- UD—damage resulting from the loss of the original shape of wooden elements;
- UP—damage to wooden elements attacked by biological pests.

The purpose of such a conceptual and technical systematization of damage is a comprehensive diagnosis of the extent to which a building element is worn. This assessment, in turn, leads to the implication of stating under what technical conditions—good, satisfactory, average, poor, or bad—the building element was (is) maintained. It is difficult to define a fuzzy set with such a broad meaning as "average technical condition of maintenance" using one membership function. In this case, a semantic analysis of the term "technical wear of a building element" was used, which was denoted with the symbol of a fuzzy set "Z". Let the technical wear of building element Z consist of: mechanical wear of its structure and texture (fuzzy set ZM), its technical wear caused by water penetration and moisture penetration (fuzzy set ZW), technical wear resulting from the loss of its original shape (fuzzy set ZD), and technical wear caused by the attack of biological pests (fuzzy ZP harvest). This sum can then be expressed as follows:

$$Z = ZM \cup ZW \cup ZD \cup ZP \tag{1}$$

and when assuming the identity of the degree of technical wear and its visual symptom ($Z \Leftrightarrow U$)—damage to a building element that is integrated into the above-described damage sets—this expression takes the following form:

$$U = UM \cup UW \cup UD \cup UP \tag{2}$$

If technical wear was assumed in the observed states with its measure—the degree of wear as a fuzzy set with no crisp membership boundary of $\{z\} = Z$—then the visual image of this wear—global damage to a building element—should be treated as a fuzzy set, the fuzzy events of which are arguments—distinguished types of damage $\{u\} = U$. Therefore, fuzzy random events are fuzzy sets that express the degree of technical wear, for which there is no complete (measurable) certainty of membership to the II, III or IV class of the technical maintenance of an element. The question then arises: what is the probability of an element being worn, which will more or less represent its average maintenance conditions. To put it simply, what is the probability that an element is more or less (approximately) worn?

The approach of Zadeh [3,4], who defined the probabilities of fuzzy events in the form of real numbers from interval [0, 1], was used in the research. Therefore, the probability of a fuzzy event, which is the technical wear of a building element, which corresponds to satisfactory, average, and poor maintenance conditions, was defined as:

$$P(Z)\text{II, III, IV} = \sum_{i=1}^{n} p(z_i)\mu_{zi}(z_i), \text{ if } Z = \{z_i\} = \{z_1, z_2, \ldots, z_n\} \tag{3}$$

For the global damage of a structural element, which is assumed equivalently to the event of technical wear, the probability of its occurrence is expressed by the following analogous relationship:

$$P(U)\text{II, III, IV} = \sum_{j=1}^{m} p(u_j)\,\mu_{uj}(u_j), \text{ gdy } U = \{u_j\} = \{u_1, u_2, \ldots, u_m\} \tag{4}$$

The probabilities $p(u_i)$ of the occurrence of elementary damage u_i in sets II, III, IV were calculated and then presented in Table 1.

Table 1. Probability $p(u_i)$ of elementary damage u_i for 10 selected building elements.

No. of Damage	Name of Damage	Probability of Damage (u) that Corresponds to the II, III, and IV Maintenance Conditions of an Element											
		Z2–Foundations			Z3–Basement Walls			Z4–Solid Floors Above Basements			Z7–Structural Walls		
		$p(u)II$	$p(u)III$	$p(u)IV$	$p(u)II$	$p(u)III$	$p(u)IV$	$p(u)II$	$p(u)III$	$p(u)IV$	$p(u)II$	$p(u)III$	$p(u)IV$
u1	Mechanical damage												
u2	Leaks												
u3	Brick losses	0.71	0.77	0.91	0.50	0.63	0.79	0.61	0.76	0.65	0.81	0.98	1.00
u4	Mortar losses				0.50	0.63	0.88				0.63	0.95	1.00
u5	Brick decay	0.43	0.81	0.64	0.42	0.65	0.71	0.83	0.71	0.70	0.44	0.89	0.80
u6	Mortar decay				0.25	0.53	0.58				0.50	0.79	0.76
u7	Peeling off of paint coatings												
u8	Falling off of paint coatings												
u9	Cracks in bricks	0.57	0.83	0.55	0.33	0.60	0.46	0.65	0.58	0.43	0.06	0.30	0.28
u10	Cracks on plaster				0.25	0.44	0.29				0.75	0.64	0.76
u11	Scratching on walls										0.00	0.10	0.20
u12	Scratching on plaster										0.75	0.39	0.80
u13	Loosening of plaster												
u14	Falling off of plaster sheets												
u15	Dampness	0.00	0.66	1.00	0.08	0.33	1.00	0.09	0.02	0.65	0.13	0.67	0.96
u16	Weeping	0.00	0.30	0.91	0.00	0.18	0.46	0.00	0.02	0.57	0.00	0.15	0.36
u17	Biological corrosion of bricks	0.00	0.00	0.73	0.00	0.16	0.25				0.00	0.31	0.84
u18	Fungus												
u19	Mold and rot	0.00	0.00	0.18	0.00	0.00	0.13				0.00	0.02	0.16
u20	Corrosion raid of steel beams							0.13	0.60	0.70			
u21	Surface corrosion of steel beams							0.52	0.71	0.78			
u22	Deep corrosion of steel beams							0.00	0.22	0.43			
u23	Flooding with water							0.00	0.00	0.09			
u24	Dynamic sensitivity of floor beams												
u25	Deformations of wooden beams												
u26	Skewing of window joinery												
u27	Warping of window joinery												
u28	Delamination of wooden elements												
u29	Partial insect infestation of wooden elements												
u30	Complete insect infestation of wooden elements												

Table 1. *Cont.*

No. of Damage	Name of Damage	Probability of Damage (u) that Corresponds to the II, III, and IV Maintenance Conditions of an Element											
		Z8–Inter-Story Wooden Floors			Z9–Stairs			Z10–Roof Construction			Z13–Window Joinery		
		p(u)II	p(u)III	p(u)IV	p(u)II	p(u)III	p(u)IV	p(u)II	p(u)III	p(u)IV	p(u)II	p(u)III	p(u)IV
u1	Mechanical damage				0.82	0.88	0.84				0.88	0.85	1.00
u2	Leaks										0.88	0.92	1.00
u3	Brick losses				0.65	0.86	0.75						
u4	Mortar losses												
u5	Brick decay												
u6	Mortar decay												
u7	Peeling off of paint coatings												
u8	Falling off of paint coatings												
u9	Cracks in bricks												
u10	Cracks on plaster												
u11	Scratching on walls												
u12	Scratching on plaster	0.74	0.92	0.75									
u13	Loosening of plaster	0.44	0.57	0.63									
u14	Falling off of plaster sheets												
u15	Dampness	0.78	0.95	0.75				0.08	0.60	0.79	0.00	0.35	0.96
u16	Weeping	0.39	0.76	0.63	0.00	0.57	0.88	0.17	0.29	0.63	0.00	0.12	0.74
u17	Biological corrosion of bricks												
u18	Fungus	0.00	0.08	0.63									
u19	Mold and rot										0.00	0.04	0.22
u20	Corrosion raid of steel beams				0.06	0.82	0.91						
u21	Surface corrosion of steel beams				0.06	0.45	0.88						
u22	Deep corrosion of steel beams				0.00	0.06	0.28						
u23	Flooding with water												
u24	Dynamic sensitivity of floor beams	0.59	0.70	0.50									
u25	Deformations of wooden beams	0.30	0.43	0.38									
u26	Skewing of window joinery										0.41	0.81	0.93
u27	Warping of window joinery										0.35	0.69	0.56
u28	Delamination of wooden elements							0.67	0.33	0.63			
u29	Partial insect infestation of wooden elements				0.00	0.02	0.09	0.08	0.10	0.19	0.00	0.02	0.19
u30	Complete insect infestation of wooden elements	0.07	0.43	0.63				0,00	0.33	0.77	0.00	0.00	0.00

Table 1. *Cont.*

No. of Damage	Name of Damage	Probability of Damage (u) that Corresponds to the II, III, and IV Maintenance Conditions of an Element					
		Z15–Inner Plasters			Z20–Facades		
		p(u)II	p(u)III	p(u)IV	p(u)II	p(u)III	p(u)IV
u1	Mechanical damage	0.80	0.69	0.68	0.65	0.73	0.78
u2	Leaks						
u3	Brick losses						
u4	Mortar losses						
u5	Brick decay						
u6	Mortar decay	0.40	0.81	0.95	0.59	0.86	0.96
u7	Peeling off of paint coatings	0.90	0.76	0.86	0.59	0.91	0.96
u8	Falling off of paint coatings	0.20	0.31	0.23	0.00	0.23	0.30
u9	Cracks in bricks						
u10	Cracks on plaster	0.50	0.78	0.91	0.41	0.86	1.00
u11	Scratching on walls						
u12	Scratching on plaster	0.30	0.78	0.86	0.35	0.59	1.00
u13	Loosening of plaster	0.00	0.29	0.95	0.00	0.14	0.89
u14	Falling off of plaster sheets	0.00	0.00	0.36	0.00	0.05	0.22
u15	Dampness	0.00	0.00	0.23	0.00	0.00	0.93
u16	Weeping	0.00	0.00	0.14	0.00	0.00	0.41
u17	Biological corrosion of bricks						
u18	Fungus	0.00	0.00	0.00	0.00	0.00	0.04
u19	Mold and rot	0.00	0.00	0.00	0.00	0.00	0.00
u20	Corrosion raid of steel beams						
u21	Surface corrosion of steel beams						
u22	Deep corrosion of steel beams						
u23	Flooding with water						
u24	Dynamic sensitivity of floor beams						
u25	Deformations of wooden beams						
u26	Skewing of window joinery						
u27	Warping of window joinery						
u28	Delamination of wooden elements						
u29	Partial insect infestation of wooden elements						
u30	Complete insect infestation of wooden elements						

It should be noted that a slightly simplified approach, in which a fuzzy number is assigned to the probability of fuzzy events, was used here. This is opposed to the *Yager* approach [5,6], in which the probabilities are fuzzy events. It is important that the study did not consider the differences between the concepts of fuzziness and randomness. It was only assumed, although these phenomena are different and described differently, that they may nevertheless occur together as two types of uncertainty.

2.2. Model of Determining the Conditional Probabilities of the Process of Technical Wear in Relation to the Occurrence of Damage

The preliminary assumption: the process of technical wear of building elements occurs when there is identifiable damage: {ZII, ZIII, ZIV} = Z $\Leftrightarrow$ U.

The technical wear of building elements, determined by a group of experts in the II, III, and IV state of their technical maintenance, takes the following argument values:

- ZII = {20, 25, 30}% = {z1, z2, z3}, moreover, z4 = 0;
- ZIII = {35, 40, 45, 50}% = {z1, z2, z3, z4};
- ZIV = {55, 60, 65, 70}% = {z1, z2, z3, z4}.

Technical inspections of the residential buildings were executed by a team of experts consisted of:

- 1 architect;
- 1 structural engineer;
- 1 mechanical/sanitary engineer;
- 1 electrical engineer;
- 2 quantity surveyors;
- 1 technician/administrator.

In order to simplify the calculations, it was assumed that the domain of sets defined as fuzzy (ZII, ZIII, and ZIV) is interval [0.2, 0.7], and each of the sets contains a sum of N arguments of z1, z2, z3, z4. Each of these arguments occurs n times in the set. Without complicating the method with operations performed on fuzzy sets, it can be assumed that the degree to which arguments z1, z2, z3, z4 belong to fuzzy sets ZII, ZIII, ZIV is equal to the frequency of their occurrence in the sets:

$$\mu_{zi} = n_i/N_k, \text{ where } i = 1, 2, 3, 4, \text{ and } k = 1,2,3 \Leftrightarrow \text{II, III, IV} \tag{5}$$

Each of the fuzzy sets ZII, ZIII, ZIV can be written with the use of the membership function as follows:

$$\text{ZII} = (\mu_{z1}/z1 + \mu_{z2}/z2 + \mu_{z3}/z3)\text{II} \tag{6}$$

$$\text{ZIII} = (\mu_{z1}/z1 + \mu_{z2}/z2 + \mu_{z3}/z3 + \mu_{z4}/z4)\text{III} \tag{7}$$

$$\text{ZIV} = (\mu_{z1}/z1 + \mu_{z2}/z2 + \mu_{z3}/z3 + \mu_{z4}/z4)\text{IV} \tag{8}$$

and when supplementing the output data with the values of their intersections:

$$\text{ZII} \bullet \text{ZIII} = (\mu_{z1}\mu_{z1}/z1 + \mu_{z2}\mu_{z2}/z2 + \mu_{z3}\mu_{z3}/z3)\text{II, III} \tag{9}$$

$$\text{ZII} \bullet \text{ZIV} = (\mu_{z1}\mu_{z1}/z1 + \mu_{z2}\mu_{z2}/z2 + \mu_{z3}\mu_{z3}/z3)\text{II, IV} \tag{10}$$

$$\text{ZIII} \bullet \text{ZIV} = (\mu_{z1}\mu_{z1}/z1 + \mu_{z2}\mu_{z2}/z2 + \mu_{z3}\mu_{z3}/z3 + \mu_{z4}\mu_{z4}/z4)\text{III, IV} \tag{11}$$

$$\text{ZII} \bullet \text{ZIII} \bullet \text{ZIV} = (\mu_{z1}\mu_{z1}\mu_{z1}/z1 + \mu_{z2}\mu_{z2}\mu_{z2}/z2 + \mu_{z3}\mu_{z3}\mu_{z3}/z3)\text{II, III, IV} \tag{12}$$

The probabilities of the occurrence of individual arguments in sets ZII, ZIII, ZIV are as follows:

$$p(z1)\text{II} = 1/3; p(z2)\text{II} = 1/3; p(z3)\text{II} = 1/3; p(z4)\text{II} = 0 \tag{13}$$

$$p(z1)\text{III} = 1/4; p(z2)\text{III} = 1/4; p(z3)\text{III} = 1/4; p(z4)\text{III} = 1/4 \tag{14}$$

$$p(z1)\text{IV} = 1/4; p(z2)\text{IV} = 1/4; p(z3)\text{IV} = 1/4; p(z4)\text{IV} = 1/4 \tag{15}$$

When using dependence (3), the degrees of membership of arguments z1, z2, z3, and z4 (6–12), and the probabilities of the occurrence of particular arguments in sets ZII, ZIII, ZIV (13–15), the partial probabilities of the occurrence of technical wear processes were calculated as fuzzy events in the satisfactory, average, and poor technical maintenance conditions of the analyzed residential buildings:

$$P(ZII) \ = \ (\sum_{i=1}^{3} p(zi) \ \mu_{zi}(zi)) II \tag{16}$$

$$P(ZIII) \ = \ (\sum_{i=1}^{4} p(zi) \ \mu_{zi}(zi)) III \tag{17}$$

$$P(ZIV) \ = \ (\sum_{i=1}^{4} p(zi) \ \mu_{zi}(zi)) IV \tag{18}$$

and their products:

$$P(ZII \bullet ZIII) \ = \ \sum_{i=1}^{3} [(p(zi) \ \mu_{zi}(zi)) II \bullet (p(zi) \ \mu_{zi}(zi)) III] \tag{19}$$

$$P(ZII \bullet ZIV) \ = \ \sum_{i=1}^{3} [(p(zi) \ \mu_{zi}(z_i)) II \bullet (p(zi) \ \mu_{zi}(zi)) IV] \tag{20}$$

$$P(ZIII \bullet ZIV) \ = \ \sum_{i=1}^{4} [(p(zi) \ \mu_{zi}(zi)) III \bullet (p(zi) \ \mu_{zi}(zi)) IV] \tag{21}$$

$$P(ZII \bullet ZIII \bullet ZIV) \ = \ \sum_{i=1}^{3} [(p(zi) \ \mu_{zi}(zi)) II \bullet (p(zi) \ \mu_{zi}(zi)) III \bullet (p(zi) \ \mu_{zi}(zi)) IV] \tag{22}$$

abilities of the occurrence of a set of damage to residential building elements in relation to the processes of their wear. It was assumed that the conditional probabilities, defined in such a way, correspond to the frequency of the occurrence of all elementary damage related to a single element in the following building maintenance conditions:

- satisfactory—P(U/ZII);
- average—P(U/ZIII);
- poor—P(U/ZIV);
- satisfactory and average—P(U/ZII•ZIII);
- satisfactory and poor—P(U/ZII•ZIV);
- average and poor—P(U/ZIII•ZIV);
- satisfactory, average and poor—P(U/ZII•ZIII•ZIV).

The above calculations of conditional and partial probabilities (16—22) allowed the probability of the occurrence of a group of damage to be determined in the middle, non-acute technical maintenance states of the analyzed residential buildings:

$$
\begin{aligned}
P(U) = &\ P(U/ZII) \bullet P(ZII) + P(U/ZIII) \bullet P(ZIII) + P(U/ZIV) \bullet P(ZIV) - P(U/ZII \bullet ZIII) \bullet P(ZII \bullet ZIII) \\
&- P(U/ZII \bullet ZIV) \bullet P(ZII \bullet ZIV) - P(U/ZIII \bullet ZIV) \bullet P(ZIII \bullet ZIV) \\
&+ P(U/ZII \bullet ZIII \bullet ZIV) \bullet P(ZII \bullet ZIII \bullet ZIV)
\end{aligned} \tag{23}
$$

In the last stage of the developed model, the Bayes formula [37–39] for a posteriori probabilities was used. It determines the conditional probabilities of fuzzy events (i.e., the processes of the technical wear of building elements) in relation to another fuzzy event, i.e., the occurrence of their damage. The Bayes formula under satisfactory, average, and poor fuzziness conditions is as follows:

$$P(ZII/U) = \frac{P(U/ZII) \bullet P(ZII)}{P(U)} \tag{24}$$

$$P(ZIII/U) = \frac{P(U/ZIII) \bullet P(ZIII)}{P(U)} \tag{25}$$

$$P(ZIV/U) = \frac{P(U/ZIV) \bullet P(ZIV)}{P(U)} \tag{26}$$

The defined conditional probabilities of fuzzy event Z = {z1, z2, z3, z4} were supplemented, using the relationships (3) and (13)–(15), with the calculations of its mean value in relation to the probabilistic measure P(Z) in classes II, III, and IV of the technical maintenance of building elements:

$$m_p(Z)II, III, IV = 1/P(Z)II, III, IV \bullet \sum_{i=1}^{4} p(zi) \, \mu_{zi}(zi)zi \tag{27}$$

The values of the conditional probabilities of the technical wear processes Z, which correspond to the II, III, and IV maintenance conditions of 10 selected elements of the analyzed buildings, in relation to the occurrence of their damage U, and with their mean values in relation to the probabilistic measure P(Z), are given in Table 2.

Table 2. Fuzzy conditional probabilities of the technical wear process in relation to the probabilistic measure and the occurrence of damage, as well as the fuzzy conditional probabilities of the occurrence of a group of damage in relation to the technical wear process.

| Group Number | Building Element | No. of Damage | Name of Damage | Conditional Probabilities of Fuzzy Events According to the Bayes Formula for: | | | | | |
| | | | | the Process of Technical wear (Z) that Corresponds to the II, III and IV Conditions of the Maintenance of an Element in Relation to its Damage (U) / Average Value of the Fuzzy Event (Z) in Relation to the Probabilistic Measure P(Z) | | | the Occurrence of the Group of Damage (U) that Corresponds to the II, III and IV Conditions of Maintenance of an Element in Relation to the Wear Process (Z) | | |
				P(ZII/U)/mp(ZII)	P(ZIII/U)/mp(ZIII)	P(ZIV/U)/mp(ZIV)	P(UII/Z)	P(UIII/Z)	P(UIV/Z)
Z2	Foundations	u3 u5 u9 u15 u16 u17 u19	Brick losses Brick decay Cracks in bricks Dampness Weeping Biological corrosion of bricks Mold and rot	0.2655/0.300	0.3957/0.441	0.5755/0.600	0.0000	0.2702	0.9586
Z3	Basement walls	u3 u4 u5 u6 u9 u10 u15 u16 u17 u19	Brick losses Mortar losses Brick decay Mortar decay Cracks in bricks Cracks on plaster Dampness Weeping Biological corrosion of bricks Mold and rot	0.2879/0.300	0.3915/0.429	0.5187/0.607	0.0000	0.5543	0.8494
Z4	Solid floors above basements	u3 u5 u9 u15 u16 u20 u21 u22 u23	Brick losses Brick decay Cracks in bricks Dampness Weeping Corrosion raid of steel beams Surface corrosion of steel beams Deep corrosion of steel beams Flooding with water	0.3846/0.270	0.3778/0.406	0.5173/0.613	0.2371	0.4577	0.6330

Table 2. *Cont.*

Group Number	Building Element	No. of Damage	Name of Damage	Conditional Probabilities of Fuzzy Events According to the Bayes Formula for:					
				the Process of Technical wear (Z) that Corresponds to the II, III and IV Conditions of the Maintenance of an Element in Relation to its Damage (U) / Average Value of the Fuzzy Event (Z) in Relation to the Probabilistic Measure P(Z)			the Occurrence of the Group of Damage (U) that Corresponds to the II, III and IV Conditions of Maintenance of an Element in Relation to the Wear Process (Z)		
				P(ZII/U)/mp(ZII)	P(ZIII/U)/mp(ZIII)	P(ZIV/U)/mp(ZIV)	P(UII/Z)	P(UIII/Z)	P(UIV/Z)
Z7	Structural walls	u3	Brick losses	0.3464/0.300	0.3999/0.431	0.5125/0.626	0.0000	0.4487	1.0000
		u4	Mortar losses						
		u5	Brick decay						
		u6	Mortar decay						
		u9	Cracks in bricks						
		u10	Cracks on plaster						
		u11	Scratching on walls						
		u12	Scratching on plaster						
		u15	Dampness						
		u16	Weeping						
		u17	Biological corrosion of bricks						
		u19	Mold and rot						
Z8	Inter-story wooden floors	u12	Scratching on plaster	0.3508/0.283	0.4401/0.435	0.6443/0.636	0.0000	0.3048	0.9615
		u13	Loosening of plaster						
		u15	Dampness						
		u16	Weeping						
		u18	Fungus						
		u24	Dynamic sensitivity of floor beams						
		u25	Deformations of wooden beams						
		u30	Complete insect infestation of wooden elements						
Z9	Stairs	u1	Mechanical damage	0.2622/0.268	0.4613/0.438	0.5842/0.600	0.2498	0.5828	0.6852
		u3	Brick losses						
		u16	Weeping						
		u20	Corrosion raid of steel beams						
		u21	Surface corrosion of steel beams						
		u22	Deep corrosion of steel beams						
		u29	Partial insect infestation of wooden elements						

Table 2. *Cont.*

Group Number	Building Element	No. of Damage	Name of Damage	Conditional Probabilities of Fuzzy Events According to the Bayes Formula for:					
				the Process of Technical wear (Z) that Corresponds to the II, III and IV Conditions of the Maintenance of an Element in Relation to its Damage (U) / Average Value of the Fuzzy Event (Z) in Relation to the Probabilistic Measure P(Z)			the Occurrence of the Group of Damage (U) that Corresponds to the II, III and IV Conditions of Maintenance of an Element in Relation to the Wear Process (Z)		
				P(ZII/U)/mp(ZII)	P(ZIII/U)/mp(ZIII)	P(ZIV/U)/mp(ZIV)	P(UII/Z)	P(UIII/Z)	P(UIV/Z)
Z10	Roof con-struction	u15 u16 u28 u29 u30	Dampness Weeping Delamination of wooden elements Partial insect infestation of wooden elements Complete insect infestation of wooden elements	0.2817/0.288	0.3510/0.437	0.6466/0.630	0.0513	0.6579	0.7999
Z13	Window joinery	u1 u2 u15 u16 u19 u26 u27 u29 u30	Mechanical damage Leaks Dampness Weeping Mold and rot Skewing of window joinery Warping of window joinery Partial insect infestation of wooden elements Complete insect infestation of wooden	0.3275/0.271	0.3717/0.423	0.5538/0.648	0.2130	0.4992	0.7592
Z15	Inner plasters	u1 u6 u7 u8 u10 u12 u13 u14 u15 u16 u18 u19	Mechanical damage Mortar decay Peeling off of paint coatings Falling off of paint coatings Cracks on plaster Scratching on plaster Loosening of plaster Falling off of plaster sheets Dampness Weeping Fungus Mold and rot	0.3303/0.290	0.3579/0.431	0.4995/0.648	0.1444	0.4578	0.7792

Table 2. *Cont.*

Group Number	Building Element	No. of Damage	Name of Damage	Conditional Probabilities of Fuzzy Events According to the Bayes Formula for:					
				the Process of Technical wear (Z) that Corresponds to the II, III and IV Conditions of the Maintenance of an Element in Relation to its Damage (U) / Average Value of the Fuzzy Event (Z) in Relation to the Probabilistic Measure P(Z)			the Occurrence of the Group of Damage (U) that Corresponds to the II, III and IV Conditions of Maintenance of an Element in Relation to the Wear Process (Z)		
				P(ZII/U)/mp(ZII)	P(ZIII/U)/mp(ZIII)	P(ZIV/U)/mp(ZIV)	P(UII/Z)	P(UIII/Z)	P(UIV/Z)
Z20	Facades	u1 u6 u7 u8 u10 u12 u13 u14 u15 u16 u18 u19	Mechanical damage Mortar decay Peeling off of paint coatings Falling off of paint coatings Cracks on plaster Scratching on plaster Loosening of plaster Falling off of plaster sheets Dampness Weeping Fungus Mold and rot	0.2854/0.253	0.3670/0.450	0.6299/0.641	0.1250	0.3623	0.9012

2.3. The Model for Determining the Conditional Probabilities of a Set of Damage in Relation to the Process of Their Technical Wear

The preliminary assumption: damage to building elements occurs when there is a process of their technical wear, which can be estimated within the range of 0–100%: {UII, UIII, UIV} = U $\Leftrightarrow$ Z.

Damage to building elements, which is identified by experts in classes II, III, and IV of their technical maintenance, is defined as being dichotomous variables that assume values "0" (damage does not occur) or "1" (damage occurs). The domain of the set of damage, defined as fuzzy UII, UIII, UIV, is binary {0}, {1}.

It was assumed that the measure of the degree of membership of a single damage μ_{uj} to the set of a group of damage U, which is the symptom of the ongoing wear processes Z, is the feature that most fully expresses the correlation between these variables. This can be a point two-series correlation coefficient r(Z) $\Leftrightarrow$ r(U), which is determined in each of the states II, III, and IV of the technical maintenance.

Each of the fuzzy sets UII, UIII, UIV can therefore be written using the membership function as follows:

$$\text{UII, III, IV} = (\sum_{j=1}^{m} r(u_j)/u_j)\text{II, III, IV, gdzie j} \rightarrow m \in [5, 12] \tag{28}$$

and when supplementing the output data with the values of their products:

$$\text{UII}\bullet\text{UIII} = (\sum_{j=1}^{m} r(u_j)\bullet r(u_j)/u_j)\text{II, III} \tag{29}$$

$$\text{UII}\bullet\text{UIV} = (\sum_{j=1}^{m} r(u_j)\bullet r(u_j)/u_j)\text{II, IV} \tag{30}$$

$$\text{UIII}\bullet\text{UIV} = (\sum_{j=1}^{m} r(u_j)\bullet r(u_j)/u_j)\text{III, IV} \tag{31}$$

$$\text{UII}\bullet\text{UIII}\bullet\text{UIV} = (\sum_{j=1}^{m} r(u_j)\bullet r(u_j)\bullet r(u_j)/u_j)\text{II, III, IV} \tag{32}$$

When using relationship (4), the degrees of memberships of individual damage $\mu_{uj} = r(u_j)$ to sets of groups of damage U (28)–(32), and by having data concerning the probabilities of individual damage in sets ZII, ZIII, ZIV (13)–(15), the partial probabilities of the damage were calculated as fuzzy events in the satisfactory, average, and poor technical maintenance conditions of the analyzed residential buildings:

$$P(\text{UII}) = (\sum_{j=1}^{m} p(u_j)r(u_j))\text{II} \tag{33}$$

$$P(\text{UIII}) = (\sum_{j=1}^{m} p(u_j)r(u_j))\text{III} \tag{34}$$

$$P(\text{UIV}) = (\sum_{j=1}^{m} p(u_j)r(u_j))\text{IV} \tag{35}$$

and their products:

$$P(\text{UII}\bullet\text{UIII}) = \sum_{j=1}^{m} [p(u_j)r(u_j))\text{II}\bullet p(u_j)r(u_j))\text{III}] \tag{36}$$

$$P(UII \bullet UIV) = \sum_{j=1}^{m} [p(u_j)r(u_j))II \bullet p(u_j)r(u_j))IV] \tag{37}$$

$$P(UIII \bullet UIV) = \sum_{j=1}^{m} [p(u_j)r(u_j))III \bullet p(u_j)r(u_j))IV \tag{38}$$

$$P(ZII \bullet ZIII \bullet ZIV) = \sum_{j=1}^{m} [p(u_j)r(u_j))II \bullet p(u_j)r(u_j))III \bullet p(u_j)r(u_j))IV] \tag{39}$$

The next stage of the created model involved the calculation of the conditional probabilities of the wear processes of the residential buildings' elements in relation to the occurrence of their damage. Due to the assumption that damage is an expression of technical wear, it was assumed, as in the case of defining the wear processes, that the conditional probabilities of the technical wear correspond to the frequency of the occurrence of all the elementary damage ($\{u_j\} = 1$) of a selected building element in the II, III, IV conditions of its maintenance: $P(Z/UII)$, $P(Z/UIII)$, $P(Z/UIV)$, $P(Z/UII \bullet UIII)$, $P(Z/UII \bullet UIV)$, $P(Z/UIII \bullet UIV)$, $P(Z/UII \bullet UIII \bullet UIV)$.

The above calculations of the conditional and partial probabilities (33–39) enabled the probability of the occurrence of technical wear processes to be determined in the middle, non-acute technical maintenance states of the analyzed residential buildings:

$$\begin{aligned} P(Z) = &\ P(Z/UII) \bullet P(UII) + P(Z/UIII) \bullet P(UIII) + P(Z/UIV) \bullet P(UIV) \\ &- P(Z/UII \bullet UIII) \bullet P(UII \bullet UIII) - P(Z/UII \bullet UIV) \bullet P(UII \bullet UIV) - P(Z/UIII \bullet UIV) \bullet P(UIII \bullet UIV) \\ &+ P(Z/UII \bullet UIII \bullet UIV) \bullet P(UII \bullet UIII \bullet UIV) \end{aligned} \tag{40}$$

In the last stage of the developed model, the Bayes formula for a posteriori probabilities was used again, which determines the conditional probabilities of fuzzy events (i.e., the occurrence of damage to building elements) in relation to another fuzzy event (i.e., the processes of their technical wear) [38]. The Bayes formula under satisfactory, moderate, and poor fuzziness conditions is as follows:

$$P(UII/Z) = \frac{P(Z/UII) \bullet P(UII)}{P(Z)} \tag{41}$$

$$P(UIII/Z) = \frac{P(Z/UIII) \bullet P(UIII)}{P(Z)} \tag{42}$$

$$P(UIV/Z) = \frac{P(Z/UIV) \bullet P(UIV)}{P(Z)} \tag{43}$$

In this case, the mean value $m_p(U)$ of fuzzy event $U = \{u\}$ in relation to the probabilistic measure $P(U)$ is a constant value equal to one, because only the cases in which the dichotomous variable occurred were taken into account.

The values of the conditional probabilities of the occurrence of a group of damage, which correspond to the II, III, and IV maintenance conditions of 10 selected elements of the analyzed buildings, in relation to the processes of their technical wear, are presented in Table 2.

3. Results

The results of research concerning the impact of damage to building elements on their technical wear in the Bayes conditional probability domain (for damage and technical wear as fuzzy events) led to the following conclusions (within two aspects A and B—Table 2):

A.   the probability of the conditional process of the technical wear, which corresponds to the three middle states of maintenance of the building elements, with regards to damage—P (Z/U) II, III, IV—is as follows:

- the conditional probability of the technical wear of an element in relation to its damage increases with the deterioration of the maintenance conditions of the building (this is an exceptionally steady increase, even in the case of different building elements);
- the probability of such a conditionally defined fuzzy event indicates the state of the technical wear for which the fuzzy damage occurs with the highest intensity, and it amounts, for the following elements of the tested residential buildings in their average maintenance condition P(ZIII/U), to:
  - ○ for foundations: dampness of foundations 0.40
  - ○ for basement walls: crack in bricks 0.39
  - ○ for solid floors above basements: dampness of floors 0.38
  - ○ for structural walls: cracks of plaster 0.40
  - ○ for wooden inter-storey floors: weeping on floors 0.44
  - ○ for internal stairs: weeping on stairs 0.46
  - ○ for roof constructions: delamination of beams 0.35
  - ○ for window joinery: mold and rot on windows 0.37
  - ○ for inner plasters: scratches on plaster 0.36
  - ○ for facades: scratches on plaster 0.37

The above values are therefore a fuzzy value of the probability of the degree of the technical wear, which was determined as an average degree, i.e., within the range of 35–50%—in the case of the occurrence of a fuzzy damage to the building element;

- this probability is characterized by a low standard deviation and a narrow range of the results of various elements within each of the considered building maintenance conditions—satisfactory (0.2622–0.3846), average (0.3510–0.4613) and poor (0.4995–0.6466). A similar remark concerns the mean value of this probability in relation to its probabilistic measure;

B. the conditional probability of a group of damage, which corresponds to the three middle states of maintenance of building elements, in relation to the process of their technical wear—P(U/Z) II, III, IV—is as follows:

- the conditional probability of damage to the element in relation to its technical wear increases with the deterioration of the building maintenance conditions;
- the probability of such a conditionally defined fuzzy event is indicated by the damage that most intensely affects the technical wear of the following elements of the tested residential buildings, and it amounts in their average maintenance condition P(UIII/Z) to:
  - ○ for foundations: dampness of foundations 0.27
  - ○ for basement walls: crack in bricks 0.55
  - ○ for solid floors above basements: dampness of floors 0.46
  - ○ for structural walls: cracks of plaster 0.45
  - ○ for wooden inter-storey floors: weeping on floors 0.31
  - ○ for internal stairs: weeping on stairs 0.58
  - ○ for roof constructions: delamination of beams 0.66
  - ○ for window joinery: mold and rot on windows 0.50
  - ○ for inner plasters: scratches on plaster 0.46
  - ○ for facades: scratches on plaster 0.36

The above values are therefore a fuzzy value of the probability of damage to a building element, but only in the case that its fuzzy technical wear is determined to be an average degree, i.e., within the range of 35–50%;

- the irregularity of this increase and the too-high coefficients of variation indicate only a partial identity of the fuzzy event defined within aspect B with the reverse event; the fuzzy event determined within aspect A is characterized by a much greater consistency of the obtained results.

4. Discussion and Conclusions

Quantitative damage analysis, which was carried out using empirical methods of assessing the technical condition of a building, indicates the type and size of damage to the building's elements, which are characteristic of the appropriate maintenance conditions. Research concerning the cause–effect relationships ("damage-technical wear") in fuzzy calculus allowed for a numerical approach to the impact of building maintenance conditions on the degree of technical wear of its elements. The analysis of fuzzy cause–effect relationships ("damage-technical wear") created the possibility of determining conditional probabilities of these dependencies that are treated as fuzzy events. The fuzzy conditional probabilities of the technical wear process in relation to the occurrence of damage (with a probabilistic measure), as well as conditional probabilities of the occurrence of a group of damage in relation to the process of technical wear, were determined.

The research methodology has been prepared in such a way that allowed the previously prepared qualitative model to be transformed into a quantitative model. Therefore, the diagnosis of the impact of the maintenance of the residential buildings on the amount of their technical wear was executed using quantitative methods in fuzzy set categories, and also by using the authors' own model that was created in the conditions of fuzziness. The model allowed for the determination of the conditional probabilities of the process of technical wear, and also the set of damage according to both Bayes formulas applied to fuzzy sets operations.

The research procedure was developed in a way that allowed for the transition of a previously prepared qualitative model into a quantitative model. The diagnosis of the impact of the maintenance of residential buildings on the amount of their technical wear was carried out using quantitative methods in the categories of fuzzy sets, and also by using the authors' own model of determining the mutually dependent probabilities created in the conditions of fuzziness. The model enabled the conditional probabilities of the process of the technical wear, as well as the set of damage, to be determined according to probabilistic Bayes formulas. Moreover, it also allowed the fuzzy approach of Zadeh to be combined with the Yager approach. In such a multi-criteria fuzzy technical assessment of residential buildings, a simplified approach was used. In this approach, the probability of fuzzy events was assigned to a fuzzy measure, as opposed to the Yager approach, in which the probabilities are fuzzy events. The differences between the concepts of fuzziness and randomness were not considered in the study. It was assumed that these phenomena are different and described differently, however, they may—as two types of uncertainty—occur together.

The methods and results of the research presented in the article indicated a way that allows for the transition of the previously prepared qualitative model into a quantitative model. The diagnosis of the impact of the maintenance of residential buildings on the amount of their technical wear was carried out using quantitative methods in the categories of fuzzy sets, and also using the authors' own models created in fuzzy conditions. The key question from the subjective expert assessment of the technical condition of the evaluated residential buildings was answered: what is the probability of the wear of an element that may be more or less represented by its average maintenance conditions? Therefore, the probability that the element is more or less worn was determined. It was proven that the conditional probability of the technical wear of an element in relation to its failure increases with the deterioration of the maintenance conditions of the building, and this increase is extremely regular, even in the case of different building elements. This probability is characterized by a low standard deviation and a narrow range of the dispersion of the results in the case of various elements within each of the considered building maintenance conditions.

Author Contributions: Conceptualization, J.K., T.S.; methodology, J.K.; software, T.S.; validation, J.K.; formal analysis, J.K., T.S.; investigation, J.K., T.S.; resources, J.K., T.S.; writing—original draft preparation, J.K.; writing—review and editing, J.K., T.S.; supervision, J.K. All authors have read and agreed to the published version of the manuscript.

Funding: This research received no external funding.

Institutional Review Board Statement: Not applicable.

Informed Consent Statement: Not applicable.

Data Availability Statement: No new data were created or analyzed in this study. Data sharing is not applicable to this article.

Conflicts of Interest: The authors declare no conflict of interest.

References

1. Kacprzyk, J. *Fuzzy Sets in the System Analysis*; PWN: Warsaw, Poland, 1986.
2. Konior, J. Decision assumptions on building maintenance management. Probabilistic methods. *Arch. Civ. Eng.* **2007**, *53*, 403–423.
3. Zadeh, L.A. Fuzzy sets. *Inf. Control.* **1965**, *8*, 338–353. [CrossRef]
4. Zadeh, L.; Aliev, R. *Fuzzy Logic Theory and Applications*; World Scientific Publishing Co Pte Ltd.: Singapore, 2018.
5. Yager, R.R. A note on probabilities of fuzzy events. *Inf. Sci.* **1979**, *18*, 113–129. [CrossRef]
6. Yager, R.R. On the fuzzy cardinality of a fuzzy set. *Int. J. Gen. Syst.* **2006**, *35*, 191–206. [CrossRef]
7. Nowogońska, B. Diagnoses in the aging process of residential buildings constructed using traditional technology. *Buildings* **2019**, *9*, 126. [CrossRef]
8. Nowogońska, B. Intensity of damage in the aging process of buildings. *Arch. Civ. Eng.* **2020**, *66*, 19–31. [CrossRef]
9. Nowogońska, B.; Korentz, J. Value of technical wear and costs of restoring performance characteristics to residential buildings. *Buildings* **2020**, *10*, 9. [CrossRef]
10. Nowogońska, B. The Method of Predicting the Extent of Changes in the Performance Characteristics of Residential Buildings. *Arch. Civ. Eng.* **2019**, *65*, 81–89. [CrossRef]
11. Nowogońska, B. Proposal for determing the scale of renovation needs of residential buildings. *Civ. Environ. Eng. Rep.* **2016**, *22*, 137–144. [CrossRef]
12. Lee, S.; Lee, S.; Kim, J. Evaluating the Impact of Defect Risks in Residential Buildings at the Occupancy Phase. *Sustainability* **2018**, *10*, 4466. [CrossRef]
13. Plebankiewicz, E.; Wieczorek, D.; Zima, K. Life cycle cost modelling of buildings with consideration of the risk. *Arch. Civ. Eng. LXII* **2016**, *62*, 149–166. [CrossRef]
14. Plebankiewicz, E.; Wieczorek, D.; Zima, K. Quantification of the risk addition in life cycle cost of a building object. *Tech. Trans.* **2017**, *5*, 35–45.
15. Plebankiewicz, E.; Wieczorek, D.; Zima, K. Model estimation of the whole life cost of a building with respect to risk factors. *Technol. Econ. Dev. Econ.* **2019**, *25*, 20–38.
16. Wieczorek, D. Fuzzy risk assessment in the life cycle of building object—Selection of the right defuzzification method. In *AIP Conference Proceedings*; AIP Publishing: Melville, NY, USA, 2018; Volume 1978, p. 240005.
17. Ibadov, N. Fuzzy Estimation of Activities Duration in Construction Projects. *Arch. Civ. Eng. LXI* **2015**, *61*, 23–34. [CrossRef]
18. Ibadov, N.; Kulejewski, J. Construction projects planning using network model with the fuzzy decision node. *Int. J. Environ. Sci. Technol.* **2019**, *16*, 4347–4354. [CrossRef]
19. Ibadov, N.; Kulejewski, J. The assessment of construction project risks with the use of fuzzy sets theory. *Tech. Trans.* **2014**, *1*, 175–182.
20. Ibadov, N. The alternative net model with the fuzzy decision node for the construction projects planning. *Arch. Civ. Eng.* **2018**, *64*, 3–20. [CrossRef]
21. Jain, K.K.; Bhattacharjee, B. Application of Fuzzy Concepts to the Visual Assessment of Deteriorating Reinforced Concrete Structures. *J. Constr. Eng. Manag.* **2012**, *138*, 399–408. [CrossRef]
22. Andrić, J.M.; Wang, J.; Zou, P.X.; Zhang, J.; Zhong, R. Fuzzy Logic–Based Method for Risk Assessment of Belt and Road Infrastructure Projects. *J. Constr. Eng. Manag.* **2019**, *145*, 238–262. [CrossRef]
23. Marzouk, M.; Amin, A. Predicting Construction materials prices using fuzzy logic and neural networks. *J. Constr. Eng. Manag.* **2013**, *139*, 1190–1198. [CrossRef]
24. Knight, K.; Robinson, R.; Fayek, A. Use of fuzzy logic of predicting design cost overruns on building projects. *J. Constr. Eng. Manag.* **2002**, *128*, 503–512. [CrossRef]
25. Sharma, S.; Goyal, P.K. Fuzzy assessment of the risk factors causing cost overrun in construction industry. *Evol. Intell.* **2019**, 1–13. [CrossRef]
26. Al-Humaidi, H.M.; Hadipriono, T.F. Fuzzy logic approach to model delays in construction projects using rotational fuzzy fault tree models. *Civ. Eng. Environ. Syst.* **2010**, *27*, 329–351. [CrossRef]
27. Ammar, M.; Zayed, T.; Moselhi, O. Fuzzy-based life-cycle cost model for decision making under subjectivity. *J. Constr. Eng. Manag.* **2012**, *139*, 556–563. [CrossRef]
28. Chan, K.Y.; Kwong, C.K.; Dillon, T.S.; Fung, K.Y. An intelligent fuzzy regression approach for affective product design that captures nonlinearity and fuzziness. *J. Eng. Des.* **2011**, *22*, 523–542. [CrossRef]

29. Nasirzadeh, F.; Afshar, A.; Khanzadi, M.; Howick, S. Integrating system dynamics and fuzzy logic modelling for construction risk management. *Constr. Manag. Econ.* **2008**, *26*, 1197–1212. [CrossRef]
30. Konior, J. Technical assessment of old buildings by fuzzy approach. *Arch. Civ. Eng.* **2019**, *65*, 129–142. [CrossRef]
31. Konior, J. Technical Assessment of old buildings by probabilistic approach. *Arch. Civ. Eng.* **2020**, *66*, 443–466. [CrossRef]
32. Konior, J. Maintenance of apartment buildings and their measurable deterioration. *Tech. Trans. Czas. Tech.* **2017**, *6*, 101–107. [CrossRef]
33. Konior, K. Bi-serial correlation of civil engineering building elements under constant technical deterioration. *J. Sci. Gen. Tadeusz Kosiuszko Mil. Acad. L. Forces* **2016**, *179*, 142–150.
34. Konior, J. Intensity of defects in residential buildings and their technical wear. *Tech. Trans. Civ. Eng.* **2014**, *111*, 137–146.
35. Konior, J.; Sawicki, M.; Szóstak, M. Intensity of the Formation of Defects in Residential Buildings with Regards to Changes in Their Reliability. *Appl. Sci.* **2020**, *10*, 6651. [CrossRef]
36. Konior, J.; Sawicki, M.; Szóstak, M. Influence of Age on the Technical Wear of Tenement Houses. *Appl. Sci.* **2020**, *10*, 6651. [CrossRef]
37. Konior, J.; Sawicki, M.; Szóstak, M. Damage and Technical Wear of Tenement Houses in Fuzzy Set Categories. *Appl. Sci.* **2021**, *11*, 1484. [CrossRef]
38. Multi-Author Work under Kapliński, O. *lead. Research Models and Methods in Construction Projects Engineering*; PAN KILiW: Warsaw, Poland, 2007.
39. Multi-Author Work under Czapliński, K. lead. Assessment of Wroclaw Downtown Apartment Houses' Technical Conditions. Reports series "U"; Building Engineering Institute at Wroclaw University of Science and Technology: Wroclaw, Poland, 1984-96.
40. Hellwig, Z. *Elements of Probability Calculus and Mathematical Statistics*; PWN: Warsaw, Poland, 2001. (In Polish)
41. Morrison, D. *Multivariate Statistical Methods*, 4th ed.; Duxbury Press: London, UK, 2004.
42. Witte, R.; Witte, J. *Statistics*, 11th ed.; Wiley: New York, NY, USA, 2017.

applied sciences

MDPI

Article

Damage and Technical Wear of Tenement Houses in Fuzzy Set Categories

Jarosław Konior *, Marek Sawicki and Mariusz Szóstak

Department of Building Engineering, Faculty of Civil Engineering, Wroclaw University of Science and Technology, 50-370 Wrocław, Poland; marek.sawicki@pwr.edu.pl (M.S.); mariusz.szostak@pwr.edu.pl (M.S.)
* Correspondence: jaroslaw.konior@pwr.edu.pl; Tel.: +48-71-320-23-69

Abstract: The results and conclusions of the research presented in the article concern the topic of the technical maintenance and wear of traditionally erected residential buildings. The cause and effect relations between the occurrence of damage to the elements of tenement houses, which are treated as an expression of their maintenance conditions, and the size of the technical wear of these elements were determined in a representative and purposefully selected sample of 102 apartment houses built in the second half of the 19th and early 20th centuries in the Wroclaw, Poland downtown district "Srodmiescie". Recognition of the impact of the maintenance of residential buildings on the level of their technical wear was carried out using quantitative methods from fuzzy set categories, and also with the use of the authors' own model. The created model, based on the Zadeh function, was created in fuzzy conditions for the purpose of assessing the degree of damage to selected building elements. The treatment of the problem with regard to fuzzy criteria allowed for the synthesis of elementary criteria, which give the greatest approximations at the technical research stage of a residential building, into a global assessment of the degree of the wear of its elements. Moreover, it also significantly reduced the subjective factor of this assessment, which had a significant impact on the results of the research obtained in the case of good, medium and poor conditions of tenement houses. It was proven that the conditions of maintenance and use of buildings determine the amount of technical wear of their elements. The state of exploitation of the examined tenement houses is reflected in the mechanical damage to the internal structure of the elements (determined in fuzzy categories). This damage has a significant frequency and cumulative effects, and is characteristic for buildings with satisfactory and average maintenance.

Keywords: tenement houses; technical wear; damage; maintenance; fuzzy sets

Citation: Konior, J.; Sawicki, M.; Szóstak, M. Damage and Technical Wear of Tenement Houses in Fuzzy Set Categories. *Appl. Sci.* **2021**, *11*, 1484. https://doi.org/10.3390/app11041484

Academic Editor: Asterios Bakolas
Received: 18 January 2021
Accepted: 4 February 2021
Published: 6 February 2021

Publisher's Note: MDPI stays neutral with regard to jurisdictional claims in published maps and institutional affiliations.

1. Introduction

1.1. Source Literature

The aim of the research was to identify the impact of the processes associated with the broadly understood maintenance of old residential buildings with a traditional construction on the size and intensity of the wear of their elements. The degree of technical wear of residential building elements is a parameter of fundamental importance in the comprehensive assessment of their technical condition, regardless of the approach that was used in the test method. The aim of the research was achieved through the analysis of the symptoms of the technical wear process—understanding the mechanism of the phenomenon of damage and identifying the size and intensity of damage to the elements of the evaluated buildings.

Essential research of tenement houses aims to undertake a qualitative analysis of detected defects and identify all particular defects of their elements. Therefore, the "reason–effect" model is applied as follows:

$$[\text{ REASONS}] \rightarrow [\text{observed} \xleftrightarrow{\text{SYMPTOMS}} \text{measured}] \rightarrow [\text{EFFECTS }]$$

107

Commonly used mathematical methods and the broadly understood system analysis deal with real tasks in which the basic goal is the possibility of including all the types of indeterminacy among modeled quantities and the relationships between them. Every indeterminacy has traditionally been equated with the uncertainty of a random type, which has enabled known probabilistic and statistical tools to be used. In practice, however, there are many cases in which the indeterminacy of the type of inaccuracy, ambiguity and imprecision of meanings can be found. However, these situations are not of a random nature, and therefore traditional probabilistic models may not be adequate [1–7]. When assessing the possibility of random and/or fuzzy events occurring in construction investment projects, apart from immeasurable (qualitative) criteria, measurable (quantitative) criteria are also used. These quantitative criteria are expressed in a mathematical model that describes multiple phenomena of construction engineering processes. Only some of these criteria are strictly defined concepts—boundary, extreme. Most of these criteria are approximate. Their value is determined using descriptive methods, e.g., "good quality", "short term", "low budget". Therefore, concepts of this type cannot be adequately represented as a conventional set. To overcome this difficulty, in 1965, *Lotfi A. Zadeh* of the University of California in Berkeley introduced the concept of a fuzzy set with its membership function [8–10].

Zadeh [8,9], when developing the foundations of fuzzy set theory, formulated the following principle: "in general, complexity and precision are inversely related to each other in the sense that if the complexity of the problem under consideration increases, the possibility of its precise analysis decreases". *Yager* [11,12] independently came to a similar conclusion when examining the uncertainty in probability. However, people can cope with situations in which all attempts at the mathematical formalization of a task and its solutions are unsuccessful due to the fact that, e.g., it is impossible to build an exact mathematical model, or it would take too long to solve it. *Zadeh* saw the reasons for this in the ability of the human mind to think in approximate categories, which microprocessors do not have. Due to this, a person can process approximate and ambiguous data, create models of the most complex processes, determine approximate solutions, etc. Fuzzy set theory, according to *Zadeh* and *Yager*, is therefore a tool used to formalize this approximate reasoning in vague and ambiguous terms.

For a long time, "uncertainty" and "ambiguity" have been used as synonyms for a lack of knowledge, which is decreasing as research progresses. Relatively recently, starting from the 1970s, these terms began to be treated as a reflection of reality, without the previous clearly negative meaning. It was then that the first major works in the field of multiple applications of fuzzy sets occurred, including *Zadeh* [8,9], *Yager* [11,12] and *Sanchez* [13]. Summing up, among the formal apparatuses that led to the development of fuzzy set theory, the first place is occupied by multi-valued logics. Previously, since ancient times, almost the entire development of logic could have been equated with two-valued logic, in which a statement can only be either true or false. The fact of such a polarization of truth and falsehood was considered as an essential feature of any "logical" reasoning. Many logicians, represented especially by *Lukaszewicz* (a co-founder of the Polish School of Logic), were already aware of the mismatch between such "rigid" logic and reality. The explosion of interest in multi-valued logics also aroused a significant increase in the interest of fuzziness and its origins, which was widely described in the later works of *Zadeh* [9], *Yager* [12], *Sanchez* [13] and *Kasprzyk* [14].

The stage preceding the main scope of the work was the conducting of a qualitative analysis of damage to the elements of the tested residential buildings [1,15]. The technical characteristics and typological ordering of this damage, understood as an expression of the quality of maintenance of residential buildings, enabled the exploitation conditions of the considered objects to be identified.

A number of works by *Nowogońska* [16–20] were used in the methodical approach to the technical assessment of tenement houses, and the fuzzy calculus presented in the publications of *Plebankiewicz, Wieczorek* and *Zima* [21–26] was used in the assessment of the whole service life of a building object. The works of *Ibadov* [27–30] and other authors [31–39], which concerned the construction investment process with the fuzzy phase, allowed for the practical application of uncertain and subjective events when determining the degree of damage to the tested tenement houses.

1.2. Subject of Study

A group of old tenement houses (that is, those erected before the First World War) takes an important place in Polish building resources. This group includes about 10.1% of the whole number of urban flats. What is more, the importance of this type of building relies on the fact that it takes part in creating an urban environment. At present, an action needs to be directed to the repair of the old land development. Doubtless, cultural aspects motivate all this action. To estimate its technical and economic justification, the degree of the technical wear of the old land development must be recognized and calculated.

This paper is a result of technical research and analyses on the old apartment houses in Wrocław, Poland [40]. The aim of the analysis is to provide information, which should help to direct an action, connected with this group of residential buildings. They are the apartment houses which were built at the turn of the nineteenth and twentieth centuries. The buildings are situated in the part of the city which (as a district from very few ones) was not completely destroyed by the war activities. The apartment houses are three- or four-storey buildings, made of bricks, erected in longitudinal, usually three-row, structural systems. Apart from the floors over the basement, which are solid ones, all the inter-storey floors represent typical wooden floors. All the buildings are covered with wooden rafter framing, usually a purlin–collar one. The staircases are composed of wooden or steel structural elements with wooden flights of steps.

1.3. Research Problem

While appraising building elements' technical wear—apart from applying the measurable (qualitative) criteria—the immeasurable (quantitative) criteria representing symptoms (pinpointed defects) of their deterioration have been taken into account. Only very few of these criteria can be classified at a high level of probability. There are symptoms of extreme characters, described by extreme dichotomic divisions. It is, however, agreed that between, e.g., a total pest attack to wooden elements and a lack of pests, the mid-states appear. Their value is often appreciated in a verbal way, e.g., "substantially", considerably", "significantly", "partially", "hardly" and it is always used in a description of detected defects as a result of a building object's technical inspections.

When assessing the degree of technical wear of building elements, apart from measurable (quantitative) criteria, immeasurable (qualitative) criteria are also used. They are expressed in the analysis of symptoms, i.c., damage, which lowers the technical condition and utility value of building elements. Only some of these criteria can be quantified with a big approximation. These are the symptoms with an extreme character, e.g., inter-story ceilings that are replaced with new elements that are not damp. It can then be assumed that the damage, and the technical wear it causes, take a value of zero.

In turn, flooding of the floors above basements does not raise doubts regarding the occurrence of the total dampness, and therefore the degrees of damage and technical wear caused by moisture take values equal to one within the variability interval of [0, 1]. Most of these criteria, however, are qualitative. Their value is determined verbally, e.g., as "significant", "poor", "strong", "almost not at all", "partial" or "complete", and it always appears in the description of damage phenomena. The interpretation of the effects of these phenomena, which is performed according to qualitative (i.e., subjective) premises, leads to the indiscriminate categorization of the technical maintenance conditions for buildings and their elements, i.e., good, satisfactory, average, poor or bad. Therefore, can a building element with a degree of technical wear of, e.g., 15%, be considered good or satisfactory from the point of view of the technical maintenance quality? Does significant biological contamination of wooden floor beams determine their 100% wear?

Striving for a quantification of criteria that are inherently qualitative (and therefore immeasurable), and trying to determine the relations between them, led to the use of the category of fuzzy sets with regard to this issue. Their properties enable damage to building elements, as well as the conditions of their technical maintenance, to be described within an unambiguous quantitative (measurable) aspect.

Therefore, the research led towards looking at the problem from this angle, which allowed the description of naturally qualitative (immeasurable) variables and the determination of existing relations between them in fuzzy set categories [15,25–39]. The advantages of fuzzy theory made it possible to describe the defects, representing three middle states (II, III, IV) of conditions of the building elements' maintenance, in a clear quantitative (measurable) aspect. Doubtless, fuzzy conditions are fully represented in these mid-states.

2. Research Methodology

2.1. Fuzzy Set Theory

The basic concept of the theory that was used in this paper is the concept of a fuzzy set [8,9,11–14]. The definition of a fuzzy set can be formulated as follows: a fuzzy set is set A, the x elements of which are characterized by the lack of a clear boundary between the membership and non-membership of x to A. The degree of the membership of element x to fuzzy set A is described by function $\mu A(x)$, which is called the membership function. The $\mu A(x)$ function takes values from the interval of [0, 1], where:

$$\mu A(x) = 0, \text{ which means that x is not a member of A;}$$

$$\mu A(x) = 1, \text{ which means that x is a full member of A.}$$

Fuzzy set A in a certain space (in this paper, it is the area of considerations concerning the observed states) $X = \{x\}$, which is written as $A \subseteq X$, is called the set of pairs:

$$A = \{(\mu A(x), x)\}, \forall\, x \in X.$$

Therefore, two basic fuzzy sets can be distinguished in a problem (each one is described in the three following observed states—II, III, IV):

- a fuzzy set of the technical wear of building elements $A \subseteq Ze \Leftrightarrow Z$ (to simplify the designations): $Z = \{(\mu Z(z), z)\}, \forall\, z \in Z$;
- a fuzzy set of damage to building elements $B \subseteq U$:

$U = \{(\mu U(u), u)\}, \forall\, u \in U.$

The basic operations performed on the fuzzy sets defined in the article are presented below:

- the absolute complement of the fuzzy set $A \subseteq X$, denoted as $-A$:

$$\mu_{-A}(x) = 1 - \mu_A(x), \forall\, x \in X \tag{1}$$

- the multiple sum of fuzzy sets $A, B \subseteq X$, denoted as $A \cup B$:

$$\mu_{A \cup B}(x) = \mu_A(x) \vee \mu_B(x), \forall\, x \in X \text{ (symbol } \vee \text{ denotes „max")} \tag{2}$$

- the intersection of fuzzy sets $A, B \subseteq X$, denoted as $A \cap B$:

$$\mu_{A \cap B}(x) = \mu_A(x) \wedge \mu_B(x), \forall\, x \in X \text{ (symbol } \wedge \text{ denotes „min")} \tag{3}$$

- the k-th power (k > 0) of fuzzy set $A \subseteq X$, denoted as Ak:

$$\mu_A{}^k(x) = (\mu(x))^k, \forall\, x \in X \tag{4}$$

Special cases of exponentiation include:

- the concentration of fuzzy set A $\subseteq$ X, denoted as CON (A):

$$\mu_{CON(A)}(x) = (\mu_A(x))^2, \forall\, x \in X \tag{5}$$

- the dilution of fuzzy set A $\subseteq$ X, denoted as DIL (A):

$$\mu_{DIL(A)}(x) = (\mu_A(x))^{0.5}, \forall\, x \in X \tag{6}$$

All these operations, which are of great importance in linguistic semantics, are interpreted as:

- $-A \Leftrightarrow$ "not A";
- $A \cup B \Leftrightarrow$ "A or B";
- $A \cap B \Leftrightarrow$ "A and B";
- CON (A) $\Leftrightarrow$ "strong A" (crispens the fuzzy set);
- DIL (A) $\Leftrightarrow$ "more or less, likely A" (flattens the fuzzy set).

When visually assessing the technical wear of building elements that inspected tenement houses consist of, the symptoms of their damage are taken into account, i.e., individual damage that can be categorized into the following groups of damage:

- UM—mechanical damage to the structure and texture of building elements;
- UW—damage to building elements caused by water penetration and moisture penetration;
- UD—damage resulting from the loss of the original shape of wooden elements;
- UP—damage to wooden elements attacked by biological pests.

The purpose of such a conceptual and technical systematization of damage is a comprehensive diagnosis of the extent to which a building element is worn. This assessment, in turn, leads to the implication of stating under what technical conditions—good, satisfactory, average, poor or bad—the building element was (is) maintained. The terms "good technical condition of maintenance", "satisfactory technical condition of maintenance", etc., can be considered as fuzzy sets with regard to semantic (qualitative) and technical (quantitative) aspects.

It is difficult to define a fuzzy set with such a broad meaning as "average technical condition of maintenance" using one membership function. In this case, a semantic analysis of the term "technical wear of a building element" was used, which was denoted with the symbol of a fuzzy set "Z". Let the technical wear of building element Z consist of: mechanical wear of its structure and texture (fuzzy set ZM), its technical wear caused by water penetration and moisture penetration (fuzzy set ZW), technical wear resulting from the loss of its original shape (fuzzy set ZD) and technical wear caused by the attack of biological pests (fuzzy ZP harvest). This sum can then be expressed as follows:

$$Z = ZM \cup ZW \cup ZD \cup ZP \tag{7}$$

and when assuming the identity of the degree of technical wear and its visual symptom ($Z \Leftrightarrow U$)—damage to a building element that is integrated into the above-described damage sets (Expression (7))—it takes the following form:

$$U = UM \cup UW \cup UD \cup UP \tag{8}$$

2.2. Research Model

The aim of the proposed model is to assess the technical wear of a building element with regard to the overriding criterion, i.e., "slightly worn, worn, significantly worn". The concepts defined in this way at the basic level best describe the behavior of a building element in its three middle maintenance states. It is in them, after rejecting the extreme states (i.e., good and bad) that have the most reliable evaluation principles from the technical point of view, that the fuzzy conditions are most fully represented. The basic principles of fuzzy logic and approximate reasoning were applied [8,9,11–15], and the fuzzy state was described as follows: its damage means that it can be classified as being in a satisfactory (II), average (III) and poor (IV) technical condition of maintenance. Therefore, in each maintenance state, the total damage to a building element is a multiplicity sum of the sets of damage, and it is expressed by Formula (9):

$$U(II, III, IV) = UM(II, III, IV) \cup UW(II, III, IV) \cup UD(II, III, IV) \cup UP(II, III, IV) \qquad (9)$$

where each set of damage, in each of the three maintenance states (II, III, IV), is a set of basic damage u_j, which represents elementary lower order criteria:

- $UM = \{u_1, u_2, \ldots, u_{14}\}$;
- $UW = \{u_{15}, u_{16}, \ldots, u_{23}\}$;
- $UD = \{u_{24}, u_{25}, \ldots, u_{28}\}$;
- $UP = \{u_{29}, u_{30}\}$.

Multiplication sum (9) can be written in each of the three maintenance states (II, III, IV) using the membership function:

$$\mu_U = \mu_{UM} \vee \mu_{UW} \vee \mu_{UD} \vee \mu_{UP} \qquad (10)$$

There is an intermediate stage between identifying damage at the elementary level, which occurs in everyday construction practice, and merging it into sets of damage in terms of their similarity regarding the wear processes. This stage involves the selection of damage of the same type but of different intensity (e.g., pitting corrosion, surface corrosion, deep corrosion of steel beams), or damage occurring to complex elements (e.g., structural walls—decay of brick or mortar). This method of combining elementary damage was used in the research, which led to the obtaining of greater possibilities of using operations of system analysis in fuzzy sets. In the considered sample of downtown tenement houses, this division is as follows:

- $\{u_1, u_2\} = U1 \Leftrightarrow$ mechanical damage and leaks;
- $\{u_3, u_4\} = U2 \Leftrightarrow$ brick and mortar losses;
- $\{u_5, u_6\} = U3 \Leftrightarrow$ brick and mortar decay;
- $\{u_7, u_8\} = U4 \Leftrightarrow$ peeling off and decomposing of the paint coatings;
- $\{u_9, u_{10}\} = U5 \Leftrightarrow$ cracks in brick and plaster;
- $\{u_{11}, u_{12}\} = U6 \Leftrightarrow$ scratching on walls and plaster;
- $\{u_{13}, u_{14}\} = U7 \Leftrightarrow$ loosening and falling off of plaster sheets;
- $\{u_{15}, u_{16}, u_{23}\} = U8 \Leftrightarrow$ dampness, weeping and flooding with water;
- $\{u_{17}, u_{18}, u_{19}\} = U9 \Leftrightarrow$ brick corrosion, fungus and mold;
- $\{u_{20}, u_{21}, u_{22}\} = U10 \Leftrightarrow$ pitting corrosion, surface corrosion and deep corrosion of steel beams;
- $\{u_{24}, u_{25}\} = U12 \Leftrightarrow$ dynamic sensitivity and deformation of floor beams;
- $\{u_{26}, u_{27}, u_{28}\} = U13 \Leftrightarrow$ torsional buckling and distortion of window joinery and wood elements;
- $\{u_{29}, u_{30}\} = U14 \Leftrightarrow$ touchwood and biological infestation of wooden elements.

In each of the damage types distinguished in this way, there is an intersection of two or three elementary fuzzy sets. Between them, as is the case between sets of damage, there is a multiple sum of the fuzzy sets that are defined above. All these dependencies can be described by the general formula for assessing the degree of damage to the elements of the analyzed residential buildings in their middle maintenance states which, when using the membership function, is as follows:

$$\mu_U = (\mu_{u1} \wedge \mu_{u2}) \vee (\mu_{u3} \wedge \mu_{u4}) \vee (\mu_{u5} \wedge \mu_{u6}) \vee (\mu_{u7} \wedge \mu_{u8}) \vee (\mu_{u9} \wedge \mu_{u10}) \vee$$
$$\vee (\mu_{u11} \wedge \mu_{u12}) \vee (\mu_{u13} \wedge \mu_{u14}) \vee (\mu_{u15} \wedge \mu_{u16} \wedge \mu_{u23}) \vee (\mu_{u17} \wedge \mu_{u18} \wedge \mu_{u19}) \vee \qquad (11)$$
$$\vee (\mu_{u20} \wedge \mu_{u21} \wedge \mu_{u22}) \vee (\mu_{u24} \wedge \mu_{u25}) \vee (\mu_{u26} \wedge \mu_{u27} \wedge \mu_{u28}) \vee (\mu_{u29} \wedge \mu_{u30})$$

Due to the fact that the greatest approximations of the observed states can be obtained at the level of elementary criteria, the degrees of membership of damage $u_1 \div u_{30}$ to fuzzy sets UM, UW, UD, UP were calculated at the stage of the basic comparative analysis, in which the fundamental probabilistic measure is the probability of the occurrence of a single damage $p(u_j)$ in the II, III and IV maintenance states. The probability of $p(u_j)$ is therefore a feature that determines the membership to elementary sets $u_1 \div u_{30}$. It would not be a mistake to simply identify the probabilities $p(u_j)$ with the degrees of memberships μ_{uj}, which are described linearly by the membership function operating on the domain [0, 1]. However, in order to present the properties of fuzzy sets more closely, the function used by *Zadeh* [8–10] was chosen for intensifying the contrast of the fuzzy set $A \subseteq X$:

$$\mu_{INT(A)}(x) = \begin{cases} 2(\mu_A(x))^2, \forall x : \mu_A(x) < 0.5 \\ 1 - 2(1 - \mu_A(x))^2, \forall x : \mu_A(x) \geq 0.5 \end{cases} \tag{12}$$

Therefore, the intensification of contrast increases the membership degrees that are greater than or equal to 0.5, while reducing the membership degrees that are lower than 0.5 (Figure 1).

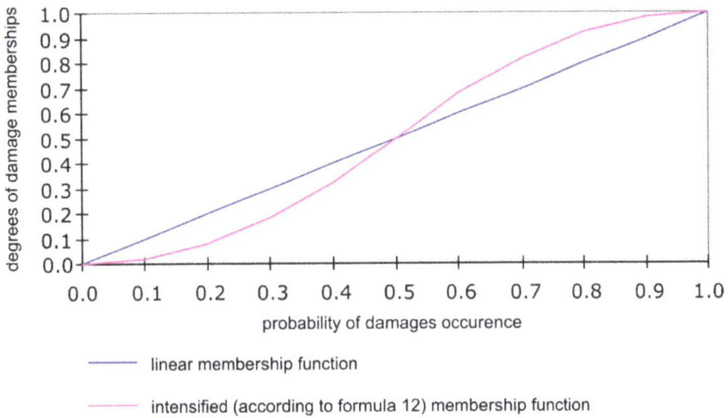

Figure 1. The effect of the contrast intensification of the degrees of damage memberships.

The final stage of the created model for assessing the technical wear (damage degree) of selected building elements in the three middle states of their technical maintenance is to estimate the size of the impact of elementary damage on the total damage. The study of the observed states and the conclusions from the proposed method of associating the occurring damage with the occurrence of the process of technical wear indicate a significant range of the strength of this relation within one building element in the maintenance states II, III and IV [1–7]. None of the values of the bi-serial correlation coefficient r(Z), which is a measure of this relationship, reaches a value of 1 in domain [0, 1]. Therefore, when taking the extreme value from this range as a reference point, it can be assumed that none of the values of the correlation coefficient r(Z) concentrates the fuzzy set U, while each of them—to a different degree—dilutes it. Considerations regarding the relationship between these dependencies and the analysis of the effects of the dilution process of fuzzy sets have led to the determination of the weights of the degrees of membership of elementary damage uj as a function of the correlation coefficient r(Z):

$$\mu_{uj} = [f(\mu_{uj})]^{1/r(Z)} \tag{13}$$

The result of the proposed Formula (13) is the following change in the membership function:

- $r(Z) \rightarrow 0 \Rightarrow \mu_{uj} \rightarrow 0$;
- $r(Z) \rightarrow 1 \Rightarrow \mu_{uj} \rightarrow \mu_{uj}$.

The application of the original procedures of the intensification and dilution of membership functions, according to Formulas (12) and (13), to the general Formula (11) of the model for assessing the degree of damage to the elements of the analyzed tenement houses in terms of fuzzy sets allowed for the transition from the data recorded using non-measurable variables to results defined by measurable values. The proposed model gives a numerical answer to the question of to what extent is a building element damaged. The total degrees of damage to the ten selected elements of the analyzed buildings S(U) in the maintenance states II, III and IV are presented in Table 1.

Table 1. The degree of fuzzy damage to building elements in their middle maintenance states. (grey backgroud is necessary to distinguish extreme values)

Group Number	Building Element	Damage Number	Damage Description	Degree of Fuzzy Damage Set S(U) Corresponding to the Maintenance States II, III and IV		
				S(U)II	S(U)III	S(U)IV
Z2	Foundations	u_3	brick losses	0.24	0	0
		u_5	brick decay	0	0.59	0
		u_9	brick cracks	0	0	0
		u_{15}	dampness of foundations	0	0	0
		u_{16}	weeping on foundations	0	0	0.97
		u_{17}	biological corrosion of bricks	0	0	0
		u_{19}	mold and rot on foundations	0	0	0
Z3	Basement walls	u_3	brick losses	0.05	0.25	0.67
		u_4	mortar losses	0	0	0
		u_5	brick decay	0	0	0
		u_6	mortar decay	0	0	0
		u_9	cracks in bricks	0	0	0
		u_{10}	cracks in mortar	0	0	0
		u_{15}	dampness of walls	0	0	0
		u_{16}	weeping on walls	0	0	0
		u_{17}	biological corrosion of bricks	0	0	0
		u_{19}	mold and rot on walls	0	0	0
Z4	Solid floors above basements	u_3	brick losses	0.01	0.22	0
		u_5	brick decay	0	0	0
		u_9	cracks in bricks	0	0	0
		u_{15}	dampness of floors	0	0	0
		u_{16}	weeping on floors	0	0	0.50
		u_{20}	corrosion raid on steel beams	0	0	0
		u_{21}	surface corrosion of steel beams	0	0	0
		u_{22}	deep corrosion of steel beams	0	0	0
		u_{23}	flooding of floors with water	0	0	0

Table 1. *Cont.*

Group Number	Building Element	Damage Number	Damage Description	Degree of Fuzzy Damage Set S(U) Corresponding to the Maintenance States II, III and IV		
				S(U)II	S(U)III	S(U)IV
Z7	Structural walls	u_3	brick losses	0	0	1.00
		u_4	mortar losses	0.34	0.93	0
		u_5	brick decay	0	0	0
		u_6	mortar decay	0	0	0
		u_9	cracks in bricks	0	0	0
		u_{10}	cracks on plaster	0	0	0
		u_{11}	scratching on walls	0	0	0
		u_{12}	scratching on plaster	0	0	0
		u_{15}	dampness of walls	0	0	0
		u_{16}	weeping on walls	0	0	0
		u_{17}	biological corrosion of bricks	0	0	0
		u_{19}	mold and rot on walls	0	0	0
Z8	Inter-story wooden floors	u_{12}	scratching on the plaster of the ceiling	0	0	0
		u_{13}	peeling of ceiling plaster	0	0	0
		u_{15}	dampness of floors	0	0	0
		u_{16}	weeping on floors	0.01	0.64	0
		u_{18}	fungus on floors	0	0	0.49
		u_{24}	dynamic sensitivity of floor beams	0	0	0
		u_{25}	deformations of wooden beams	0	0	0
		u_{30}	complete insect infestation of wooden beams	0	0	0
Z9	Stairs	u_1	mechanical damage	0.26	0.56	0
		u_3	brick losses	0	0	0
		u_{16}	weeping on stairs	0	0	0.95
		u_{20}	corrosion raid on steel beams	0	0	0
		u_{21}	surface corrosion of steel beams	0	0	0
		u_{22}	deep corrosion of steel beams	0	0	0
		u_{29}	partial insect infestation of wooden elements	0	0	0
Z10	Roof construction	u_{15}	dampness of truss	0	0	0
		u_{16}	weeping on wooden elements	0	0.43	0.53
		u_{28}	delamination of beams	0.03	0	0
		u_{29}	partial insect infestation of wooden elements	0	0	0
		u_{30}	complete insect infestation of wooden beams	0	0	0

Table 1. *Cont.*

Group Number	Building Element	Damage Number	Damage Description	Degree of Fuzzy Damage Set S(U) Corresponding to the Maintenance States II, III and IV		
				S(U)II	S(U)III	S(U)IV
		u_1	mechanical damage	0	0.85	0
		u_2	window leaks	0.89	0	1.00
		u_{15}	dampness of windows	0	0	0
		u_{16}	stains on windows	0	0	0
		u_{19}	mold and rot on windows	0	0	0
Z13	Window joinery	u_{26}	skewing of window joinery	0	0	0
		u_{27}	warping of window joinery	0	0	0
		u_{29}	partial insect infestation of window joinery	0	0	0
		u_{30}	complete insect infestation of window joinery	0	0	0
		u_1	mechanical damage to plaster	0.40	0	0
		u_6	plaster decay	0	0.85	0
		u_7	peeling off of paint coatings	0	0	0
		u_8	falling off of paint coatings	0	0	0
		u_{10}	cracks in plaster	0	0	0.95
Z15	Inner plasters	u_{12}	scratching on plaster	0	0	0
		u_{13}	loosening of plaster	0	0	0
		u_{14}	flaking off of sheets of plaster	0	0	0
		u_{15}	dampness of plaster	0	0	0
		u_{16}	weeping on plaster	0	0	0
		u_{18}	fungus on plaster	0	0	0
		u_{19}	mold and rot on plaster	0	0	0
		u_1	mechanical damage to plaster	0	0	0
		u_6	plaster decay	0.43	0	0
		u_7	peeling off of paint coatings	0	0	0
		u_8	falling off of paint coatings	0	0	0
Z20	Facades	u_{10}	cracks in plaster	0	0.94	0
		u_{12}	scratching on plaster	0	0	1.00
		u_{13}	loosening of plaster	0	0	0
		u_{14}	flaking off of sheets of plaster	0	0	0
		u_{15}	dampness of plaster	0	0	0
		u_{16}	weeping on plaster	0	0	0
		u_{18}	fungus on plaster	0	0	0
		u_{19}	mold and rot on plaster	0	0	0

3. Results

The analysis of the results of the research concerning the impact of damage to building elements on their technical wear with regard to fuzzy sets leads to the following conclusions (Table 1):

a. in the field of assessing the degree of fuzzy damage to elements of downtown tenement houses—S(U) II, III, IV:

- the development of the model presented in the article allowed the fundamental question of to what extent a building element is worn (damaged), when knowing that it is (more or less) satisfactorily, moderately or poorly maintained, to be answered;
- the use of simple operations in the fuzzy set calculus enabled the influence of both elementary damage that occurs with a specific frequency (probability) and the measure of its interdependence (correlation) on the observed technical wear of building elements to be considered;
- as a result of the proposed model, which is based on fuzzy set theory, it was possible to identify the elementary damage that determines the degree of destruction of the building's elements;

b. when determining the degree of damage of 10 selected building elements according to fuzzy criteria, it was indicated that there is a need for an individual approach to each of the elements (especially structural) during the process of their technical assessment. However, several regularities can be identified:

- the degree of damage to the element increases with the deterioration of its maintenance conditions (although not proportionally to the maintenance conditions and not equally for different types of elements). For instance, degrees of fuzzy damage set S(U) corresponding to the maintenance states II, III and IV grow in the following way: Z3—basement walls—u_3—brick losses: 0.05; 0.25; 0.67. It most often differs from the observed values of the degree of the technical wear that was determined using the probabilistic approach [1]—in particular, in poor conditions of building maintenance, the degree of damage exceeds 70% of its technical wear threshold;
- elementary damage that determines the degree of destruction of an element comes much more often from group I (mechanical damage to the structure and texture of elements) than was the case in the analysis of the observed states. Only under poor conditions of building maintenance does the analysis of the observed random [1] and fuzzy [15] phenomena show a great similarity—the decisive damage is the destruction of the element caused by water penetration and moisture penetration (group II);
- at the level of the greatest detail, the type of damage and the degrees of fuzzy damage to the elements of the downtown tenement houses were determined. In the most representative, i.e., average/satisfactory condition of maintenance—S (U) III—the degrees were as follows:

  - for foundations: brick decay 0.59
  - for basement walls: brick decrements 0.25
  - for solid floors above basements: brick decrements 0.22
  - for structural walls: mortar decrements 0.93
  - for wooden inter-storey floors: weeping 0.64
  - for internal stairs: mechanical damage 0.56
  - for roof constructions: weeping on wooden elements 0.43
  - for window joinery: mechanical damage 0.85
  - for inner plasters: plaster decay 0.85
  - for facades: cracks on plaster 0.94

4. Summary and Discussion

At the beginning, general methodological conclusions were formulated. They resulted from the modeling of the impact of the maintenance of tenement houses on the technical wear of their elements in fuzzy conditions. Such an approach gives much greater possibilities of studying cause and effect relationships than the probabilistic analysis [1]:

a. the use of simple operations in the fuzzy set calculus enables the simultaneous recognition of the impact of elementary damage that occurs with a specific frequency (probability), and also the measure of its interdependence (correlation) on the observed technical wear of building elements;

b. in the effect of fuzzy transformations, it is possible to identify the elementary damage that determines the degree of destruction of the building element. The result of the cumulative effects of frequently occurring mechanical damage to the structure and texture of elements indicates that this type of damage is no less important in the process of the technical wear of elements of downtown tenement houses;

c. consideration of the problem with regard to fuzzy phenomena allows for the synthesis of elementary criteria. This gives the greatest approximations (at the stage of the technical investigation of a residential building) for the global assessment of the degree of wear of the building's elements. In addition, it significantly reduces the subjective factor of this assessment, which has the greatest impact on the result of research conducted for the middle maintenance states of buildings.

The consequence of systematizing the most important processes that influence the loss of functional properties of residential buildings was the creation of the authors; own qualitative model and its transformation into a quantitative model. This, in turn, enabled a multi-criteria quantitative analysis of the cause–effect phenomena—"damage–technical wear"—of the most important elements of downtown residential buildings to be conducted in the so-called conventional and fuzzy sets. In conventional sets, in which attempts were made to describe the observed (empirical) states with the use of theoretical formulas, the probabilistic side of the problem and its random nature were considered [1]. In turn, in fuzzy sets, the observed states of cause–effect phenomena in the fuzzy conditions [15] (i.e., uncertainty as to the very fact of their occurrence) were analyzed.

The fact that the membership function of a fuzzy set assumes values from interval [0, 1] leads to the hasty conclusion that fuzziness is a hidden form of randomness, and therefore fuzzy set theory is basically nothing new in relation to probability. The differences between fuzziness and randomness, however, concern both their nature and the formal differences between probabilistic calculus and fuzzy sets. The nature of these phenomena lies in the problem of the uncertainty of the type of randomness and fuzziness. In the case of randomness, the event is strictly defined, while its occurrence is uncertain. Therefore, randomness can be equated with the uncertainty regarding an element's membership or non-membership. This is not the case with fuzziness, which concerns the very degree of membership of an element to a set, and therefore an event is no longer strictly defined. Such events are the ones analyzed in the paper—the occurring damage of building elements and the processes of their technical wear. Their nature, in the authors' opinion, is more fuzzy than random.

The differences between randomness and fuzziness can be presented with regard to the following three points of view:

- level of uncertainty;
- number of decision makers;
- number of steps in the decision process.

Regarding "the degree of uncertainty", the following decision-making situations, with an increasing degree of uncertainty, can be distinguished:

- Certainty: all the information that describes the issue of decision making is deterministic;
- Risk: information that describes the decision-making issue is probabilistic, i.e., the data have appropriate probability distributions;

- Uncertainty: even the probabilities are not known. Making decisions is usually reduced to using a minimax strategy;
- Fuzziness: uncertainty not only relates to the occurrence of an event, but also to its meaning in general, and this can no longer be considered using probabilistic methods. Of course, further extensions, such as adding risk to fuzziness, are also possible.

The sense of a fuzzy set can therefore be used to formally determine and quantitatively express ambiguous concepts that are always present in the programming and analysis of a construction process. Thus, fuzzy set theory is a theory of classes in which the transition from membership to non-membership does not have a jumping character, as is the case in a conventional set, but instead it is gradual. Striving for a quantification of criteria that are inherently qualitative (and therefore immeasurable), and trying to determine the relations between them, led to the use of the category of fuzzy sets with regard to this issue. Their properties enable elementary construction processes to be mathematically described as fuzzy events within an unambiguous quantitative (measurable) aspect.

To sum up, the approach of the creator of fuzzy set theory [8–10] (*Lofti Zadeh*, who, unlike *Yager* and *Kaufmann* [11,12], assumed the fuzzy set as a random event) was consciously used by the authors. This enabled the question of what is the probability that a building element is more or less (approximately) worn to be answered. Therefore, the differences between the concepts of fuzziness and randomness were not considered. It was only assumed that although these phenomena are different and described differently, they may nevertheless occur together as two types of uncertainty.

Author Contributions: Conceptualization, J.K., M.S. (Marek Sawicki) and M.S. (Mariusz Szóstak); methodology, J.K.; software, J.K. and M.S. (Mariusz Szóstak); validation, J.K., M.S. (Marek Sawicki) and M.S. (Mariusz Szóstak); formal analysis, J.K., M.S. (Marek Sawicki) and M.S. (Mariusz Szóstak); investigation, J.K. and M.S. (Marek Sawicki); resources, J.K. and M.S. (Marek Sawicki); writing—original draft preparation, J.K.; writing—review and editing, J.K. and M.S. (Mariusz Szóstak); supervision, M.S. (Marek Sawicki). All authors have read and agreed to the published version of the manuscript.

Funding: This research received no external funding.

Institutional Review Board Statement: Not applicable.

Informed Consent Statement: Not applicable.

Data Availability Statement: No new data were created or analyzed in this study. Data sharing is not applicable to this article.

Conflicts of Interest: The authors declare no conflict of interest.

References

1. Konior, J. Technical Assessment of old buildings by probabilistic approach. *Arch. Civ. Eng.* **2020**, *66*, 443–466. [CrossRef]
2. Konior, J. Decision assumptions on building maintenance management. Probabilistic methods. *Arch. Civ. Eng.* **2007**, *53*, 403–423.
3. Konior, J. Maintenance of apartment buildings and their measurable deterioration. *Tech. Trans. Czas. Tech.* **2017**, *6*, 101–107. [CrossRef]
4. Konior, K. Bi-serial correlation of civil engineering building elements under constant technical deterioration. *J. Sci. Gen. Tadeusz Kosiuszko Mil. Acad. Land Forces* **2016**, *179*, 142–150.
5. Konior, J. Intensity of defects in residential buildings and their technical wear. *Tech. Trans. Civ. Eng.* **2014**, *111*, 137–146.
6. Konior, J.; Sawicki, M.; Szóstak, M. Intensity of the Formation of Defects in Residential Buildings with Regards to Changes in Their Reliability. *Appl. Sci.* **2020**, *10*, 6651. [CrossRef]
7. Konior, J.; Sawicki, M.; Szóstak, M. Influence of Age on the Technical Wear of Tenement Houses. *Appl. Sci.* **2020**, *11*, 297. [CrossRef]
8. Zadeh, L. Fuzzy sets. *Inf. Control* **1965**, *8*, 338–353. [CrossRef]
9. Zadeh, L. Fuzzy sets and systems. *Int. J. Gen. Syst.* **1990**, *17*, 129–138. [CrossRef]
10. Zadeh, L.; Aliev, R. *Fuzzy Logic Theory and Applications*; World Scientific Publishing Co Pte Ltd.: Singapore, 2018.
11. Yager, R.R. *A Note on Probabilities of Fuzzy Events Information and Science*; Iona College: New Rochelle, NY, USA, 1979; Volume 18.
12. Yager, R.R. On the fuzzy cardinality of a fuzzy set. *Int. J. Gen. Syst.* **2006**, *35*, 191–206. [CrossRef]
13. Sanchez, E. Resolution of composite fuzzy relation equations. *Inf. Control* **1976**, *30*, 38–48. [CrossRef]
14. Kacprzyk, J. *Fuzzy Sets in the System Analysis*; PWN: Warsaw, Poland, 1986.

15. Konior, J. Technical assessment of old buildings by fuzzy approach. *Arch. Civ. Eng.* **2019**, *65*, 129–142. [CrossRef]
16. Nowogońska, B. Diagnoses in the aging process of residential buildings constructed using traditional technology. *Buildings* **2019**, *9*, 126. [CrossRef]
17. Nowogońska, B. Intensity of damage in the aging process of buildings. *Arch. Civ. Eng.* **2020**, *66*, 19–31. [CrossRef]
18. Nowogońska, B.; Korentz, J. Value of technical wear and costs of restoring performance characteristics to residential buildings. *Buildings* **2020**, *10*, 9. [CrossRef]
19. Nowogońska, B. The Method of Predicting the Extent of Changes in the Performance Characteristics of Residential Buildings. *Arch. Civ. Eng.* **2019**, *65*, 81–89. [CrossRef]
20. Nowogońska, B. Proposal for determing the scale of renovation needs of residential buildings. *Civ. Environ. Eng. Rep.* **2016**, *22*, 137–144. [CrossRef]
21. Plebankiewicz, E.; Karcińska, P. Creating a Construction Schedule Specifying Fuzzy Norms and the Number of Workers. *Arch. Civ. Eng.* **2016**, *62*, 149–166.
22. Plebankiewicz, E.; Wieczorek, D.; Zima, K. Life cycle cost modelling of buildings with consideration of the risk. *Arch. Civ. Eng.* **2016**, *62*, 149–166. [CrossRef]
23. Plebankiewicz, E.; Wieczorek, D.; Zima, K. Quantification of the risk addition in life cycle cost of a building object. *Tech. Trans.* **2017**, *5*, 35–45.
24. Plebankiewicz, E.; Wieczorek, D.; Zima, K. Model estimation of the whole life cost of a building with respect to risk factors. *Technol. Econ. Dev. Econ.* **2019**, *25*, 20–38.
25. Plebankiewicz, E.; Wieczorek, D. Fuzzy risk assessment in the service life of building objects. *Mater. Bud.* **2016**, *6*, 59–61.
26. Wieczorek, D. Fuzzy risk assessment in the life cycle of building object—Selection of the right defuzzification method. In Proceedings of the AIP Conference Proceedings, Maharashtra, India, 5–6 July 2018; AIP Publishing: Melville, NY, USA, 2018; Volume 1978, p. 240005.
27. Ibadov, N. Fuzzy Estimation of Activities Duration in Construction Projects. *Arch. Civ. Eng.* **2015**, *LXI*, 23–34. [CrossRef]
28. Ibadov, N.; Kulejewski, J. Construction projects planning using network model with the fuzzy decision node. *Int. J. Environ. Sci. Technol.* **2019**, *16*, 4347–4354. [CrossRef]
29. Ibadov, N.; Kulejewski, J. The assessment of construction project risks with the use of fuzzy sets theory. *Tech. Trans.* **2014**, *1*, 175–182.
30. Ibadov, N. The alternative net model with the fuzzy decision node for the construction projects planning. *Arch. Civ. Eng.* **2018**, *64*, 3–20. [CrossRef]
31. Kamal, K.J.; Jain, B.B. Application of Fuzzy Concepts to the Visual Assessment of Deteriorating Reinforced Concrete Structures. *J. Constr. Eng. Manag.* **2012**, *138*, 399–408.
32. Andrić, J.M.; Wang, J.X.W.; Zou, P.; Zhang, J. Fuzzy Logic–Based Method for Risk Assessment of Belt and Road Infrastructure Projects. *J. Constr. Eng. Manag.* **2019**, *145*, 238–262. [CrossRef]
33. Marzouk, M.; Amin, A. Predicting Construction materials prices using fuzzy logic and neural networks. *J. Constr. Eng. Manag.* **2013**, *139*, 1190–1198. [CrossRef]
34. Knight, K.; Robinson, R.; Fayek, A. Use of fuzzy logic of predicting design cost overruns on building projects. *J. Constr. Eng. Manag.* **2002**, *128*, 503–512. [CrossRef]
35. Sharma, S.; Goyal, P.K. Fuzzy assessment of the risk factors causing cost overrun in construction industry. *Evol. Intell.* **2019**, 1–13. [CrossRef]
36. Al-Humaidi, H.M.; Hadipriono, T.F. Fuzzy logic approach to model delays in construction projects using rotational fuzzy fault tree models. *Civ. Eng. Environ. Syst.* **2010**, *27*, 329–351. [CrossRef]
37. Ammar, M.T.; Zayed, T.; Moselhi, O. Fuzzy-based life-cycle cost model for decision making under subjectivity. *J. Constr. Eng. Manag.* **2012**, *139*, 556–563. [CrossRef]
38. Chan, K.Y.; Kwong, C.K.; Dillon, T.S.; Fung, K.Y. An intelligent fuzzy regression approach for affective product design that captures nonlinearity and fuzziness. *J. Eng.* **2011**, *22*, 523–542. [CrossRef]
39. Nasirzadeh, F.; Afshar, A.; Khanzadi, M.; Howick, S. Integrating system dynamics and fuzzy logic modelling for construction risk management. *Constr. Manag. Econ.* **2008**, *26*, 1197–1212. [CrossRef]
40. Czapliński, K. *Assessment of Wroclaw Downtown Apartment Houses' Technical Conditions*; Reports series "U" of Building Engineering Institute; Wroclaw University of Science and Technology: Wroclaw, Poland, 1984.

Article

Intensity of the Formation of Defects in Residential Buildings with Regards to Changes in Their Reliability

Jarosław Konior, Marek Sawicki and Mariusz Szóstak *

Department of Building Engineering, Faculty of Civil Engineering, Wroclaw University of Science and Technology, 50-370 Wrocław, Poland; jaroslaw.konior@pwr.edu.pl (J.K.); marek.sawicki@pwr.edu.pl (M.S.)
* Correspondence: mariusz.szostak@pwr.edu.pl; Tel.: +48-71-320-23-69

Received: 27 August 2020; Accepted: 19 September 2020; Published: 23 September 2020

Abstract: Defining the basic determinants of the level of reliability with regards to the use of residential buildings and determining the function of the intensity of their characteristic defects are important issues concerning renovation strategy. The distribution of the exploitation time of residential buildings, the function of their reliability, and the distribution of the defect intensity of examined buildings are interdependent terms. Therefore, it can be assumed that the defect intensity of an object will be higher with an increase in its exploitation time. However, it is neither an increase reflecting the length of the building's service life nor the value directly proportional to its age. The article presents a model and method of testing the defects and reliability of a representative group of traditional downtown residential buildings, which were erected in Wroclaw, Poland at the turn of the 19th and 20th centuries. A basic conclusion was drawn regarding the mechanism of damage of residential buildings: for the period of using the facility, in which the time of correct operation until failure has an exponential distribution, the average remaining time of failure-free operation is unchanged at any time. It was confirmed that the tested residential buildings, after a certain period of failure-free operation, fulfil their functions, just like new buildings. The optimal moment of renovation occurs after the end of the second period of operation, before the period of rapid wear. The study of the course of the damage intensity function over time reflects the wear process of a residential building in a representative sample of downtown residential buildings that were erected using traditional methods. Defining the average duration of the correct failure-free operation of an object by the reliability function, which determines the probability with which the correct operation time of an object will be longer than its age, has a practical application in the exploitation of a residential building and its components.

Keywords: residential buildings; defects; intensity; reliability; technical wear

1. Introduction

1.1. Damage to Building Objects

Damage is an event that involves the loss of serviceability of an element or building [1–10]. It is related to them reaching their limit state. The exceeding of the limit state that is appropriate for the subsequent utility functions of individual elements of a residential building reduces their exploitation potential [11]. An element loses various utility functions when it reaches the serviceability limit state and enters the state that is defined in reliability theory as being defective (but fit for use). This state lasts until all of its functions exceed this limit state. The element then becomes unusable. The serviceability limit state is a contractual value, which depends on the adopted criteria.

The stimuli that cause the limit states to be exceeded by successive element functions may be sudden (random damage), gradual (aging damage), or have a nature of relaxation extortion (gradual aging of an element and its sudden transition to a state of being unfit for use occur together) [12]. Aging processes, which gradually occur, are usually caused by damage of a deterministic nature (predictable in a given time) [13]. Random damage occurs suddenly (breakdowns, catastrophes), or it is caused by accelerated wear (sporadic defects and technological defects) [14]. Defects (sporadic defects) are typical damage that result from the poor performance of executive works, the poor quality of the used construction materials, or they are caused by both of these causes simultaneously [15]. On the other hand, erroneous design assumptions and defective design and material solutions cause technological (chronic) defects are damage. Poor workmanship and the low-quality of built-in materials only exacerbate this problem [16].

If a technical element contains a sporadic or chronic defect, the limit state of its individual functions is reached faster. Subsequently, there is a clear reduction in the resistance of the material to external stimuli. Sudden damage causes unexpected changes in the essential physical parameters that determine the performance of the element's basic operational tasks. During a failure, the limit values of the safety functions of the element's structure are not exceeded, while, during a disaster, the parameters change beyond the values that are permissible by the requirements. Gradual damage is the result of aging activities. Aging processes are associated with irreversible structural changes in the materials that are used in building components. They result from physicochemical reactions that occur over time due to the operation of destructive stimuli on a macro and micro scale.

The rate of aging of materials depends on:

- the resistance of material to destructive stimuli; and,
- the intensity of the impact of destructive stimuli [17–21].

The accumulation of the effects of these interactions causes structural changes in material. The result of these external and internal destructive processes is the reduction of the material's resistance to damage that occurs during various periods of operation [22]. Consequently, there is a gradual increase in wear and a loss of functional properties of the element, which leads to its inoperability, and later to it being unfit for use. Exceeding the serviceability limit state by the individual operational functions of an element does not mean its full, physical destruction. Full physical destruction (and, at the same time, complete technical, social, and economic wear) occurs when the technical features of fundamental importance for the appropriate performance of the element's operational functions do not meet the parameters that guarantee safe operation [23]. The issue of safety is defined here by appropriate standards, technological guidelines, conditions of admission to use, approvals, and technical certifications [24]. Therefore, the criterion of safe operation is absolutely essential in the exploitation process, regardless of the nature of the causes of damage to a building's elements [25].

The so-called Lorenz curve illustrates the typical course of the wear process of building elements during their operation [26–28]. In this process, we can distinguish the following three basic intervals of a building's age t and the corresponding intervals of technical wear Zt—Figure 1:

- a warranty and post-warranty period of up to about 0.15 of a building's age t, in which the object "adjusts" and shows technical wear Zt at a level of 0.2,
- a period of normal exploitation of up to around 0.75 of a building's age t, in which the facility is properly maintained and shows technical wear Zt at a level of 0.5, and
- a period of planned exploitation of up to 1.0 of a building's age t, which is equal to its expected durability T, and in which the object should be renovated/modernized until it reaches a level of technical wear of 1.0.

The last period of planned exploitation and the period of unplanned use of residential buildings, the age of which exceeds their literature service life [29], is the subject of research and analysis regarding the intensity of damage and the change in the reliability state of buildings that qualified for the targeted research sample that is described below.

Figure 1. A typical course of the wear process of residential buildings during their exploitation.

1.2. Literature Review

During exploitation, construction objects are subjected to continuous destructive processes of various courses. With the passage of time, their functional properties decrease, and their partial restoration occurs as a result of repairs [30]. Therefore, during the use of buildings, it becomes necessary to carry out renovation works [31], which, according to the principle of sustainable development, should be included in the life cycle costs of construction objects [32–34].

Deciding which repair solutions to choose is a difficult and complex task. To this end, many models and methods have been developed in order to support policymakers and building administrators. They include the computer decision model for selecting repair options [35]; a simplified method of estimating technical degradation, which uses artificial neural networks [36]; and, the feasibility assessment of works using fuzzy stochastic networks [37].

In the proposed models and methods, an important element is the correct assessment of the size and intensity of defects to structural elements. A significant problem in the discussed issue is an increase of damage and partial defects [38], which is, the change in the building's reliability state during its operation. The analysis of the technical condition and intensity of damage requires appropriate modelling. The Rayleigh distribution [39,40] or Weilbull distribution [41,42] can be used for modeling this phenomenon.

The Weilbull exponential distribution is often used to evaluate the distribution of normal operation time [43], because it assumes that failures are only caused by external random events. However, in reality, there is no such exponential model of reliability distribution. Significant approximations, in which a negligible influence of wear processes is assumed, are made in the exponential distribution. A special example of the Weibull distribution is the Rayleigh distribution. This distribution, in turn, occurs when the wear of an element increases over time, i.e., it is the main cause of failure over time. The appropriate modelling of exploitation scenarios helps to select the optimal planning of renovation works for a building.

The aim of the research was to determine, while using the example of over 102 tested residential buildings, how the intensity of damage affects the reliability of construction objects.

2. Research Method

2.1. Research Sample

The subject of the research [44–48] involves tenement houses in a separate part of the downtown district in Wroclaw, Poland. The buildings are situated along downtown streets of secondary importance in an urban layout that has remained unchanged for years. They are front buildings, and also outbuildings with a modest architectural design and economical functional standard. The facilities were built of brick in longitudinal, usually three-bay, structural systems.

102 tenement houses were mainly erected in the second half of the nineteenth century, until the outbreak of World War I. However, three of them are 170 years old. It is difficult to determine with certainty the type of building development due to the enormous scale of war damage that took place in this region in 1945; it can be assumed that, at the time of the examination, almost 2/3 of the buildings were built in compact developments, 1/5 in semi-compact developments, and 1/6 as free-standing buildings. The number of storeys varies from 2 to 5: 9% are two-story buildings, 10% are three-story buildings, 39% are four-story buildings, and 42% are five-story buildings. The vast majority of tenement houses (84%) have a basement under the entire building, 9% under a part of the building, and 7% have no basement at all. With the exception of three buildings, all of them have a usable attic. 83% of the attics are used as a drying room and 17% have been converted into apartments.

The apartments were designed without sanitary installations. Water intake points, as well as sinks and toilets (c.c.), were later installed on the staircase landings and even in the kitchens of the apartments. Furnaces heat most of the apartments, and only a few have central heating made by the residents themselves. Electrical installations, originally designed as surface-mounted, after the unprofessional modifications of tenants, are placed under the plaster. Gas installations were gradually introduced, with the development of the city network, to almost all apartments.

The term "tenement houses" defines the above-described downtown residential buildings with construction and material solutions that are typical for the turn of the 19th and 20th centuries, similar functions and standards, and a specific form of ownership (the so-called pre-war "tenement houses") in all parts of this article.

The research sample, covering 102 technically assessed residential buildings from Wroclaw's Srodmiescie district, was selected from a group of 160 examined buildings. The overriding criterion for the selection of the sample was the obtaining of a comparable group of objects. Mutual comparability of downtown tenement houses meant:

- age coherence, i.e., a similar period of erection, maintenance and exploitation with regards to historical and social aspects;
- compact development in the urban layout that has remained unchanged for years;
- similar location along downtown street routes with an urban, but not representative, character;
- construction and material homogeneity, especially regarding the load-bearing structure of buildings; and,
- identical functional solutions, which are understood as the standard of apartment amenities and furnishings in force at that time, and also a specific standard of living of residents.

A method of selecting the research sample at the level of greater detail was based on the mutual similarity of all technical solutions of downtown tenement houses.

The selected research sample, according to the criteria presented above, is representative with regards to one of the concepts (specific for the adopted purpose of the study) of representativeness [49,50]. It contains all the values of the variables, which could be recreated from previous research that had a different objective function than the one that was adopted in this study. However, these values were compiled and processed in such a way that it is possible to make conclusions about the cause–effect relationships between them in the general population. Thus, the typological representativeness of the sample into which the desired types of homogeneous variables are classified can be assumed. Because of the fact that the structure of the population and its properties were well recognized earlier, such a selection of the research sample can also be considered to be deliberate. It should be noted that the sample may not be representative in terms of the distributions of the examined variables, which may—for the adopted significance level—not correspond to the analogous distributions in the general population. It is also not known—at this stage of the research—whether the selected sample is representative due to the correspondence between its variables and the identically defined variables in the entire set of downtown residential buildings.

2.2. Research Model

The general scheme of the cause-and-effect model—"defect–technical wear of building elements"—is the result of a synthesis of the results of visual studies of a selected sample of tenement houses in Wroclaw's Srodmiescie district. The scheme of the considered model at the level of greatest generalization is as follows:

$$[\text{CAUSES}] \rightarrow \left[\text{observed} \overset{\text{SYMPTOMS}}{\leftrightarrow} \text{measured}\right] \rightarrow [\text{EFFECTS}]$$

or in a more detailed elaboration—Figure 2:

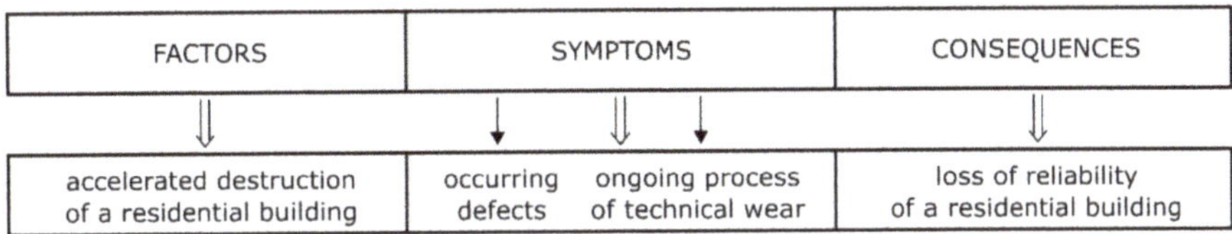

Figure 2. General diagram of the cause and effect model—"damage–technical wear of building elements".

The theoretical model of the technical wear of elements of residential buildings is a function of time t and their assumed durability T. The comparative analysis of the observed and theoretical wear shows that it is practically impossible to determine the exact form of the dependence between the size of wear of an element and its age. This difficulty results from the influence of many factors, which are individual for each residential building and can only be described by a complex mathematical model. In this situation, low complexity models should be selected and their compliance with empirical observations should be assumed as the selection criterion when the task of the researcher is to determine the trend of the phenomenon. Therefore, the research was limited to the search for trend functions from among linear, power (multiplicative), as well as exponential and hyperbolic relationships [51–57].

At any time during the assessment of the technical condition of any selected group of residential buildings and their elements, the group of experts acts at the intermediate stage of the proposed model—the analysis of the symptoms of the observed states. It cannot measure the causes (factors). However, it may take their impact into account when carrying out the assessment. In the case of effects (consequences), short-term effects (e.g., loss of utility values) and intentional effects (concerning decisions about the future of a residential building) can be distinguished. Further steps depend on adopting one of the multi-criteria decision making methods, e.g., according to [58]. The more reliable and meaningful the research on the symptoms of damage to a residential building's elements in the observed states, the more reliable the reasons regarding further decision-making analysis.

The key element of the technical examination of residential buildings should be the assessment of the size and intensity of the damage (symptoms) of their structural elements, which is carefully prepared in advance in terms of methodology. This assessment, which is supported by the theoretical recognition of the failure mechanism with regards to the reliability of technical facilities, leads to the determination of the causes of damage, and it enables a decision that is based on numerical evidence concerning the future of residential buildings to be made.

2.3. Research Method

The concept of the reliability of a residential building is always associated with the performance of exploitation tasks [7,59–63]. The performance of a task by a residential building involves its correct fulfilment of certain functions under certain operating conditions and within a specified time. If this function is denoted by φ, the building's working conditions by χ, and the building's operation time by t, then the task to be performed by the facility can be written as an ordered triple [φ,χ,t]. By knowing the function that the building needs to perform, it t is possible to establish such a set of requirements (ωφ) for the features of a residential building (characterized by a number of essential and auxiliary technical-operational, economic, and other parameters that are important in the process of exploitation and maintenance) that their fulfilment is a necessary and sufficient condition for the correct implementation of the assigned functions (φ) by the building. It has been assumed, with some simplifications, that the assessed residential building is, from the point of view of exploitation theory, a two-state object. This means that it may be fit to perform its function (assuming the actual state that is characterized by meeting the requirements (ωφ)), or unfit to perform its function (assuming the physical state that is characterized by a failure to meet the requirements (ωφ)). The task of the facility, which is then understood as an event (Z) (e.g., with regards to the provision of housing services), is written as the following ordered triple: [ωφ,χ,t]. It was further assumed that the requirements regarding a residential building and its maintenance conditions are known, i.e., the pair [ωφ,χ] is fixed and, consequently, it was assumed that the reliability of residential buildings could be assessed as a function of time (t).

Thus, the concept of the reliability of a residential building is defined, as follows: the reliability of a residential building is its property, which is seen as its ability to meet the requirements (ωφ) within the designated limits of being fit and unfit under certain given maintenance conditions (χ) and exploitation time (t).

The above considerations allowed for the reliability measure to be defined according to general formula [30,61]:

$$R(t) = P\{\tau > t\} \tag{1}$$

where:

τ—time of the failure-free operation of an element; and,

$R(t)$—the function of reliability that describes the probability of the failure-free operation of an element during the time period t (it assumes values from the interval [0, 1], where $R(0) = 1$, and $R(\infty) = 0$), which can be also expressed as $lim\{t \to \infty\} R(t) = 0$.

More precisely, the reliability function R(t) denotes the probability of the correct operation of an object in the interval [0, t]. For the considered residential buildings with repairable elements, the index (1) characterizes their reliability until the first defect. The course of the reliability curve coincides with the change in the serviceability value due to the fact that they are also complex objects.

The function $R(t)$ is also a transformed distribution function, the form of which is as follows:

$$F(t) = P\{\tau > t\} = 1 - R(t) \tag{2}$$

$F(t)$ determines the probability with which the time of the correct operation of a residential building will be shorter than its expected service life (t). If it is assumed that such an event occurs at time $t = 0$, i.e., an object is fit for use at the moment of putting it into operation, then it can be assumed that it will also perform this task at any time t_i (from interval $0 < t_i \leq t$, where $i = 1, 2, \dots$). The general reliability of a residential building is then equal to the product of reliability:

$$R = R(t)R(0) = R(\omega\varphi, \chi, t)\Delta P\{Z(\omega\varphi, \chi, 0)\} \tag{3}$$

The symbol $R(0)$ denotes the so-called initial reliability of an element, i.e., the probability that it will be fit at the moment of starting the task $(t = 0)$. Therefore, the definition of the concept of reliability was used and, as a result, the following was obtained:

$$R(0) = P\{Z(\omega\varphi, \chi, t)\} \tag{4}$$

and then:

$$R(t) = P\{Z(\omega\varphi, \chi, t_i)\},\ 0 < t_i < t/Z(\omega\varphi, \chi, 0) \tag{5}$$

It was further assumed that event $Z(\omega\phi,\ \chi,\ 0)$ is certain, i.e., it is certain that, if a residential building adopted from the investment process to the exploitation process meets all the specified functions, then it will fulfil all the operational tasks assigned to it and it can also be inhabited by residents. In this case, $R(0) = 1$, and expression (5) takes the following form:

$$R(t) = P\{Z(\omega\varphi, \chi, t_i)\},\ 0 < t_i \le t \tag{6}$$

Ultimately, the overall reliability of a residential building can be expressed by the $R(t)$ function, which may have different distributions in different periods of the building's operation. Most often, variable (t) can be treated as a random variable of the continuous type and then the density function is a derivative of the distribution function $F(t)$:

$$f(t) = F'(t) = -R'(t) \tag{7}$$

and then:

$$F(t) = \int_0^\infty f(t)dt = \int_0^\infty \left(-R'^{(t)}dt \right) \tag{8}$$

These are important relationships in the process of analysing building structures.

Note that the reliability function is a decreasing function. This means that e.g., $R(t_i) < R(t_{i-1})$, and, in extreme cases for $t_i = 0$ and $t_i = \infty$, the reliability function takes the values $R(0) = 1$ and $R(\infty) = 0$. This is contrary to the distribution function $F(t)$, also called the unreliability function, for which t_i: $F(0) = 0$ and $F(\infty) = 1$ for the same values.

Further considerations were based on the definition of durability (T), which can be formalized as a function of the following quantities [61,64]: the reliability of a residential building $R(t)$, which is considered to be an event of randomly reaching the limit state by its element; the flux of physical aging extortions of a building and its elements $W(t)$ (sometimes in the form of step stimuli); and, the level of resistance of elements to the effects of extortions $D(t)$. When presenting the problem with regards to reliability, it can be defined as the functional:

$$R(t) = \Psi\{T(t), W(t), D(t)\} \tag{9}$$

in which reliability is expressed by the durability and flux of changes in the level of aging and, thus, the processes of physical wear (if the problem is simplified by not taking the processes of social wear into account). The wear processes are manifested by damage to a residential building and its elements, as a result of which the building is unfit for use (loses its serviceability value) and requires renovation (if it is technically possible and economically justified). With exploitation time, a residential building becomes unfit for use more often, as it is subjected to increasingly frequent damage (aging processes). For the sake of simplifying the considerations, distinguishing between the state of operability and inoperability was omitted. However, the subsequent state of being fit for use, after renovation, already represents a level of serviceability that is lower than the previous one. Therefore, it is possible to talk about a greater intensity of damage to the object with an increase in its operation time, although this is neither an increase reflecting the length of the building's service life nor an increase directly proportional to its age.

An important issue in renovation strategy is to define the basic determinants of the level of reliability concerning the use of residential buildings. The basic element of these studies is the determination of the defect intensity function, which is itself an important issue with regards to the assessment of renovation decisions regarding residential buildings.

When considering any two renovation intervals, it can be assumed that the random variable τ, which defines the time of failure-free operation of a building, takes values from the following interval [61]:

$$\left[\tau_{Ri} < \tau \leq \tau_{Ri+1}\right] \tag{10}$$

or, in general:

$$\left[t_i < \tau \leq t_i + \Delta t_i\right] \tag{11}$$

If it is assumed that in the interval $[0,t_i]$ no damage was found and, in the interval $[t_i,t_i+\Delta t_i]$, damage could occur, the expression $R(t_i+\Delta t_i)/R(t_i)$ is called the conditional probability of such an event, which assumes that there will be no damage in the interval $[t_i+\Delta t_i]$ if there is no damage in the interval $[0,t_i]$. It is a relationship with a domain defined on the interval $[0,1]$, also known as the Bayesian formula concerning conditional probability [37,48]:

$$P\{t_i; t_i + \Delta t_i\} = \frac{R(t_i + \Delta t_i)}{R(t_i)} \tag{12}$$

In extreme cases, damage may occur at the moment of $t_i+\Delta t_i$, and then the probability that is determined by equation (12) will assume value 1 (because $R(t_i+\Delta t_i) = R(t_i)$). If the damage occurs at the moment of t_i, then $P\{t_i; t_i+\Delta t_i\} = 0$ (and then $R(t_i+\Delta t_i) = 0$). Formula (12) indicates the conditional probability of the duration of the correct and fault-free operation of the object.

In subsequent steps, the following transformations were made, which allowed the concepts derived from expression (12) to be defined:

- relationship (12) was subtracted from unity:

$$1 - P\{t_i; t_i + \Delta t_i\} = U\{t_i; t_i + \Delta t_i\} \tag{13}$$

and the resulting complement $U\{t_i;t_i+\Delta t_i\}$ was interpreted as the probability of a faulty and incorrect operation of the object;

- the obtained expression (13) was divided by Δt_i, and the average value of the probability of damage in the object's operating time interval with the length Δt_i was obtained:

$$\frac{U\{t_i; t_i + \Delta t_i\}}{\Delta t_i} \tag{14}$$

- the limit (probability) of this transformed expression for $\Delta t_i \to 0$ was then calculated and the sought function of damage intensity $\lambda(t_i) = \lambda(t)$ was obtained:

$$\lim_{\Delta t_i \to 0} \frac{U\{t_i; t_i + \Delta t_i\}}{\Delta t_i} = \lambda(t) \tag{15}$$

- the obtained relationship (15) was further transformed through appropriate integration and differentiation operations until a more convenient mathematical form was obtained. It expressed the relationship between the function of damage intensity and reliability function:

$$R(t) = e^{-\int_0^t \lambda(t)dt} \tag{16}$$

Relationship (16) is the basic formula in the theory of exploitation, which is used under the name of the Wiener formula.

The distribution of the service life (exploitation) of a residential object $f(t)$, the reliability function $R(t)$, and the distribution of damage intensity $\lambda(t)$ are interdependent terms. The function of damage intensity itself depends—just like the reliability function (because it is a set of its arguments)—on many factors. These include the physical and chemical properties of the elements of residential buildings, the aging and wear processes $W(t)$, the requirements for these objects ($\omega\phi$), and the maintenance conditions of these objects (χ). The damage intensity function $\lambda(t)$ can take various forms. It may be monotonically increasing or decreasing (possibly with a few extreme points) or it may be constant over time. The time course of the damage intensity function $\lambda(t)$ reflects the course of the wear processes of a residential building throughout its service life. Table 1 shows the average values of the damage intensity function $\lambda(t)$ for the ten most important elements of the examined downtown tenement houses, as well as the indication of the lack of damage intensity $\lambda(t) < 0.12$, the damage tendency $0.12 < \lambda(t) < 0.20$, and a strong damage intensity $\lambda(t) > 0.20$.

The statistical form of this reliability measure is also used apart from the probabilistic (based on the probability calculus) approach for determining the failure intensity function in the exploitation theory [61]:

$$\overline{\lambda(t)} = \frac{n(t + \Delta t) - n(t)}{N(t)\Delta t} \tag{17}$$

where:

$N(t)$—the number of objects fit for use until time t;

$N(t + \Delta t)$—the number of damaged objects until time $t + \Delta t$; and,

$\overline{\lambda(t)}$—a statistical measure of damage intensity.

$\overline{\lambda(t)}$ is therefore the share of the number of defects in the analysed time unit (Δt) in interval $[t,t+\Delta t]$ and in the number of objects fit for use at the beginning of this interval, i.e., at time t.

For the renovation strategy of a residential building, an important issue is to find—with the assumed reliability level—a rational moment of renovation τR, i.e., to determine the most technically and economically advantageous average time of exploitation of the object $\tau 0$ from one renovation to the next one. The measure of the average inter-repair time (or the value of the average random variable τ, i.e., the variable that determines the time of the correct operation of the facility) is:

$$\tau_0 = E(\tau) = \int_0^{\tau_R} R(t)dt \tag{18}$$

It should be assumed that, in the period of building adaptation, i.e., for objects damaged during this period, the distribution of the damage intensity function is usually explained by various functions that have a monotonic and decreasing character of their course (more or less, depending on the nature of the damage intensity). The processes of the adaptation period are associated with the loss of serviceability, which is caused by the exceeding of the limit states. After a fairly long-term effect of loads, when they are continuously distributed and cyclically repeated in the process of exploitation, the Weibull function can be used, the density of which has the form of:

$$f(t) = \frac{\beta}{\alpha} t^{\beta-1} e^{\frac{t\beta}{\alpha}} \tag{19}$$

where: β—shape parameter, α—scale parameter.

Because the damage intensity function can be written using the parameters α and β in the form:

$$\lambda(t) = \frac{\beta}{\alpha} t^{\beta-1} \tag{20}$$

then:

$$R(t) = e^{-\int_0^{\infty} \frac{\beta}{\alpha} t^{\beta-1} dt} \tag{21}$$

Table 1. Average values of the damage intensity function λ(t) for the ten most important elements of residential buildings.

Average Values of Intensity Defects Formation for 10 Selected Most Critical Elements of Residential Buildings		Foundations	Walls of Basement	Solid Floor over Basement	Main Walls	Inter - Storey Wooden Floors	Stairs	Roof (Rafter Framing)	Window Joinery	Inner Plasters	Facades
defect No	type of defect	λ(t)2	λ(t)3	λ(t)4	λ(t)5	λ(t)6	λ(t)7	λ(t)8	λ(t)9	λ(t)10	λ(t)11
d1	Mechanical defects						0.05		0.29	0.09	0.28
d2	Leaks								0.26		
d3	Mechanical decrements of bricks	0.13	0.23	0.08	0.19		0.03				
d4	Mechanical decrements of mortar		0.28		0.30						
d5	Decrements caused by rotten bricks	0.14	0.07	0.00	0.17						
d6	Decrements caused by rotten mortar		0.05		0.09					0.47	0.48
d7	Paint coating's peeling off									0.15	0.55
d8	Paint coating's falling off									0.25	0.57
d9	Craks of bricks	0.05	0.01	0.05	0.11						
d10	Craks of plaster		0.03		0.03					0.30	0.63
d11	Scratching of walls				0.21						
d12	Scratching of plaster				0.12	0.05				0.18	0.63
d13	Loosening of plaster					0.09				0.67	0.81
d14	Plaster's falling off									0.57	0.50
d15	Signs of permanent damp	0.70	0.74	0.58	0.56	0.07		0.43	0.83	0.70	0.84
d16	Weeping	0.64	0.52	0.67	0.46	0.27	0.59	0.50	0.74	0.61	0.79
d17	Biological corrosion of bricks	0.36	0.31		0.67						
d18	House fungus					0.45				0.38	0.60
d19	Mould & decay	0.49	0.43		0.34				0.49	0.41	0.56
d20	Localized corrosion of steel beams			0.42				0.54			
d21	Surface corrosion of steel beams			0.29				0.61			

Table 1. *Cont.*

d22	Pitting corrosion of steel beams			**0.55**			**0.53**				
d23	Flooding of foundation			**0.45**							
d24	Wooden beams of floor sensitiveness to dynamic activity of human's weight					0.00					
d25	Deformation of wooden beams					0.12					
d26	Skewing of joinery									**0.42**	
d27	Warp of joinery									0.04	
d28	Stratification of wooden elements							0.07			
d29	Partial deterioration of wooden elements pest attacked						**0.38**	**0.28**	**0.45**		
d30	Total deterioration of wooden elements pest attacked					**0,43**		0.57	0.42		
	number of cases:	100	93	93	100	100	100	100	100	97	100

Damage intensity function $\lambda(t)$ appears in three following ranges: $\lambda(t)$—lack of intensity defects formation; $\lambda(t)$—tendency to intensity defects formation; $\lambda(t)$—strong intensity defects formation.

A typical distribution for objects subjected to damage in the first period of exploitation may be the gamma distribution in the form:

$$f(t) = \frac{\lambda(\lambda(t))^{\beta-1}e^{-\lambda(t)}}{\int_0^\infty e^{-t}t^{\beta-1}dt} \tag{22}$$

in which the denominator is the so-called Euler's integral. This distribution represents the processes of loss of serviceability well, which are based on the cumulative effects of external factors. Ultimately, as a result of the elimination of all the defects and faults covered by the warranty, it is assumed that gradual and sudden defects (the latter ones that result from the step operation of stimuli in the conditions of accumulation of wear) are completely eliminated. This characterizes the end of the adaptation period of a residential building and the beginning of its normal operation period. The shape of the intensity function becomes "smooth". Hence, value 1 for the shape parameter (β) can be assumed for both the Weibull distribution and gamma distribution. It is a very characteristic period of a residential building's service life, in which the risk function takes a constant value ($\lambda(t) = \lambda$). This results from the following transformations of the mentioned functions:

- for the Weibull function:

$$\lambda(t) = \frac{\beta}{\alpha}t^{\beta-1} = \frac{1}{\alpha}t^{1-1} = \frac{1}{\alpha} = \lambda \tag{23}$$

- for the gamma function (using a shortcut for long calculations):

$$\lambda(t) = \frac{f(t)}{R(t)} \tag{24}$$

$$f(t) = \lambda e^{-\lambda t} \tag{25}$$

$$F(t) = 1 - e^{-\lambda t} \tag{26}$$

and using dependence (2):

$$R(t) = e^{-\lambda t} \tag{27}$$

ultimately:

$$\lambda(t) = \frac{\lambda e^{-\lambda t}}{e^{-\lambda t}} = \lambda(= const) \tag{28}$$

If the risk function has a domain defined by segment ti of the exploitation time ($0 < t_i \leq t$), then $\lambda(t)$ = const (for $\lambda > 0$ and $t > 0$) and the time of correct operation of a residential building has an exponential distribution. Finally, after several mathematical transformations of Equation (18), it is possible to determine the most technically and economically advantageous average service life of the object $\tau 0$ from the end of the warranty period:

$$\tau_0 = \frac{1}{\lambda} \tag{29}$$

The last dependence leads to an extremely important conclusion for the mechanism of the occurrence of defects in residential buildings: for the period of using a facility, in which the time of correct operation to damage has an exponential distribution, the average remaining time of failure-free operation is unchanged at any time. Therefore, after a certain time of failure-free operation, residential buildings fulfil their functions, just like new ones, and after exceeding the planned exploitation time beyond the assumed service life ($t > T$), they show "over-durability"—Figure 3:

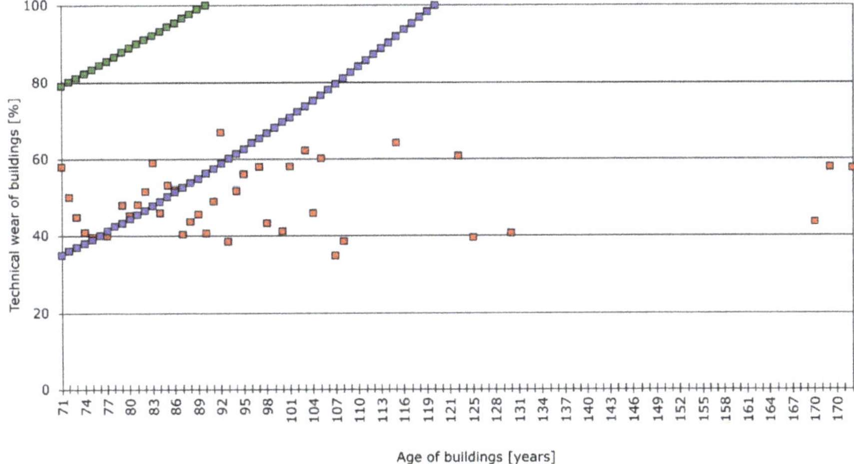

Figure 3. A diagram of the identified "over-durability" of the examined downtown tenement houses with regards to the charts of theoretical technical wear during very good and bad maintenance conditions.

It is clearly seen in Figure 3 that the observed technical wear of examined buildings is within the area that is determined by theoretical wear at poor and good maintenance by the age of 85 years. Within the range of $85 < t < 120$, the theoretical curves do not match the observed ones (the older, the worse). Over the age of 120 years, the inspected buildings should head for "technical death" ($t = T$), whereas a trend of "over-durability" ($t > T$) is noted on regular basis.

3. Conclusions

The adopted model and method of testing a representative group of downtown residential buildings with a traditional construction, which were erected at the turn of the 19th and 20th centuries, indicate that the age of the elements of old residential buildings is of secondary importance in the process of the intensity of loss of their serviceability value. No more than 30% of the element's damage can be explained by the passage of time if we assume that the coefficient of determination is the measure of the adjustment of the mathematical models (as a function of the technical wear of building elements over time), which are tested in the nonlinear regression method. Therefore, it is not age that determines the course of the technical wear of the analysed building components.

The analysis of the exploitation processes of residential buildings and the transformations of the basic dependencies of the reliability theory indicate that, for the service life of an object, in which the time of correct operation to failure has an exponential distribution (it is basically the service life corresponding to the length of operation of the considered residential buildings), the average remaining time of failure-free operation is unchanged at any time. Theoretically, residential buildings fulfil their functions, just like new ones, after a certain period of failure-free operation. The optimal moment of renovation occurs after the end of the second period of operation, before the period of rapid wear. Expressing the average duration of the correct failure-free operation of an object ($\tau 0$) by the reliability function $R(t)$, which determines the probability with which the correct operation time of an object will be longer than t_i, has a practical application in the exploitation of a residential building and its components.

The study of the course of the damage intensity function $\lambda(t)$ over time reflects the wear process of a residential building in a representative sample of downtown residential buildings erected while using traditional methods. It authorizes the formulation of the following conclusions—Table 1:

- there is measurable the damage intensity function in interval $[0, t]$ for all 10 tested building elements, but the damage intensity force shows a significant span (from 0.00 to 0.84);
- as a rule, damage that is caused by water penetration and moisture penetration is of the highest intensity -0.54 on average;
- the technical condition of each of the tested elements also shows the intensity of defects that are characteristic for their design and material solutions, e.g.,:

  - damage to wooden parts of elements (ceiling beams, stair treads, roof trusses, window joinery), which are attacked by biological pests;
  - mechanical damage to the structure and texture, the intensity of which applies only to those elements in which the damage may cause the intensification of the impact of subsequent (cumulative) defects, e.g., construction walls underground and aboveground, as well as internal and external plasters (but not foundations or massive cellar ceilings); and,

- damage that is manifested by the loss of the original shape of wooden elements can be considered as not very intense; an exception is the torsion of window joinery (with an intensity of 0.42), for which this damage determines a significant decrease in its serviceability value.

4. Summary and Discussion

The practical approach to the problem of the intensity of the formation of defects in the tested residential buildings with regards to the change in their reliability condition enables summing up the following findings:

- generally used normative definitions of the reliability of buildings facilitate the study and interpretation of the course of exploitation processes of residential buildings;
- for the purpose of a comprehensive assessment of changes in the reliability level of residential buildings, various reliability characteristics should be used, in which the damage intensity function is of key importance, as it enables the construction of other reliability indicators; and,
- using the characteristics of the reliability of a residential building in renovation decisions allows for a rational renovation strategy to be determined by e.g., the determination of maintenance intervals on the basis of established damage intensity distributions.

The methodological approach to the technical assessment of buildings, and their durability and reliability, has been known and presented in the literature for many years, especially in the papers of Arendarski [5], Zaleski [8,9], Thierry [21], and Tymiński [61]. Their works are used as manuals for managers and administrators regarding the use of buildings. Nowogońska [11,18,28,30,31,39,40] deals with the study of the impact of the maintenance of residential buildings on both the degree of damage and the reliability function. The reliable results of these studies are presented with a division into building elements with the greatest share and significance for the proper functioning of the examined buildings. Such a division is particularly important in the last period of the building's "service life", when its operation time is approaching its expected durability.

A similar approach, as presented in previous publications, was adopted by the authors of this article, who conducted the technical assessment of tenement houses under the supervision of Marcinkowska and Czapliński [44–48,58]. This methodology of diagnosing residential buildings has been presented for many years by the researchers of the so-called "German school"—Deutschmann [6] and Zimmermann [10]. They indicated the methods of measuring technical wear, damage size and the aging process of the load-bearing structure of engineering objects. Plebankiewicz, Zima, and Wieczorek [32,33] deal with the life cycle of a building object with regards to the risk and cost of its restoration as a result of renovation activities, which—in the opinion of the authors of this publication—should be a secondary feature: cost versus technical. After all, increasing cultural, historical, and humanistic aspects prevail over material ones when making decisions about the so-called "technical death" of facilities that are located in the centres and suburbs of many cities. Therefore, the priority seems to be to study the damage intensity of residential buildings in the context of changes in their reliability state.

It is worth noting that the discussed quantitative data can provide the basis for programming the size and structure of specialized construction companies that are involved in the maintenance and renovation of residential buildings. These data are included in the technical information that is necessary for managing buildings and designing the organization of these maintenance activities for residential buildings, which, in turn, determines the quality of broadly understood housing maintenance conditions.

Finally, attention should be paid to the individual nature of the results of the study, which was based on research on a homogeneous coherent group of downtown tenement houses. The transfer of the results of the technical assessment to a different population of residential buildings with a traditional construction should be conducted with great caution and with the necessity to perform surveys. Undoubtedly, such studies should be preceded by the careful, purposeful selection of a typological sample that is representative for the general population. Such a sample may contain a much smaller number of objects, but it is extremely important that the decisive selection criterion for the technical assessment involves the elements (or only parts of them) that are essential for the structure (load-bearing structure) of the building. This division is especially important when examining composite and complex elements. It can then be assumed that, from the point of view of the ultimate limit states of elements, the degree of their technical wear (while maintaining the safe and reliable operating conditions of the object) is equal to 75%.

The methodological aspects of the reliability of the quantitative results of the technical assessment should also aim to minimize the subjectivity of expert judgment in the process of technical examinations of residential buildings by specifying the type of the predicted random impacts, and by determining the variability of at least some of them. It should also be remembered that the issue of technical tests of buildings (especially residential buildings) needs to be updated with full recognition of the forms of their immaterial wear—social and economic. It is a sign of recent times that it is the psychological aspects of the perception of the process of decline in the serviceability value of flats by their users, being supported by the analysis of the profitability of replacing entire buildings, which play a fundamental role when making decisions regarding the future of downtown housing developments.

Therefore, the results of the research should be treated as an exploratory study, the main aim of which was to model the solution of cause–effect relationships. These dependencies indicate the type and size of damage that shows the impact of the maintenance conditions of downtown residential buildings on the technical wear of their components. Like any exploratory solution, it should be treated as a multi-criteria recognition of the mechanism of the occurrence and effects of phenomena that are encountered by an adjudicator at each stage of the technical assessment of an engineering object. However, this assessment, in its nature, includes an unmeasurable (partly subjective) aspect. The construction of a new model of the technical inspection of residential buildings, which is based on the assumptions and conclusions resulting from the study, will allow the burden of the results of the technical assessment to be considered as more quantitative than qualitative. The intention of the authors is that further work related to the broadly understood diagnosis of technical objects should go in this direction.

Author Contributions: Conceptualization, J.K., M.S. (Marek Sawicki) and M.S. (Mariusz Szóstak); methodology, J.K.; software, J.K. and M.S. (Mariusz Szóstak); validation, J.K., M.S. (Marek Sawicki) and M.S. (Mariusz Szóstak); formal analysis, J.K., M.S. (Marek Sawicki) and M.S. (Mariusz Szóstak); investigation, J.K. and M.S. (Marek Sawicki); resources, J.K. and M.S. (Marek Sawicki); writing—original draft preparation, J.K.; writing—review and editing, J.K. and M.S. (Mariusz Szóstak); supervision, M.S. (Marek Sawicki). All authors have read and agreed to the published version of the manuscript.

Funding: This research received no external funding.

Conflicts of Interest: The authors declare no conflict of interest.

References

1. Kim, B.; Ahn, Y.; Lee, S. LDA-based model for defect management in residential buildings. *Sustainability* **2019**, *11*, 7201. [CrossRef]
2. Rotimi, F.; Tookey, J.; Rotimi, J. Evaluating defect reporting in new residential buildings in New Zealand. *Buildings* **2015**, *5*, 39–55. [CrossRef]
3. Sandeford, P. Planning for durability and life cycle maintenance of buildings during design. In Proceedings of the International Conference on Performance-based and Life-cycle Structural Engineering, Brisbane, QLD, Australia, 9–11 December 2015; University of Queensland Library: Brisbane, Australia, 2016; pp. 1094–1103.
4. Šnīdere, L.; Geipele, I.; Stāmure, I. Case study of standard multi-storey residential building owners and tenants' perception of building technical conditions and renovation issues. *Balt. J. Real Estate Econ. Constr. Manag.* **2017**, *5*, 6–22. [CrossRef]
5. Arendarski, J. *Durability and Reliability of Residential Buildings Erected Using Industrialized Methods*; Arkady: Warsaw, Poland, 1978. (In Polish)
6. Deutschmann, E. *Development of Measuring Techniques In Situ for Residential Buildings*; Wissenschaftliche Zeitschrift der Technischen Universitaet: Dresden, Germany, 1987.
7. Merminod, P. *Investigation of Wear and Damages of Old Buildings*; Chantiers Revue du Batiment du Genie civil et de la Securite: France, 1984.
8. Zaleski, J. *Renovations and Modernizations of Residential Buildings*; Arkady: Warsaw, Poland, 1995. (In Polish)
9. Zaleski, J. *Renovation of Residential Buildings, Guide*; Arkady: Warsaw, Poland, 1995. (In Polish)
10. Zimmermann, G. *Impairment of Structures-Building Failure, Damages, Wear and Tear and Aging*; Deutsches Architektenblatt, Ausgabe Baden: Wuerttemberg, Germany, 1976.
11. Nowogońska, B. Prognosis of the technical condition of masonry walls in residential buildings. *Bud. Archit.* **2014**, *13*, 27–32. [CrossRef]
12. Macarulla, M.; Forcada, N.; Casals, M.; Gangolells, M.; Fuertes, A.; Roca, X. Standardizing housing defects: Classification, validation, and benefits. *J. Constr. Eng. Manag.* **2013**, *139*, 968–976. [CrossRef]
13. Chen, C.; Juan, Y.; Hsu, Y. Developing a systematic approach to evaluate and predict building service life. *J. Civil Eng. Manag.* **2017**, *23*, 890–901. [CrossRef]
14. Chong, W.; Low, S. Assessment of defects at construction and occupancy stages. *J. Perform. Constr. Facil.* **2005**, *19*, 283–289. [CrossRef]
15. Pan, W.; Thomas, R. Defects and their influencing factors of posthandover new-build homes. *J. Perform. Constr. Facil.* **2015**, *29*, 04014119. [CrossRef]
16. Kim, Y.; Oh, S.; Cho, Y.; Seo, J. A PDA and wireless web-integrated system for quality inspection and defect management of apartment housing projects. *Autom. Constr.* **2008**, *17*, 163–179. [CrossRef]
17. Jernberg, P.; Lacasse, M.; Haagenrud, S.; Sjöström, C. *Guide and Bibliography to Service Life and Durability Research for Buildings and Components*; Joint CIB W80/RILEM TC 140–TSL Committee on Service Life of Building Materials and Components CIB Report, Publication 295; CIB: Rotterdam, The Netherlands, 2004.
18. Nowogońska, B. Diagnoses in the aging process of residential buildings constructed using traditional technology. *Buildings* **2019**, *9*, 126.
19. van Breugel, K.; Koleva, D.; van Beek, T. *The Ageing of Materials and Structures*; Springer International Publishing: Delft, The Netherlands, 2018.
20. Thier, H. *Cases of Damage and Their Analysis*; Deutscher Verlag fuer Schweisstechnik: Duesseldorf, Germany, 1985.
21. Thierry, J.; Zaleski, J. *Renovation of Buildings and Strengthening of Structures*; Arkady: Warsaw, Poland, 1982. (In Polish)
22. Silva, A.; de Brito, J.; Gaspar, P. *Methodologies for Service Life Prediction of Buildings with a Focus on Facade Claadings*; Springer International Publishing: Lisbon, Portugal, 2016.
23. Thomsen, A.; van Der Flier, K. Understanding obsolescence: A conceptual model for buildings. *Build. Res. Inf.* **2011**, *39*, 352–362. [CrossRef]

24. Hoła, B.; Nowobilski, T. Analysis of the influence of socio-economic factors on occupational safety in the construction industry. *Sustainability* **2019**, *11*, 4469. [CrossRef]

25. Grzyl, B.; Apollo, M.; Miszewska-Urbańska, E.; Kristowski, A. Management of exploitation in terms of life cycle costs of built structures. *Acta Sci. Pol. Archit.* **2017**, *16*, 85–89.

26. Forster, A.; Kayan, B. Maintenance for historic buildings: A current perspective. *Struct. Surv.* **2009**, *27*, 210–229. [CrossRef]

27. Sugier, J.; Anders, G. Modelling and evaluation of deterioration process with maintenance activities. *Maint. Reliab.* **2013**, *15*, 305–311.

28. Nowogońska, B.; Korentz, J. Value of technical wear and costs of restoring performance characteristics to residential buildings. *Buildings* **2020**, *10*, 9. [CrossRef]

29. Ramuhalli, P.; Henager, C.; Griffin, J.; Meyer, R.; Coble, J.; Pitman, S.; Bond, L. Material aging and degradation detection and remaining life assessment for plant life management, Third International Conference on Nuclear Power Plant Life Management. In Proceedings of the International Conference, Salt Lake City, UT, USA, 14–18 May 2012.

30. Nowogońska, B. Intensity of damage in the aging process of buildings. *Arch. Civil Eng.* **2020**, *66*, 19–31.

31. Nowogońska, B. Proposal for determing the scale of renovation needs of residential buildings. *Civil Environ. Eng. Rep.* **2016**, *22*, 137–144. [CrossRef]

32. Plebankiewicz, E.; Zima, K.; Wieczorek, D. Original model for estimating the whole life costs of buildings and its verification. *Arch. Civil Eng.* **2019**, *65*, 163–179. [CrossRef]

33. Plebankiewicz, E.; Zima, K.; Wieczorek, D. Life cycle cost modelling of buildings with consideration of the risk. *Arch. Civil Eng.* **2016**, *65*, 149–166. [CrossRef]

34. Grant, A.; Ries, R. Impact of building service life models on life cycle assessment. *Build. Res. Inf.* **2013**, *41*, 168–186. [CrossRef]

35. Bucoń, R.; Sobotka, A. Decision-making model for choosing residential building repair variants. *J. Civil Eng. Manag.* **2015**, *21*, 893–901. [CrossRef]

36. Knyziak, P. Estimating the technical deterioration of large-panel residential buildings using artificial neural networks. *Procedia Eng.* **2014**, *91*, 394–399. [CrossRef]

37. Radziszewska-Zielina, E.; Śladowski, G. Proposal of the use of a fuzzy stochastic network for the preliminary evaluation of the feasibility of the process of the adaptation of a historical building to a particular form of use. *Iop. Conf. Ser. Mater. Sci. Eng.* **2017**, *245*, 1–11. [CrossRef]

38. Grądzki, R.; Lindstedt, P. Method of assessment of technical object aptitude in environment of exploitation and service conditions. *Maint. Reliab.* **2015**, *17*, 54–63. [CrossRef]

39. Nowogońska, B. Preventive services of residential buildings according to the Pareto principle. *Iop. Conf. Ser. Mater. Sci. Eng.* **2019**, *471*, 112034. [CrossRef]

40. Nowogońska, B. The life cycle of a building as a technical object. *Period. Polytech. Civil Eng.* **2016**, *60*, 331–336. [CrossRef]

41. Zaidi, A.; Bouamama, B.; Tagina, M. Bayesian reliability models of weibull systems: State of the art. *Int. J. Appl. Math. Comput. Sci.* **2012**, *22*, 585–600. [CrossRef]

42. Cordeiro, G.; Ortega, E.; Lemonte, A. The exponential-Weibull lifetime distribution. *J. Stat. Comput. Simul.* **2014**, *84*, 2592–2606. [CrossRef]

43. Khelassi, A.; Theilliol, D.; Weber, P. Reconfigurability analysis for reliable fault-tolerant control design. *Int. J. Appl. Math. Comput. Sci.* **2011**, *21*, 431–439. [CrossRef]

44. Konior, J. Technical assessment of old buildings by fuzzy approach. *Arch. Civil Eng.* **2019**, *65*, 129–142. [CrossRef]

45. Marcinkowska, E. Technical and functional requirements of building deterioration estimation. *Arch. Civil Eng.* **2002**, *48*, 125–134.

46. Marcinkowska, E.; Gawron, K. Down-town tenement-houses in Wroclaw as a subject of real estate sales. *Pr. Nauk. Inst. Budownictwa Politech. Wrocławskiej. Studia I Mater.* **2006**, *87*, 315–324. (In Polish)

47. Marcinkowska, E. Analyses of techical and functional requirements of building modernization. *Arch. Civil Eng.* **2003**, *49*, 53–62.

48. Multi-Author Work under K. *Czapliński Lead, Assessment of Wrocław Downtown Apartment Houses' Technical Conditions*; Reports Series "U" of Building Engineering Institute; Wrocław University of Technology: Wrocław, Poland, 1984; Volume 96.

49. Govindarajulu, Z. *Elements of Sampling Theory and Methods*; Pearson: Upper Saddle River, NJ, USA, 1999.
50. Barnett, V. *Elements of Sampling Theory*; PWN: Warsaw, Poland, 1982. (In Polish)
51. Baer, D. *An Introduction to Latent Variable Models in the Social Sciences*; University of Western Ontario: London, ON, Canada, 1993.
52. Gerstenkorn, T.; Śródka, T. *Combinatorics and Probability Calculus*; PWN: Warsaw, Poland, 1980. (In Polish)
53. Greń, J. *Models and Tasks of Mathematical Statistics*; PWN: Warsaw, Poland, 1978. (In Polish)
54. Hellwig, Z. *Elements of Probability Calculus and Mathematical Statistics*; PWN: Warsaw, Poland, 1993. (In Polish)
55. Stuart, A. Rank Correlation Methods. *By M. G. Kendall. Br. J. Stat. Psychol.* **1956**, *9*, 68. [CrossRef]
56. Morrison, D. *Multivariate Statistical Methods*, 4th ed.; Thomson Brooks/Cole: New York, NY, USA, 2004.
57. Zając, K. *Outline of Statistical Methods*; PWE: Warsaw, Poland, 1994. (In Polish)
58. Marcinkowska, E. *Decision-Making Problems in the Design of Buildings and Construction Processes*; Monograph of the Wroclaw University of Science and Technology: Wroclaw, Poland, 1986.
59. Sognen, O. The Scandinavian approach to maintenance and modernization, International Symposium on Management, Maintenance and Modernisation of Building Facilities. In Proceedings of the Strategies and Technologies for Maintenance and Modernisation of Building, Tokyo, Japan, 26–28 October 1994.
60. Torregrossa, J. *Liquidated Damages: Three Myths Exposed*; Construction-Specifier: London, UK, 1987.
61. Tymiński, J. *Elements of the Theory of Reliability of Residential Buildings*; Wydawnictwo WSGK w Kutnie: Kutno, Poland, 2001. (In Polish)
62. Weil, G. *Detecting the Defects*; Civil Engineering ASCE: London, UK, 1989.
63. Pihlajavaara, S. *Background and Principles of Long-Term Performance of Building Materials, Durability of Building Materials and Components*; Sereda, P., Litvan, G., Eds.; ASTM International: West Conshohocken, PA, USA, 1980; pp. 5–16.
64. Wolff, R.; Dicke, P. *Maintenance and Modernisation*; Beton: Germany, 1988.

Article

Risk Assessment of Large-Scale Infrastructure Projects—Assumptions and Context

Jana Korytárová * and Vít Hromádka

Faculty of Civil Engineering, Brno University of Technology, 60200 Brno, Czech Republic; hromadka.v@fce.vutbr.cz
* Correspondence: korytarova.j@fce.vutbr.cz; Tel.: +420-733-164-369

Featured Application: The results of the presented research extend the methodology of economic analysis and risk assessment of large infrastructure projects.

Abstract: This article deals with the partial outputs of large-scale infrastructure project risk assessment, specifically in the field of road and motorway construction. The Department of Transport spends a large amount of funds on project preparation and implementation, which however, must be allocated effectively, and with knowledge of the risks that may accompany them. Therefore, documentation for decision-making on project financing also includes their analysis. This article monitors the frequency of occurrence of individual risk factors within the qualitative risk analysis, with the support of the national risk register, and identifies dependent variables that represent part of the economic cash flows for determining project economic efficiency. At the same time, it compares these dependent variables identified by sensitivity analysis with critical variables, followed by testing the interaction of the critical variables' effect on the project efficiency using the Monte Carlo method. A partial section of the research was focused on the analysis of the probability distribution of input variables, especially "the investment costs" and "time savings of infrastructure users" variables. The research findings conclude that it is necessary to pay attention to the setting of statistical characteristics of variables entering the economic efficiency indicator calculations, as the decision of whether or not to accept projects for funding is based on them.

Keywords: CBA; investment project; probability distribution; sensitivity analyses; risk assessment

check for **updates**

Citation: Korytárová, J.; Hromádka, V. Risk Assessment of Large-Scale Infrastructure Projects—Assumptions and Context. *Appl. Sci.* **2021**, *11*, 109. https://dx.doi.org/10.3390/app1101 0109

Received: 1 December 2020
Accepted: 22 December 2020
Published: 24 December 2020

Publisher's Note: MDPI stays neutral with regard to jurisdictional claims in published maps and institutional affiliations.

1. Introduction

Transport infrastructure projects are important carriers and supporters of economic growth for national economies. Implementation of investment projects, in addition to the direct benefits for which they are implemented, brings growth potential for the national economy; they reduce unemployment, increase the sales of design and implementation companies, and thus create revenue capacity on the demand side for purchases of goods and services. Implementation of investment projects will also be a key factor in alleviating the current COVID-19 pandemic effect in all national economies; e.g., the draft of the state budget of the Czech Republic brings record investments for the future, which have been increased by CZK 178 billion for 2021 (€6.7 billion). Even so, the supply of funds for project implementation is limited. Therefore, it is always necessary to choose for financing only those projects that are efficient. The efficiency of projects to be implemented is assessed in the ex-ante period, on the basis of feasibility study data, which is addressed in the form of a cost–benefit analysis (CBA).

The authors of this article have been carrying out research into development in economic efficiency assessment of public transport infrastructure projects for a long time. In the present article they focused on the analysis of the economic outputs of road infrastructure projects, motorways, and class I roads via CBA. CBA has the largest explanatory power [1–4], which is based on the determination of cost-effectiveness against the total

Appl. Sci. **2021**, *11*, 109. https://dx.doi.org/10.3390/app11010109

societal benefits. Generally, four criteria are solved and monetized in large-scale transport infrastructure project appraisals: travel time savings, travel and operational costs, safety, and environmental cost, from different perspectives. In ref. [5] based on the modeling of economic cash flows determined by these variables, the following economic efficiency indicators were established: economic net present value (ENPV), economic internal rate of return (ERR), and benefit cost ratio (BCR) [6,7]. The values of the economic indicators were tested for critical variables and the switching values of indicators (threshold value of the indicator in terms of efficiency, e.g., ENPV = 0, ERR = discount rate) were determined. In the following step, a quantitative risk analysis using the Monte Carlo method was performed for the identified critical variables. At the same time, a qualitative risk analysis, which considered potential risk factors using a risk register [7], was performed. It monitored the project risk impact, the occurrence probability, and deduced the risk relevance for the implementation and operation of the project. Individual projects that demonstrated a positive evaluation from all perspectives examined are ready for funding, and further phases of their life cycle can be launched for them. The research question addressed by the implemented research team was which variables are risky, how strong is their influence on economic efficiency, and whether and how the projects are resilient; robust to the potential risk interaction. This concerns questions of the connection of the qualitative and quantitative risk analysis, which dependant variables are resulting from the qualitative analysis, and if they are also considered in the quantitative analysis. In the case of the important critical variable it was the objective to test the changes of the efficiency of projects while using different probability distributions.

Investors aim, not only to prevent project failure, but also to select the best alternatives among the available investment projects, so as to gain more benefits and achieve better results [8]. In the investment decision-making process of large-scale projects, many risk factors can cause decision failure [9]. This is also why decision-support systems are of high importance for investors in the construction industry [10].

2. Materials and Methods

The aim of the research described in this paper was to find the relations between the outputs of the qualitative risk analysis, sensitivity analysis, and quantitative risk analysis, which were performed in the evaluation of the economic efficiency of transport infrastructure projects, as part of the modeling of economic Cash Flow (CF) of their life cycle. For the case study, a set of projects being prepared for realization in the Czech Republic was chosen. The authors of the paper have many years of experiences in the evaluation of projects in Czech transport infrastructure, and during these years they were able to collect a large amount of input data. However, the authors would like to emphasize that for the presented procedures, and partly also for the results, it is possible, respecting individual specifics of economic evaluation in other countries, to relate them to projects carried out abroad. The research sample consisted of 20 large-scale transport infrastructure projects from the Czech Republic, which were the pre-investment phase in the 2018–2020 period, and with proven economic efficiency. Only those projects that could be compared with each other due to the fact that they were processed according to the same methodological procedure, e.g., according to the Departmental Methodology valid since 2017 [7], were included in the research sample.

Net cash flow (NCF) for the calculation of economic ratios consisted of the savings in the costs of the suggested (investment) variant related to the zero variant (without investment). The calculation formula consists of four types of particular benefits; socio-economic savings. They are savings in travel and operating costs, savings in travel time costs, reduction in accident costs, and savings in exogenous costs. The time value of money determining the amount of the discount rate for the calculation of the ENPV indicator was set at 5% for the Czech Republic in the EU programming period 2014–2020.

The research presented in this article examined project risk frequency, and the impact on their economic efficiency and robustness of economic efficiency indicators, using sensitivity analysis, and finally involved confirmation or refusing the robustness of projects according to the previous step by determining the cumulative probability of achieving project economic efficiency using the Monte Carlo method. To assess the real risk of failure associated with the investment, changes in the values of economic performance indicators deriving from the simultaneous change of several project variables had to be identified [11]. As stated by [12], one of the risk assessment tools is the Monte Carlo method, which combines and develops both sensitivity analysis, and scenario analysis, methods. In the resource material, Ref. [13] focused on the Monte Carlo method used in the case of the earned value management methodology. Bowers also provided a broader view of the issue of project risk assessment [14].

2.1. Data

Table 1 presents the research sample projects with their basic characteristics. It states the undiscounted economic investment costs (i.e., investment costs excluding VAT reduced by a conversion coefficient 0.807), economic internal rate of return (ERR), economic net present value (ENPV), and cost benefit ratio (BCR), which was calculated according to the following relation:

$$\frac{B}{C} = 1 + \frac{ENPV}{IC} \tag{1}$$

where:

BCR: Cost Benefit Ratio
ENPV: Economic Net Present Value
IC: Discounted Investment Costs

Table 1. Basic economic data on research sample projects.

No.	Name of the Project	IC €	ERR %	ENPV €	BCR
P1	Vestec connection	73,655,517	13.15%	134,141,506	2.90
P2	I/22 Draženov-Horažďovice	253,477,033	5.67%	25,929,610	1.11
P3	I/27 Kaznejov, bypass	91,192,128	9.50%	74,002,422	1.83
P4	I/13 Ostrov-Smilov, right bank	141,082,434	5.88%	19,811,383	1.15
P5	I/13 Ostrov-Smilov, left bank	116,820,770	7.52%	50,193,343	2.01
P6	I/26 Horšovský Týn	50,375,269	5.60%	4,578,849	1.09
P7	D0 Březiněves-Satalice var. 1	371,886,072	39.45%	2,576,573,157	8.28
P8	D0 Březiněves-Satalice var. 2	434,933,917	30.46%	2,395,820,591	6.92
P9	D0 Březiněves-Satalice var. 3	757,919,450	17.89%	1,934,644,942	3.81
P10	I11– Hradec Králové, tangent	111,776,135	17.24%	336,621,090	4.15
P11	I/18 Příbram-bypass var. 1	28,417,453	14.20%	54,410,634	2.96
P12	I/18 Příbram-bypass var. 2	49,497,029	13.21%	74,973,161	2.61
P13	I/50 Bučovice	78,579,450	7.56%	32,937,152	1.44
P14	I/36 Trnová-Fablovka-Dubina	53,652,370	19.20%	190,286,624	4.73
P15	I/11 Nové Sedlice-Opava Komárov	91,436,523	5.52%	7,834,232	1.09
P16	I/26 Holysov, bypass	56,624,471	9.19%	42,452,457	1.80
P17	D10 Praha-Kosmonosy	361,367,050	5.72%	35,994,616	1.11
P18	I/67 Bohumín-Karviná	83,937,876	5.33%	4,067,671	1.05
P19	D43 Bořitov-Staré Město	56,624,471	9.19%	42,452,457	1.80
P20	D27 Přeštice-Klatovy	128,638,259	5.12%	22,333,326	1.02

Source: Feasibility Studies of Investment projects, The State Fund for Transport Infrastructure SFDI, authors' own processing.

Qualitative risk analysis is generally based on expert opinions on the risks that threaten a particular investment project. Lists of risks are usually created, based on the knowledge of the issues addressed, which contain risks that are relevant and common for the given type of projects. A risk register was created in the Czech Republic for the purposes of risk assessment of the road infrastructure projects specified above [7]. The list of risks according to the risk register is given in Table 2.

Table 2. Risk register according to the Departmental Methodology of the Czech Republic.

No.	Risk Description
	Demand-related risks
R1	Different development of demand than expected
	Risks related to the project design
R2	Inadequate surveys and inquiries in the given locality
R3	Inadequate estimates of project work costs
	Administrative and public procurement risks
R4	Delays in awarding
R5	Building permit
	Risks related to the land purchase
R6	Land price
R7	Delays in land purchase
	Risks related to construction
R8	Exceeding investment costs
R9	Floods, landslides, etc.
R10	Archaeological findings
R11	Risks related to the contractor (bankruptcy, lack of resources)
	Operational risks
R12	Higher maintenance costs than expected
	Regulatory risks
R13	Environmental requirement change
	Other risks
R14	Public opposition

Source: Departmental methodology of the Ministry of Transport [7].

2.2. Methods

The methodological procedure was based on collection, analysis, and examination of relevant data concerning the economic efficiency assessment of individual investment projects. The outputs were aimed at answering research questions concerning the interconnectedness of individual analyses of future project uncertainties.

2.2.1. Qualitative Analysis

The significance of project risks (R) was divided into four categories: very high (VH), high (H), medium (M), and low (L). This was determined on the basis of the product of the project risk impact intensity (I) and its occurrence probability (p), with a five-interval scale of both variables, according to the following relation:

$$R = I \times p \tag{2}$$

The probability (value) and the impact intensity had the determined ranges presented in following Tables 3 and 4.

Table 3. Scale of risk occurrence probability (p).

Classification	Verbal Description	Percentage Expression
A	Very improbable	0–9%
B	Improbable	10–32%
C	Neutral	33–65%
D	Probable	66–89%
E	Very probable	90–100%

Source: Departmental methodology of the Ministry of Transport [7].

Table 4. Scale for risk impact intensity (I).

Category	Name	Verbal Description
I	Imperceptible	no significant effect on expected social benefits of the project
II	Mild	long-term project benefits are not affected but corrective measures are needed
III	Medium	loss of expected social benefits of the project, mostly financial loss and in medium- and long-term time horizon, corrective measures may solve the problem
IV	Critical	large loss of expected social benefits of the project, occurrence of adverse effects causes a loss of the project's primary function; corrective measures, even if taken on a large scale, are not sufficient to prevent major losses
V	Catastrophic	significant to complete loss of function of the project, project objectives cannot be achieved even in the long term

Source: Departmental methodology of the Ministry of Transport [7].

Table 5 shows the occurrence frequency of very high, high, and medium risks in the researched sample of projects, according to the risk register (see Table 1). In addition to the risk frequency, the table also shows the dependent variable, which enters the economic CF of the projects as a basis for the calculation of economic efficiency indicators.

It is clear from the overview given in Table 5 that the most significant risks for transport infrastructure projects identified in the pre-investment phase lie in the estimation of future demand for new infrastructure use (R1), design and preparatory work (R2), (R3), delays in obtaining construction permits (R5), land purchase (R7), and excess of project costs (R8).

The R1 risk is related to the demand, which affects the income part of the projects in the operational phase of their life cycle by a possible reduction in their expected socio-economic benefits.

The influence of other risks has a direct impact on investment costs, which thus become a significant variable in the economic assessment.

Table 5. Risk frequency according to their significance, including the dependent variable identification.

Risk No.	VH and H Risks	M Risk	Total	Dependent Variable
R1	3	5	8	Revenues alias operating phase savings
R2	5	8	12	Investment costs, beginning of the construction
R3	4	6	10	Investment costs
R4	0	5	5	Beginning of the construction
R5	0	9	9	Beginning of the construction
R6	0	2	2	Investment costs
R7	12	2	14	Beginning of the construction
R8	8	5	13	Investment costs
R9	0	1	1	Investment costs, extension of construction, delay/shortening of the operational phase for evaluation
R10	0	1	1	Investment costs, extension of construction, delay/shortening of the operational phase for evaluation
R11	0	2	2	Investment costs, extension of construction, delay/shortening of the operational phase for evaluation
R12	0	0	0	Operating costs, reduction of benefits under "Infrastructure operating costs" item
R13	0	0	0	Changes in benefits under "Externalities" item
R14	0	0	0	Influence on the beginning of construction

Source: Feasibility Studies of Investment projects, SFDI, authors' own processing.

2.2.2. Sensitivity Analysis

The outputs of the sensitivity analysis (elasticity coefficients and switching values of economic efficiency indicators) were investigated for individual projects in the following phase of the research in order to determine project resilience to changes in variables potentially affected by risks. The elasticity coefficients were determined both for investment costs and for all relevant socio-economic benefits, which as a total amount, form the income part of the economic CF (following the R1 risk).

It can be seen from the data in Table 6 that variables such as accident rate, externalities, and/or total operating costs generally have low elasticity coefficients, and are not in most cases identified as critical variables. Investment costs and the time savings of infrastructure users already showed that they very often become critical variables (EC > 1). For this reason, occurrences of switching values (i.e., ENPV = 0), which show the influence of these critical variables, were investigated in the following phase of the research. Outputs were divided into the interval of changes up to 10%, up to 30%, and over 30%. It can be clearly seen from Table 7 that the projects showed a relatively high efficiency robustness; about 70% of projects met a limit of efficiency when changing one of these critical variables up to 30%.

Table 6. Frequency of elasticity coefficient (EC) values.

Variable	$0 \leq EC < 0.5$	$0.5 \leq EC < 1$	$1 \leq EC < 1.5$	$EC \geq 1.5$
Total investment costs	5	4	4	5
Vehicle operating costs	16	1	1	0
User time costs	1	7	5	5
Accident rate	13	3	0	2
Other externalities	13	2	0	3

Table 7. Switching values of project efficiency.

Variable/Switching Value	$0 \leq$ PH < 10%	10% $\leq$ PH < 30%	PH $\geq$ 30%
Total investment costs	3	3	13
Time savings of users	2	3	14

The outputs of the sensitivity analysis and qualitative risk analysis showed that the total investment costs and time savings of transport infrastructure users represented fundamental risk variables that affected the efficiency of the investment projects. For this reason, these independent variables were tested by subsequent quantitative analysis, which was carried out by the Monte Carlo method, using Crystal Ball software [15].

In the case of the quantitative analysis, a relative index BCR was chosen, because it allows comparing the efficiency of projects of different sizes (investment demanding), and it shows the benefit of one invested currency unit. The utilization of the BCR index as one of the criterial indicators for the evaluation of the economic efficiency of public projects is methodically described in references [6,7]. The authors focused on comparing two assumptions of the probability distribution of the investment costs critical variable. The simulations were therefore performed in two variants, in the first variant the beta-PERT probability distribution was chosen for the investment costs, in the second variant a triangular asymmetric probability distribution was used. In order to be able to correctly compare the impact of the use of partial probability distributions of investment costs on the overall project results, an equally normal distribution was used for the second critical variable "time savings of infrastructure users" for both simulation variants.

The parameters of the probability distribution of investment costs in the case of the beta-PERT probability distribution assumption were therefore chosen as follows:

Minimum	project value reduced by 10%,
Most likely	project value,
Maximum	project value increased by 50%.

The parameters of the probability distribution of investment costs in the case of the asymmetric triangular probability distribution assumption were, in accordance with the recommendations arising from the background source [9], set with parameters comparable with the beta-PERT probability distribution, i.e., as follows:

Minimum	project value reduced by 10%,
Most likely	project value,
Maximum	project value increased by 50%.

Probability distribution for the time savings of infrastructure users was chosen as a normal probability distribution, where the mean value corresponded to the project value of time savings and standard deviation 10%.

3. Results

The performance of the quantitative analysis can be demonstrated on one of the projects of the tested set. The D10 Prague-Kosmonosy project, with a total investment cost of CZK 9,272,678,497 (€361,367,050), was used as an example. Simulation results when the beta-PERT probability distribution of total investment costs and the normal probability distribution for time savings of the infrastructure users were chosen, are shown in Table 8 and Figure 1. The simulated quantity dependent variable was cost-effectiveness (BCR).

The outputs of all projects showed a normal distribution of the BCR indicator. The research in [11] came to the same results, where an experiment which was identified as a pseudo-random number sequence as normally distributed was carried out.

In the interpretation of results it is necessary to respect certain limits connected with the elaborated analysis. As mentioned above, in this paper is presented the case study elaborated using projects being prepared for realization in the Czech Republic. Even if the original methodical steps used in this paper are generally accepted and used, it is necessary to respect certain national specificities in the evaluation of public investment projects. The next limit, which it is necessary to consider, is the definition of probability distributions for the simulation. In the presented analysis it was for the random variable "investment costs", and the triangle and beta-PERT probability distributions were alternatively used, which is in harmony with the present state in the references, and opinions of other experts. However, it is not possible to exclude that the real probability distribution of investment costs of partial projects will be different. However, for the correct evaluation, and the identification of the influence of the selected probability distribution on the results of the evaluated projects it was necessary to uniformly use the chosen probability distributions. In a similar limitation, it is necessary to also note the probability distributions of the random variable "time savings of infrastructure users". In this case it was uniformly selected for both variants of the simulation normal probability distribution, even if the real probability distribution of this variable can be, for partial projects, slightly different.

4. Discussion

It can be concluded from the above-stated calculations that one of the important settings of the input variables is their assumed probability distribution. From the available literature research and the authors' own expert opinion, it can be assumed that the investment costs variable tends to have a rather asymmetric probability distribution. This was also confirmed by the CBA guide [6], which considers an asymmetric triangular probability distribution in the range −5% to 20%. Makovšek [16], who dealt with a long-term analysis of cost over-runs of road constructions in Slovenia, addressed this issue in detail. Two fundamental conclusions emerged from his analysis: the fact that cost over-runs are systematic (not randomly distributed around zero) and that cost over-runs appear constantly over a time period of several decades and do not decrease (and thus do not show signs of improved forecasting tools and methods). A conclusion can also be drawn from these deductions, that the probability distribution of investment costs tends to be rather asymmetric.

An interesting comparison was published by Emhjellen [17], who dealt with the difference of values when setting different limits of normal distribution and their effect on the resulting values. Kumar [18] noted that the concessionaire aims to bear minimal cost, so maximum probability occurs at lower cost values, and hence it followed a lognormal probability distribution. Jakiukevicius [19,20] worked with normal and triangular distributions, for which he set theoretical parameters which he, based on simulations, converted to log logistics parameters. Kumar [18] adhered to a lognormal distribution of project costs. Gorecki [21] used a triangular distribution. The Czech author Hnilica [22] worked with the beta-PERT distribution, which he considered to be smoother, with possible values more concentrated around the most probable value, and the probability decreases towards the limit values faster than linearly. The authors of this article believe that the beta-PERT distribution best fits an expert estimate of the investment costs behaviour in comparing their values in the ex-ante and ex-post phases. The authors of this article carried out project simulations as mentioned above, assuming both a probability distribution of beta-PERT, and an asymmetric triangular one, and state that the results of the outputs in the expected value of "BCR-mean" ranged up to 7% for all of the projects. The outputs of all projects in both variants of solutions proved the normal distribution of the BCR indicator. The authors of the background research [4] reached the same results, where they stated that an experiment which identifies a pseudo-random number sequence as normally distributed was carried out. The reading of the frequency distribution of the evaluation indicator provides information of extreme importance, as regards the riskiness of the investment project [23].

5. Conclusions

It is clear from the above-stated findings that attention must be paid to the setting of statistical characteristics of variables which enter into the calculations of economic efficiency indicators, and on the basis of which it is decided whether or not to accept projects for financing. At present, data on post-audits of major transport infrastructure projects are beginning to be collected and analysed in the Czech Republic, and it is expected that the analyses will make possible, among other things, reaching more precise assumptions.

Although the projects proved efficient, a combination of negative changes to both variables can already bring projects with a certain value of probability into negative results. Based on the analysis of the research sample, it is clear that it cannot be clearly established for projects that a certain value of the BCR ratio predicts 100% stability of the project under the action of several critical variables. It is obvious from the mean value simulations determining the expected BCR value that projects with BCR < 1.1 show, at a certain percentage of probability, and at the critical variable limits specified above, that they shall not be 100% effective. However, the variance of the results obtained was large. Project P10 also showed an interesting result; a relatively high mean BCR ratio showed with a 5% probability that it will not be effective.

The results of the research point to the fact that it is always necessary to perform a quantitative analysis, since the results of the combination of the interaction of critical variables cannot be derived from the partial results of the sensitivity and qualitative analyses. The result will always depend on the absolute values of the critical variables of each unique project.

Author Contributions: Conceptualization. J.K. and V.H.; methodology. J.K. and V.H.; validation. J.K. and V.H.; formal analysis. J.K. and V.H.; investigation. J.K. and V.H.; resources. J.K. and V.H.; data curation. J.K. and V.H.; writing—original draft preparation. J.K. and V.H.; writing—review and editing. J.K. and V.H.; visualization. J.K.; supervision. J.K.; project administration. J.K.; funding acquisition. J.K. Both authors have read and agreed to the published version of the manuscript.

Funding: This research was funded by project Brno University of Technology No. FAST-S-20-6383 Selected Economic and Managerial Aspects in Construction Engineering.

Acknowledgments: This paper has been worked out under the project of Brno University of Technology no. FAST-S-20-6383 Selected Economic and Managerial Aspects in Construction Engineering.

Conflicts of Interest: The authors declare no conflict of interest.

References

1. Demart, S.; Roy, B. The uses of cost-benefit analysis in public transportation decision-making in France. *Transp. Policy* **2009**, *16*, 200–212. [CrossRef]
2. Hyard, A. Cost-benefit analysis according to Sen: An application in the evaluation of transport infrastructures in France. *Transp. Res. Part A Policy Pract.* **2012**, *46*, 707–719. [CrossRef]
3. Jones, H.; Moura, F.; Domingos, T. Transport Infrastructure Project Evaluation Using Cost-Benefit Analysis. *Procedia Soc. Behav. Sci.* **2014**, *111*, 400–409. [CrossRef]
4. Mackie, P.; Worsley, T.; Eliasson, J. Transport Appraisal Revisited. *Res. Transp. Econ.* **2014**, *47*, 3–18. [CrossRef]
5. Korytárová, J.; Papežíková, P. Assessment of Large-Scale Projects Based on CBA. *Procedia Comput. Sci.* **2015**, *64*, 736–743. [CrossRef]
6. Sartori, D. Guide to Cost-benefit Analysis of Investment Projects. Economic appraisal tool for Cohesion Policy 2014–2020. In *Directorate-General for Regional and Urban Policy*; European Commission: Brussels, Belgium, 2014; ISBN 978-92-79-34796-2.
7. Ministry of Transport of the Czech Republic (MoT CZ). Departmental Guideline for the Evaluation of Economic Effectiveness of Transport Construction Projects. 2017. Available online: https://www.sfdi.cz/soubory/obrazky-clanky/metodiky/2017_03_departmental-methodology-full.pdf (accessed on 25 August 2020).
8. Mokhtari, H.; Kiani, K.; Tahmasebpoor, S. Economic evaluation of investment projects under uncertainty: A probability theory perspective. *Sci. Iran.* **2020**, *27*, 448–468. [CrossRef]

9. Liu, Y.; Ting-Hua, Y.; Cui-Qin, W. Investment decision support for engineering projects based on risk correlation analysis. *Math. Probl. Eng.* **2012**, *2012*. [CrossRef]
10. Marović, I.; Androjić, I.; Jajac, N.; Hanák, T. Urban road infrastructure maintenance planning with application of neural networks. *Complexity* **2018**. [CrossRef]
11. Nesticò, A.; He, S.; De Mare, G.; Benintendi, R.; Maselli, G. The ALARP Principle in the Cost-Benefit Analysis for the Acceptability of Investment Risk. *Sustainability* **2018**, *10*, 4668. [CrossRef]
12. Bilenko, D.; Lavrov, R.; Onyshchuk, N.; Poliakov, B.; Kabenok, Y. The Normal Distribution Formalization for Investment Economic Project Evaluation Using the Monte Carlo Method. *Montenegrin J. Econ.* **2019**, *15*, 161–171. [CrossRef]
13. Acebes, F.; Pajares, J.; Galán, J.M.; López-Paredes, A. A new approach for project control under uncertainty. Going back to the basics. *Int. J. Proj. Manag.* **2014**, *32*, 423–434. [CrossRef]
14. Bowers, J.; Khorakian, A. Integrating risk management in the innovation project. *Eur. J. Innov. Manag.* **2014**, *17*, 25–40. [CrossRef]
15. Software Oracle Crystal Ball, Perpetual Licencs, 2020–2021. Available online: https://www.oracle.com/cz/applications/crystalball/ (accessed on 25 August 2020).
16. Makovšek, D. Systematic construction risk cost estimation mechanism and unit price movements. *Transp. Policy* **2014**, *35*, 135–145. [CrossRef]
17. Emhjellen, K.; Emhjellen, M.; Osmundsen, P. Investment cost estimates and investment decisions. *Energy Policy* **2002**, *30*, 91–96. [CrossRef]
18. Kumar, L.; Apurva, J.; Velaga, N.R. Financial risk assessment and modelling of PPP based Indian highway infrastructure projects. *Transp. Policy* **2018**, *62*, 2–11. [CrossRef]
19. Jasiukevicius, L.; Vasiliauskaite, A. Risk Assessment in Public Investment Projects: Impact of Empirically-grounded Methodology on Measured Values of Intangible Obligations in Lithuania. *Procedia Soc. Behav. Sci.* **2015**, *213*, 370–375. [CrossRef]
20. Jasiukevicius, L.; Vasiliauskaite, A. Cost Overrun Risk Assessment in the Public Investment Projects: An Empirically-Grounded Research. *Inz. Ekon. Eng. Econ.* **2015**, *26*, 245–254. [CrossRef]
21. Gorecki, J.; Diaz-Madronero, M. Who Risks and Wins?—Simulated Cost Variance in Sustainable Construction Projects. *Sustainability* **2020**, *12*, 3370. [CrossRef]
22. Hnilica, J.; Fotr, J. *Aplikovaná Analýza Rizika ve Finančním Management a Investičním Rozhodování*; Grada Publishing: Prague, Czech Republic, 2009; ISBN 978-80-247-2560-4.
23. De Mare, G.; Nesticò, A.; Benintendi, R.; Maselli, G. ALARP approach for risk assessment of civil engineering projects. In *Lecture Notes in Computer Science (Including Subseries Lecture Notes in Artificial Intelligence and Lecture Notes in Bioinformatics)*; Springer: Berlin/Heidelberg, Germany, 2018; pp. 75–86, ISBN 978-3-319-95174-4. [CrossRef]

Article

New Aspects of Socioeconomic Assessment of the Railway Infrastructure Project Life Cycle

Vít Hromádka *, Jana Korytárová, Eva Vítková, Herbert Seelmann and Tomáš Funk

Faculty of Civil Engineering, Brno University of Technology, 602 00 Brno, Czech Republic; korytarova.j@fce.vutbr.cz (J.K.); vitkova.e@fce.vutbr.cz (E.V.); seelmann.h@fce.vutbr.cz (H.S.); tomas.funk@email.cz (T.F.)
* Correspondence: hromadka.v@fce.vutbr.cz; Tel.: +420-541-148-641

Received: 25 September 2020; Accepted: 19 October 2020; Published: 21 October 2020

Featured Application: The results of the presented research will become part of the methodological material for the economic analysis of railway infrastructure projects after the completion of a broader research task.

Abstract: The paper deals with the issue of evaluation of socioeconomic impacts of occurrences emerging from railway infrastructure. The presented research results form part of a broader research subject focusing on the evaluation of the socioeconomic benefits of projects for the implementation of measures aimed at increasing the safety and reliability of railway infrastructure. The research topic addresses a part of the evaluation of railway infrastructure project efficiency within its life cycle using the cost–benefit analysis method. The methodology is based on the description and definition of input variables that are essential for the process of evaluating socioeconomic impacts. It is followed by another important step, which is the analysis of the categories and the number of occurrences, separately, for regional and national lines, and, further, the data is sorted according to whether occurrences emerge at stations or on a wide line. The result of the presented research is an overview of the calculated values of the expected socioeconomic impacts of partial occurrences according to the categories related to the year of operation on the railway infrastructure and the unit of measure. The research team carried out an inquiry into the annual impacts of the subcategories of occurrences related to one railway station and one kilometer of wide line, e.g., for national lines, the impacts of €2922.72/station/year and €41.67/km of wide line/year were determined. The results of the presented research represent important and necessary inputs for the next phase of the research topic, i.e., the evaluation of the socioeconomic benefits of projects increasing the safety and reliability of railway infrastructure.

Keywords: railway infrastructure; occurrences; socioeconomic impact; economic evaluation; CBA; life cycle

---

1. Introduction

The research, the results of which are presented in the paper, is focused on the issue of the socioeconomic evaluation of projects in the field of transport, especially railway infrastructure. The issue of the economic evaluation of public projects is very broad; however, the basic principles have been known for many years. Unlike commercial projects, where profit or profit-derived cash-flow plays a key role in the economic evaluation, the evaluation of public projects is usually based on the use of cost–output methods. The most important and widely used method is the cost–benefit analysis (CBA). CBA is thoroughly described in a number of guidelines and publications (e.g., [1]). In the case of transport infrastructure projects, there is a detailed methodological guide for its elaboration [2]. However, none of these documents are detailed enough to include methodologies for evaluating all relevant costs and benefits that arise in connection with transport (and especially railway) infrastructure. The paper presents the results of applied research aimed at incorporating the benefits associated with increasing the safety and reliability of railway infrastructure as a result of the implementation of projects for the installation of new and improved security equipment. The importance of research lies mainly in the fact that, currently, the benefits associated with increasing the safety and reliability of railways are not incorporated into the socioeconomic evaluation carried out in accordance with the methodology (in the case of the Czech Republic, it is departmental methodology [2]) using CBA, although it is clear that these benefits arise from railway infrastructure project implementation. As a result, railway infrastructure projects show worse results in economic efficiency evaluation and seem to be less economically efficient. The presented paper focuses on the evaluation of the socioeconomic impacts of occurrences that emerge from railway infrastructure. The aim of the paper is to methodically describe and verify the procedure for determining the socioeconomic impacts of occurrences that emerge from railway infrastructure in a case study. The Database of Occurrences [3], which contains detailed information on occurrences emerging from the railway in the Czech Republic, managed by employees of the Railway Administration, was used for the purposes of analysis and subsequent synthesis of the obtained data into methodological steps. Occurrences from the 2011–2018 period were used for the purposes of the research. The output of the research presented in the article is to determine the values of the expected annual socioeconomic impacts of occurrences according to the categories related to railway stations or one kilometer of the track segment. The outputs of the presented research will be used in follow-up research for the purpose of determining potential savings on railway infrastructure arising from the reduction of the number of occurrences, characterized by their potential socioeconomic impact. The reduction in the number of occurrences will be achieved by implementing appropriate security measures. Their economic efficiency shall be assessed by these steps.

2. Materials and Methods

The subject of the paper is the analysis and subsequent synthesis of relevant data in order to identify the methodology for evaluating the socioeconomic impacts of occurrences emerging from railway infrastructure. The purpose of the evaluation of these socioeconomic impacts is their subsequent use in the analysis of project costs and benefits in the field of transport infrastructure. This article, from a general point of view, can be included in the issue of socioeconomic evaluation of public investment projects in the field of transport infrastructure. The principles and procedures applied in the process of cost–benefit analysis are methodically described in the Guide to CBA of Investment Projects [1], where the general rules are defined and the specifics for the process of CBA for individual types of public projects are described. The issue of public projects in the field of transport infrastructure is dealt with in detail by the Departmental Guideline of the Ministry of Transport of the Czech Republic [2], which addresses the economic evaluation of transport infrastructure projects, both in general and with a focus on individual modes of transport, i.e., projects of roads and highways, projects of railway line construction and projects in the field of transport-important waterways. The abovementioned methodology provides some procedures for evaluating the socioeconomic impacts of transport projects (e.g., user savings, transport time savings, traffic accident savings, or impact on externalities). However, some key impacts, such as the impact on transport network safety and reliability, dealt with in the research presented here, are not addressed in more detail in the methodology. One of the aims of the paper is to explore the possibilities for supplementing the methodology for the socioeconomic evaluation of investment projects in the field of railway infrastructure by assessing its impact on the safety and reliability of railway infrastructure.

Based on research into the available scientific literature, it has not been found that any of the scientific teams were directly involved in assessing the impacts associated with increasing the safety of railway infrastructure. However, the research included texts dealing with the issue of occurrences emerging from railway infrastructure and their causes, as well as technical impacts (material damage, train delays). The basic input task was the identification of occurrences emerging from railway infrastructure, their impact, prevention, and classification. Santos-Reyes [4] dealt with a general analysis of the occurrence of traffic accidents on the railway [4] and presented basic study points to be addressed in order to subsequently prevent occurrences from railway infrastructure. The methodology developed within the Dnipro National University of Railways [5] presented the definition of the categories of occurrences from railway infrastructure in relation to the amount of material damage caused. Occurrences were classified according to the severity of the consequences, which were expressed in physical quantities. The paper provides an overview of financial losses associated with subcategories of occurrences, which represent a suitable data set for its comparison with the partial outputs of the presented results.

Klockner and Toft dealt with the modeling of occurrences on the railway in their study [6]. A second significant part of the publication deals with the factors influencing the occurrence emergence from railway infrastructure. In their study, Iridiastadi and Ikatrinasari [7] presented a classification system, including subfactors, with the potential to influence the occurrence emergence. Zhou and Lei also addressed the causes of occurrences on the railway in their article [8]. In their paper, Baysari et al. [9] presented a detailed analysis of errors leading to railway occurrences. Their work was based on a set of forty reports on the investigation of occurrences in railway infrastructure in Australia. The study concluded that up to half of the occurrences were caused by equipment failure due to insufficient maintenance. The conclusions of the paper [9] are important in relation to the presented research results. The subject of a detailed examination within the presented research is mainly occurrences emerging as a result of human factor failure, not due to a technical defect. Consequently, the next scientific papers also directly focus on the influence of the human factor on occurrence emergence from railway infrastructure. The paper by Hani Tabai et al. [10], which focused on the evaluation of the relationship between engine driver demography, cognitive performance, and the risk of an occurrence emergence, where the need to pay continuous attention had been identified as one of the most important reasons for an occurrence emergence due to an engine driver's error, can serve as one of the examples. The study of Zhan [11] elicits a qualitative and quantitative analysis method to detect the human- and organization-related causes of railway accidents. The HFACS-RAs framework, based on the incident and accident data of the railway industry, is proposed in this study. Evans [12], in his paper, dealt with the influence of the speed of trains on railway infrastructure and the number and severity of occurrences and extended his own previous statistics by the influence of train speed on the severity of the occurrences. This represents an important aspect of occurrences emerging from railway infrastructure; however, this dimension was not considered within the presented research.

An important (albeit rather marginal) issue within the presented research is the causes of occurrences emerging at railway crossings [13] and as a result of suicides [14,15]. These are very important occurrences emerging from all railway infrastructure, which represent a significant part of occurrences recorded in the database used. However, they are not considered in more detail in the analysis presented in this paper. Occurrences emerging at railway crossings are the subject of a separate methodology, according to which they are further assessed, so the development of this methodology is not the subject of the presented research either. Occurrences emerging as a result of suicides represent types of occurrences that cannot be effectively prevented by the implementation of appropriate measures and, therefore, also do not fall within the scope of the research.

The last important area relevant to the scope of the presented paper is the issue of occurrence prevention in railway infrastructure. Kim et al., in their article [16], presented results of a factual analysis of the railway occurrence rate and subsequently proposed preventive care systems focused mainly on the development of a training program for railway safety and railway safety training centers. Edkins and Pollock addressed the issue of the analysis of occurrences in railway infrastructure and the subsequent proposal of preventive measures aimed at reducing the number of occurrences [17]. It comprised a retrospective analysis of 112 railway occurrences in Australia, which revealed a tendency to human error and subsequently led to the development of a railway safety checklist. Evans presented an overview of the development of statistics on rail occurrences in Europe and a proposal for preventive measures to reduce the number of occurrences [18]. Authors Cheng and Tsai dealt, in their article [19], with the issue of competencies in the case of occurrence management and the restoration of normal operations after the occurrence emergence and its resolution. The abovementioned research reveals findings explaining the emergence of occurrences and the possibilities for reducing their number, which corresponds to the findings of the research team on the Czech conditions. The conclusions of the research confirm the topicality of the issue. The presented research builds on these basic building blocks and proceeds to the further step, i.e., the quantification of impacts from the socioeconomic point of view. These are very important and inspiring findings that explain the emergence of occurrences and the possibilities for reducing their number. The conclusions of the articles confirm the need to address the issue of occurrences and to identify and quantify their impacts.

The authors of the article, following the above-listed research of foreign resources, use other surveys and analyses based on the issues in the national environment and focus on the usability of technical data for socioeconomic analysis (evaluation of economic efficiency), which can be used as a basis for the decision-making process in investment projects for the security devices. The main goal of the research is to develop and present methodological steps for the evaluation of socioeconomic impacts of occurrences emerging from railway infrastructure and to verify the methodology on the occurrences emerging from railway infrastructure in the Czech Republic. The main result of the applied research presented in this paper is the determination of the value of the annual socioeconomic impact of occurrences emerging from railway infrastructure in the Czech Republic per one kilometer of the track segment and one railway station, divided into both national and regional lines.

2.1. Data

The basic background for the research of the issue dealt with is the creation of a database of occurrences in the researched region/national economy/territorial unit. The Database of Occurrences, managed by the employees of the Railway Administration of the Czech Republic, is the key source of data for the creation of the methodology and the elaboration of the case study, where all occurrences arising in railway infrastructure of the Czech Republic in the 2009–2018 period were registered [3]. The Database of Occurrences is not a publicly accessible source of information; the database was provided to the research team by the Railway Administration of the Czech Republic, upon request, exclusively for the purpose of conducting the presented research. This represents a very large document that contains more than 500,000 items of information.

2.2. Methods

The proposed methodology is based on the collection, analysis, and use of relevant data to evaluate the societal benefits arising from the establishment of security measures. The outputs are bound for the later use of the "opportunity cost" valuation approach.

The subject of the paper is to develop and present a methodology for the quantification and evaluation of the socioeconomic impacts of occurrences emerging from railway infrastructure, divided into relevant subcategories, following the scientific literature research and basic principles of economic analysis of public investment projects.

Development of the methodology for the evaluation of the socioeconomic impacts of occurrences emerging from railway infrastructure consists of the following steps:

- Analyzing the Database of Occurrences, defining the categories, and selecting the data for further use.
- Defining the key characteristics of occurrences, including health and delay impacts on passenger and freight trains and their economic evaluation.
- Determining the expected impact of occurrences by category.

The calculation of the amount of the expected socioeconomic impacts of occurrences by the categories of per kilometer of railway line and station is performed in Section 3. The partial steps are described in detail in the following sections.

2.2.1. Analysing the Database of Occurrences, Defining the Categories, and Selecting the Data for Further Use

Occurrences emerging within the railway infrastructure can be classified into three basic groups:

- A—Serious accidents (A1–A4),
- B—Accidents (B1–B10),
- C—Incidents (C1–C21).

Subcategories of occurrences were defined within the mentioned groups of occurrences, which differ mainly in the causes of the occurrences and their impacts. For the purposes of the presented research, the following variables, which are relevant to the methodological procedure, were selected from the Database of Occurrences:

- Impact on health,

  - Death,
  - Serious injury,
  - Minor injury,

- Material damage,
- Costs,
- Number of delayed passenger trains,
- Total delay of passenger trains,
- Number of delayed freight trains,
- Total delay of freight trains, and
- The cause of the occurrence,

  - Technical,
  - Human factor,
  - Others.

These variables are listed in the Database of Occurrences for each specific occurrence.

The Database of Occurrences [3] could not be used for further research purposes in its original version without modifications as the recorded variables were developed year-on-year and the data structure was not consistent. The researchers, therefore, made the necessary adjustments for the purposes of its further use. The analysis of the Database of Occurrences was elaborated in order to address the research question of the average, minimum, and maximum annual impact values (number of occurrences, deaths, number of serious/minor injuries, material damage, and costs) of occurrences, with their possible deviation, which was monitored based on the standard deviation. It was subsequently decided, following the results of the analysis, that the follow-up research would not include occurrences due to suicides and those emerging at railway crossings. Detailed results of the above-defined analysis of the Database of Occurrences and conclusions and recommendations resulting from the analysis were published in [20].

2.2.2. Defining the Key Characteristics of Occurrences and Their Economic Evaluation

The research team defined the information and characteristics of partial occurrences within the Database of Occurrences that provided data on the partial socioeconomic impacts of the occurrences on society. An overview of this information and characteristics is presented in the previous section.

Impacts in the form of costs of the removal of material damage or other direct costs incurred are quantified for each individual occurrence and are listed directly in the occurrence database. The research, therefore, focused on the evaluation of the remaining variables. Subsequent research was thus focused on the evaluation of the following characteristics of the occurrences:

- Impact on health,
- Passenger train delays, and
- Freight train delays.

The unit impact of an occurrence in a particular category was determined using Relation (1).

$$UI_O = \sum_{i=1}^{3} AIH_i + \sum_{j=1}^{2} AITD_j + AML \tag{1}$$

where

UI_O	Unit Impact (the unit impact of the occurrence of the relevant category; CZK/Occurrence)
AIH_i	Average Impact on Health (the average health impact of the occurrence of the relevant category on health: 1—fatalities, 2—serious injuries, 3—minor injuries; CZK/Occurrence)
$AITD_j$	Average Impact on Travel Delay (the average impact of an occurrence of the relevant category on traffic delay: 1—passenger transport, 2—freight transport; CZK/Occurrence)
AML	Average Material Loss (average material damage of an occurrence of the relevant category; CZK/Occurrence)

The evaluation of the listed impacts was performed in connection with the procedures specified in the Departmental Methodology of the Ministry of Transport of the Czech Republic [2]. In the case of a health impact assessment, the following Relation (2) was used for the evaluation.

$$AIH_i = AH_i \times UIH_i \tag{2}$$

where

AIH_i	Average Impact on Health (the average health impact of the occurrence of the relevant category on health: 1—fatalities, 2—serious injuries, 3—minor injuries; CZK/Occurrence)
AH_i	Average Number (the average number of affected people by the occurrence of the relevant category: 1—fatalities, 2—serious injuries, 3—minor injuries; CZK/Occurrence)
UIH_i	Unit Impact on Health (the unit impact of the occurrence of the relevant category on the health of one person, 1—fatalities, 2—serious injuries, 3—minor injuries; CZK/Occurrence)

The input data for the purposes of the case study elaborated for the Czech Republic was taken directly from the Departmental Methodology [2], indicating the unit cost of the occurrence according to its severity. The data is shown in Table 1 below.

Table 1. Costs associated with occurrences.

Occurrence	Unit Cost €/Person
Fatality	786,457
Serious injury	190,414
Minor injury	24,581

Source: Departmental Methodology of the Ministry of Transport [2].

In the case of impacts related to passenger train delay, a calculation was carried out using Relation (3).

$$AITD_1 = PTU \times ADP \times UIDP \tag{3}$$

where

PTU	Personal Train Utilization (average passenger train occupancy; Persons/Train)
ADP	Average Delay of Personal Trains (the average total delay of passenger trains due to the occurrence of the relevant category; Train*Hour/Occurrence)
UIDP	Unit Impact of Personal Train Delay (unit cost of passenger time; CZK/Hour*Person)

The data presented in the Departmental Methodology [2] and statistical data from the Statistical Yearbook of the Czech Railways Group [21] were used for the purposes of the case study performed for the Czech Republic. Based on these sources, the expected occupancy of one passenger train was determined at 66.55 people/train and the average value of passenger time at 10.63 €/person-hour.

In the case of the evaluation of the impacts associated with the delay of freight trains, the following Relation (4) was used for the calculation:

$$AITD_2 = CTU \times ADC \times UIDC \tag{4}$$

where

CTU	Cargo Train Utilization (average freight weight of a freight train; Tons/Train)
ADC	Average Delay of Cargo Trains (the average total delay of freight trains due to an occurrence of a relevant category; Trains*Hour/Occurrence)
UIDC	Unit Impact of Cargo Train Delay (unit time cost of transported cargo; CZK/Hour)

The average freight weight of a freight train was determined at 455 t/train using statistical data [21] as part of a case study elaborated for the territory of the Czech Republic. The average value of freight transport time was subsequently determined at 0.23 €/ton using the values of freight transport time according to commodities taken from the Departmental Methodology [2] and the percentage rate of individual commodities in freight transport.

The values of the input quantities used for the calculation of Relations (1)–(4) within the case study were taken from the national sources of the Czech Republic. These sources were used in the form of official methodological documents for the socioeconomic evaluation of public projects and are based on long-term statistical data. The key methodological basis was the Departmental Methodology of the Ministry of Transport, which defines some input variables (e.g., unit impacts on health or unit cost of passenger time or cargo). The already mentioned occurrence database or statistical yearbooks of carriers or the administrator of railway infrastructure fall within the documents containing statistical data.

The quantities UIHi (Unit Impact on Health), UIDP (Unit Impact of Personal Train Delay), and UIDC (Unit Impact of Cargo Train Delay) were taken from the Departmental Guideline [2] for the purposes of the presented research. The quantities AML (Average Material Loss), AHi (Average Number), ADP (Average Delay of Personal Trains), and ADC (Average Delay of Cargo Trains) were determined using data recorded in the Database of Occurrences [18], and PTU (Personal Train Utilization) and CTU (Cargo Train Utilization) values were derived from statistical data presented in the Statistical Yearbook of Czech Railways [21].

A detailed calculation of these unit impacts of railway occurrences in railway infrastructure, including a case study to verify the functionality of the evaluation algorithm, is provided in [22].

2.2.3. Determining the Expected Impact of Occurrences by Category

Only those occurrences that emerged due to human error were considered for the calculation of the average impacts of occurrences following the conclusions of [22]. The categories of occurrences listed in Table 2 are, therefore, considered for the current research.

Table 2. Categories of occurrences.

Designation	Description
A1	Collision of railway vehicles, resulting in death or injury to at least 5 persons or large-scale damage.
A2	Derailment of a rail vehicle, resulting in death or injury to at least 5 persons or large-scale damage.
A3	Collision of a rail vehicle with an obstacle in the passage, resulting in death or injury to at least 5 persons or large-scale damage.
B1	Collision of railway vehicles, resulting in minor consequences rather than a serious accident.
B2	Derailment of a rail vehicle, resulting in minor consequences rather than a serious accident.
B3	Collision of a rail vehicle with an obstacle in the passage, resulting in minor consequences rather than a serious accident.
C1	Collision of railway vehicles, resulting in minor consequences rather than a serious accident or accident.
C2	Derailment of a rail vehicle, resulting in minor consequences rather than a serious accident or accident.
C3	Collision of a rail vehicle with an obstacle in the passage, resulting in minor consequences rather than a serious accident or accident.
C6	Unauthorized movement of the rail vehicle behind a signaling device prohibiting driving, resulting in minor consequences rather than an accident.
C12	Unsecured movement of a rail vehicle, resulting in minor consequences rather than an accident.
C16	Failure of signaling systems, resulting in minor consequences rather than an accident.
C19	Unspecified incident arising in connection with the movement of the rail vehicle, resulting in minor consequences rather than an accident.

Source: Database of Occurrences 2009–2018 [3].

Using the data contained in the Database of Occurrences [3] and the unit impacts of occurrences presented in the previous part of the text, the expected overall socioeconomic impacts of individual categories of occurrences were determined using Relation (1). The overall socioeconomic impacts include the following items:

- Impacts on health,
- Delay of passenger trains,
- Delay of freight trains, and
- Total costs.

The presented research made use of selected occurrences on two levels. The first level included occurrences taken from the Database of Occurrences [3] (adjusted for occurrences at railway crossings and suicides), whatever the cause; the second level included occurrences caused by human error. A detailed evaluation of individual categories of occurrences was described in [23]; the expected impacts of individual categories of occurrences are summarized in Table 3.

Table 3. Average total economic impacts per occurrence by category.

Category	Expected Impact Per Occurrence (EUR)
	Cause—Human Factor
A1	742,138
A2	983,734
A3	2,672,916
B1	105,200
B2	59,398
B3	74,056
C1	5880
C2	3797
C3	4202
C6	2464
C12	3325
C19	4422

The presented methodological part resulted from the current research that is focused on the evaluation of the socioeconomic impacts associated with occurrence emergence in railway infrastructure in order to consider the increase in the safety and reliability of the railway in the socioeconomic evaluation of railway infrastructure projects. These methodological steps serve as an input basis for determining the expected impact of occurrences emerging from railway infrastructure related to the network of railway lines and railway stations. The results of this research are presented in the following section.

3. Results

The key output of the research presented in this article is the determination of the expected socioeconomic impacts of individual categories of occurrences in relation to the system in which they emerge. In general, it can be stated that occurrences emerge from railway tracks or at railway stations. Railway tracks and railway stations are further divided into national and regional categories. The expected annual impact of an occurrence of the relevant category on the station or a kilometer of track is determined using the following Relation (5):

$$TI = \frac{\sum_{i=A1}^{C19}\left(O_{i,j,k} \times UI_{i,j,k}\right)}{Q_{j,k} \times t} \tag{5}$$

where

TI	Total Impact of Occurrence per year in €
O_i	Number of Occurrences per evaluated period t
i	Category of Occurrence according to Table 2
UI_i	Unit Impact of Occurrence according to Table 3
$Q_{j,k}$	Quantity of stations/tracks
j	Category of railway line (1—national, 2—regional)
k	Category of parts of railway line (1—Railway Track, 1—Railway Station)
t	Evaluated (reference) Period in years

The Number of Occurrences per evaluated period t is determined separately and further divided into occurrences emerging from the national or regional line and occurrences emerging from the tracks or at the stations. While the information on whether an occurrence emerges from a national or regional line is taken from the Database of Occurrences for the purpose of a case study of a railway line in the Czech Republic, information on whether an occurrence emerges on the tracks or at the stations had to be determined using statistical data. The calculation was performed using Relation (6).

$$O_{i,j,k} = O_{i,j} \times RO_{i,j,k} \tag{6}$$

where

$O_{i,j,k}$	Number of Occurrences of the relevant category (i) on the national (j = 1) or regional (j = 2) line emerging from the tracks (k = 1) or at the station (k = 2)
$O_{i,j}$	Number of Occurrences of the relevant category (i) on the national (j = 1) or regional (j = 2) line
$RO_{j,j,k}$	Rate of Occurrences: ratio of occurrences emerging from the tracks (k = 1) or at the stations (k = 2) on the total number of occurrences $O_{i,j}$.

The ratio of occurrences emerging from railway tracks and at the railway stations was determined using a selected set of data from the Database of Occurrences. This ratio is presented in Table 4.

Table 4. Ratio of occurrences emerging from the railway tracks and at railway stations.

Year	Occurrences in Railway Infrastructure	Occurrences on the Railway Track
2011	92.31%	7.69%
2012	92.96%	7.04%
2013	93.94%	6.06%
2014	92.86%	7.14%
2015	95.00%	5.00%
2016	95.79%	4.21%
2017	96.77%	3.23%
2018	99.21%	0.79%
Average	94.85%	5.15%

In further calculations, it is therefore assumed that the share of occurrences emerging at railway stations ($RO_{i,j,1}$) is 94.85%, and the share of occurrences emerging from the railway tracks ($RO_{i,j,2}$) is 5.15%.

As a next step, it is necessary to distinguish between occurrences emerging from the national railway lines and occurrences emerging from the regional lines. The analysis of the Database of Occurrences generally shows a larger number of occurrences on national lines and a smaller number on regional lines.

For the sake of clarity, it is appropriate to summarize which occurrences from the overall Database of Occurrences were used in the final phase of the research. Of the total number of occurrences recorded in the Database of Occurrences, occurrences emerging at railway crossings and as a result of suicides were omitted. The reasons are described in more detail in [20]. Subsequently, those categories of occurrences that, as a law, cannot be caused by the human factor were omitted. The resulting set of occurrences was subsequently reduced by occurrences objectively caused in a different way than by the failure of the human factor. Last but not least, those occurrences that emerge during handling rides, carriage shifts, or on sidings were omitted. The resulting set of occurrences was subsequently used for the final analysis.

An important input for the final calculation was also the extent of the railway network. Using the documents of the Railway Administration [24–26], the total number of railway stations on national and regional lines and the total length in kilometers of national and regional railway lines were determined. An overview of the resulting input data is given in Table 5.

Table 5. Overview of the input data for the final calculation.

Input Quantity	Value
Total length of national lines	4738 km
Total length of regional lines	4361 km
Total no. of stations on the national lines	1245
Total no. of stations on the regional lines	1380
Ratio of occurrences at stations *	94.85%
Ratio of occurrences on the tracks *	5.15%

* Values taken from Table 3.

Using the input data listed in Tables 2–4, the values of expected annual socioeconomic impacts were calculated for individual relevant categories of occurrences related to one kilometer of railway track and one railway station, separately for national and regional lines. The calculation was performed according to Equation (5), and its results are presented in Table 6.

Table 6. Calculation of the expected annual impact of subcategories of occurrences per station and per km of track.

Occurrence Category	National/ Regional	Total Number of Occurrences 2011–2018	Number of Occurrences Station	Number of Occurrences Track	Unit Impact of Occurrence in €	Expected Annual Impact in € Per Station	Expected Annual Impact in € Per km of Track
A1	national	12	11.38	0.62	757,756.47	865.98	12.35
	regional	2	1.90	0.10	757,756.47	130.21	2.24
A2	national	3	2.85	0.15	1,004,436.36	286.97	4.09
	regional	1	0.95	0.05	1,004,436.36	86.30	1.48
A3	national	2	1.90	0.10	2,729,166.45	519.82	7.41
	regional	0	0.00	0.00	2,729,166.45	0.00	0.00
B1	national	37	35.10	1.90	107,414.17	378.49	5.40
	regional	7	6.64	0.36	107,414.17	64.60	1.11
B2	national	34	32.25	1.75	60,647.57	196.38	2.80
	regional	7	6.64	0.36	60,647.57	36.48	0.63
B3	national	14	13.28	0.72	75,614.37	100.82	1.44
	regional	3	2.85	0.15	75,614.37	19.49	0.33
C1	national	75	71.14	3.86	6003.73	42.88	0.61
	regional	4	3.79	0.21	6003.73	2.06	0.04
C2	national	395	374.67	20.33	3877.39	145.86	2.08
	regional	82	77.78	4.22	3877.39	27.32	0.47
C3	national	254	240.93	13.07	4290.53	103.79	1.48
	regional	80	75.88	4.12	4290.53	29.49	0.51
C6	national	637	604.22	32.78	2516.13	152.64	2.18
	regional	86	81.57	4.43	2516.13	18.59	0.32
C12	national	104	98.65	5.35	3394.95	33.63	0.48
	regional	37	35.10	1.90	3394.95	10.79	0.19
C19	national	222	210.58	11.42	4515.41	95.47	1.36
	regional	14	13.28	0.72	4515.41	6.02	0.09

Table 5 shows the calculation and the resulting values of the expected annual socioeconomic impact on one railway station and on one kilometer of track section for individual categories of occurrences. The total expected annual socioeconomic impact of occurrences was determined by the sum of the impacts of the individual categories. The total values are given in Table 7.

Table 7. Values of the overall socioeconomic impacts of occurrences.

Part of Railway Network	Value
Station—national line	2922.72 €/station/year
Station—regional line	431.35 €/station/year
Track—national line	41.67 €/km/year
Track—regional line	7.39 €/km/year

The resulting values, given in Tables 5 and 6, represent the expected values obtained from historical data taken from the Database of Occurrences for the 2011–2018 period and from the documents of the Railway Administration defining the railway transport network. These outputs clearly point out the importance of occurrences emerging from the railway and their socioeconomic impact on society as a whole. However, the obtained results have further use. As part of follow-up research activities, these values will be used to calculate the socioeconomic benefits associated with increasing the safety and reliability of the railway line as a result of the implementation of projects that increase the level of security of the railway network.

4. Discussion

The subject of the presented research is the definition of methodological steps for determining the overall economic impacts of occurrences emerging from railway infrastructure related to a kilometer of wide line or a railway station and year. The methodological steps consist of determining the expected socioeconomic impacts for subcategories of occurrences in the form of direct financial impacts, impacts on human health, and the impact on delays of both passenger and freight trains. The partial impacts of individual categories of occurrences were subsequently related to a purpose unit, i.e., a kilometer of wide line or one railway station. The case study used to verify the functionality of the proposed procedure was based on the data obtained by the long-term monitoring of traffic on the railway infrastructure in the Czech Republic. When interpreting the calculations, it is, therefore, necessary to consider the possible specifics of railway transport in the Czech Republic in comparison with abroad. At the same time, the authors of the paper highlight the use of some input variables in values corresponding to the conditions of the Czech Republic. It mainly concerns the data taken from the Departmental Methodology of the Ministry of Transport of the Czech Republic, such as the unit impacts of traffic accidents on health, the value of passenger time, or the cost of transported cargo.

The calculations also make use of national rail transport statistics, which may also show results that are different from those of other countries. While respecting the abovementioned limitations, it is possible to proceed to the discussion on the obtained results. A very interesting partial result is the ratio between occurrences emerging on a wide line and at railway stations. As shown in Table 4, on average, in the Czech Republic, approximately 95% of occurrences emerge at railway stations and only 5% on a wide line. The conclusions resulting from these findings are commented on below. The key outputs of the research carried out are the values of the expected annual impact of the subcategories of occurrences per kilometer of wide line or per railway station, given in Table 6. Table 6 shows the values for the occurrences emerging on a national or regional line and on a wide line or at train stations. For all categories of occurrences, the dominance of national lines is evident, both in the case of emergencies occurring at railway stations as well as emergencies occurring on a wide line. These findings are clearly visible from the total values of impacts of all categories of occurrences per kilometer of wide line or one railway station, as listed in Table 7. The results shown in Table 7 clearly demonstrate the importance of occurrences emerging on national lines (almost a 6.8-times higher unit impact in the case of railway stations and a 5.6 times higher impact in the case of a wide line) compared to regional lines. Considering the above-listed results, it can be stated that while respecting the restrictions arising from the data used, as well as inputs related solely to the Czech Republic area, it is generally recommended to pay attention and use resources to eliminate occurrence emergence risk at railway stations on the national line, as, in this sector, there is a high damage occurrence risk caused by occurrence emergence from railway infrastructure. This conclusion can be useful in both planning new projects in the field of construction and the modernization of the railway infrastructure, as well as in the decision-making process on the allocation of investment funds.

Authors of the results presented in [5] arrived at methodologically similar partial results. This text provides information on the average losses associated with the occurrences in Ukraine. The categorization of occurrences used in this article is slightly different from the categories used in the presented research; a certain comparison of both is possible. However, this is not a comparison of the final results of the presented research, but a comparison of the processing of partial results listed in Table 3, i.e., the expected impacts of subcategories of occurrences. In the case of the results published in [5], the following results can be given as an example:

- Financial loss related to the occurrence (accident with serious consequences)—€38,041,
- Accident—€16,988,
- Collision of passenger or freight trains with other trains or rolling stock or the approximation of rolling stock in trains at companies and stations, which, due to their consequences, do not belong among occurrences—€11,446.

It is evident that the results presented in this paper present other values when compared with the corresponding results given in Table 3 of this article. Absolute values of quantities due to different levels of purchasing power cannot be directly compared without further adjustments. However, in part, the reason can also be seen in the fact that the overall impact, including socioeconomic damage, was determined for the purposes of the presented research results, while in the case of [5], financial loss was exclusively considered. Another reason may result from the fact that only occurrences associated with human factor failure were considered in the case of the presented research results. A slightly different classification of occurrences may also play a role. The authors also stated that while financial loss is the result of the research in [5], in the case of the presented research, this is a value obtained from long-term relevant statistics administered by a state authority.

The findings resulting from the presented research have a significant impact on the process of socioeconomic evaluation of projects of transport infrastructure, which forms an integral part of the life cycle of a public investment project. The long-term experience of the research team shows that despite the quality methodological data used to perform a socioeconomic evaluation of railway infrastructure projects, it is not possible to perform a comprehensive socioeconomic evaluation of the railway infrastructure project as the current methodological documents lack a procedure for taking into account the increase in the safety and reliability of the railway line, although, in connection with most of the evaluated projects, the increase in the safety and reliability of the railway line is very closely related to them. The presented results provide important information on the values of the socioeconomic impacts of subcategories of occurrences, and, when considering the frequency of their emergence, as derived from a detailed database of occurrences, the socioeconomic impacts related to special-purpose units are determined, in this case, for one railway station or one kilometer of railway track. In addition, all this is carried out separately for the national and regional lines, as during the research, there were significant differences in the number and severity of occurrences with regard to whether it was a regional or national line.

Follow-up research will focus on the use of the obtained data for the development of a methodology for the evaluation of benefits associated with increasing the safety and reliability of the railway. The authors assume that the results of the research will be incorporated into the existing Departmental

Methodology, which addresses the course of the socioeconomic evaluation of projects in the field of transport infrastructure in the Czech Republic.

5. Conclusions

The paper presents the results of applied research aimed at refining the evaluation of socioeconomic benefits associated with increasing the safety and reliability of the railway line as a result of the implementation of investment projects in the field of railway infrastructure within a CBA analysis. The aim of the paper was to methodically describe and verify a case study on the procedure for determining the socioeconomic impacts of occurrences that emerge from railway infrastructure. The article presents partial methodological steps that have been developed for these purposes. The research is based on a detailed Database of Occurrences emerging from the railway line in the Czech Republic, using data for the 2011–2018 period, which originally included a total of 8455 occurrences divided into subcategories. After the elimination of the occurrences connected with suicides and accidents at railway crossings, a total of 5378 occurrences was used for a more detailed analysis. The first step was to define the methodological steps for determining the socioeconomic impact of an occurrence emerging on a railway line. The overall socioeconomic impact includes material damage and costs, evaluated costs associated with health impacts, and evaluated impacts associated with delays of both passenger and freight trains. Using these documents and the entire Database of Occurrences, the expected value of the impact of an occurrence of a relevant category was subsequently determined. Combining it with additional information on the railway transport network and the ratio of occurrences emerging on the national or regional line, the expected annual impacts of occurrences of individual categories on a railway station or one kilometer of track were derived in the final phase of the research. The main goal of the paper was to present methodological steps for evaluating the socioeconomic impacts of occurrences emerging from railway infrastructure and to verify the methodology in the case of occurrences emerging from railway infrastructure in the Czech Republic area. The research presented in the paper assessed new aspects of the socioeconomic evaluation of railway projects and subsequently proposed new procedures aimed at quantifying the socioeconomic impacts of occurrences emerging from railway infrastructure. In the follow-up research, the existing process of evaluating the economic efficiency of railway infrastructure projects will be supplemented by a new dimension, i.e., considering the increase in the reliability and safety of railway infrastructure.

When interpreting the results obtained, it is necessary to accept certain limitations associated with the research carried out, as well as the data used. While the proposed methodological steps are generally applicable for the purposes defined in the research goal, the results of the case study are influenced by the fact that they were determined using data specific to the Czech Republic area. For the purposes of the case study, data from the occurrence database emerging exclusively from the Czech Republic area were used, and other input information was also taken from national methodological documents and statistics. In this regard, follow-up research on the topic is also proposed for other countries for the possibility of assessing the international applicability of the methodological steps and for comparing the situation in the field of occurrence impacts on railway infrastructure.

Follow-up research will focus on projecting the results achieved in the presented research into the final methodology to consider the benefits associated with increasing the safety and reliability of the railway network in the socioeconomic evaluation of investment projects of railway infrastructure. The aim of the authors is to incorporate the resulting methodology into the process of economic efficiency assessment of investment projects in the field of railway infrastructure in the form of an annex to the Departmental Methodology of the Ministry of Transport and the State Fund for Transport Infrastructure.

Author Contributions: Conceptualization, V.H., J.K. and E.V.; methodology, V.H., J.K., E.V., H.S. and T.F.; validation, V.H., H.S. and T.F.; formal analysis, V.H., J.K. and E.V.; investigation, V.H., J.K., E.V., H.S. and T.F.; resources, E.V., J.K., H.S. and V.H.; data curation, V.H., H.S. and T.F.; writing—original draft preparation, V.H. and J.K.; writing—review and editing, V.H., E.V., H.S. and T.F.; visualization, V.H.; supervision, J.K.; project administration, V.H.; funding acquisition, V.H. All authors have read and agreed to the published version of the manuscript.

Funding: This research was funded by the Technology Agency of the Czech Republic, grant number TL02000278.

Acknowledgments: This paper comes under the project of the Technology Agency of the Czech Republic "TL02000278 Evaluation of increased safety and reliability of railway infrastructure after its modernization or reconstruction".

Conflicts of Interest: The authors declare no conflict of interest.

References

1. Sartori, D. *Guide to Cost-Benefit Analysis of Investment Projects, Economic Appraisal Tool for Cohesion Policy 2014–2020*; European Commission, Directorate-General for Regional and Urban Policy: Brussels, Belgium, 2014; ISBN 978-92-79-34796-2.

2. Ministry of Transport of the Czech Republic (MoT CZ). Departmental Guideline for the Evaluation of Economic Effectiveness of Transport Construction Projects. 2017. Available online: http://www.sfdi.cz/pravidla-metodiky-a-ceniky/metodiky/ (accessed on 3 July 2020).

3. Czech Railway Infrastructure Administration (SŽDC). Statistika Mimořádných Událostí, Databáze 2009–2018 (Database of Occurrences 2009–2018). *Transp. Res. Procedia* **2020**, *44*, 47–52.

4. Santos-Reyes, J.; Beard, A.N.; Smith, R.A. A systemic analysis of railway accidents. In *Proceedings of the Institution of Mechanical Engineers, Part F: Journal of Rail and Rapid Transit*; SAGE Publications: London, UK, 2005; Volume 219, pp. 47–65. [CrossRef]

5. Ohar, O.M.; Rozsocha, O.V.; Shapoval, G.V.; Smachylo, Y.V. *Transport Accidents Distribution at Ukrainian Railways According to Categories Depending on Severity of Consequences. Nauka ta Progres Transportu*; Dnipro National University of Railway Transport Named after Academician V. Lazaryan: Kyiv, Ukraine, 2018; Volume 75, pp. 7–19. Available online: https://doaj.org/article/fedc1e6ccf164b6b94cbc363f5d990b4 (accessed on 7 July 2020). [CrossRef]

6. Klockner, K.; Toft, Y. Railway accidents and incidents: Complex socio-technical system accident modelling comes of age. *Saf. Sci.* **2018**, *110*, 59–66. [CrossRef]

7. Iridiastadi, H.; Ikatrinasari, Z.F. Indonesian Railway Accidents—Utilizing Human Factors Analysis and Classification System in Determining Potential Contributing Factors. *Work* **2012**, *41*, 4246–4249. [CrossRef] [PubMed]

8. Zhou, J.-L.; Lei, Y. Paths between latent and active errors: Analysis of 407 railway accidents/incidents' causes in China. *Saf. Sci.* **2018**, *110*, 47–58. [CrossRef]

9. Baysari, M.T.; McIntosh, A.S.; Wilson, J.R. Understanding the human factors contribution to railway accidents and incidents in Australia. *Accid. Anal. Prev.* **2008**, *40*, 1750–1757. [CrossRef] [PubMed]

10. Hani Tabai, B.; Bagheri, M.; Sadeghi-Firoozabadi, V.; Sze, N.N. Evaluating the impact of train drivers' cognitive and demographic characteristics on railway accidents. *Saf. Sci.* **2018**, *110*, 162–167. [CrossRef]

11. Zhan, Q.; Zheng, W.; Zhao, B.; Zhan, Q. A hybrid human and organizational analysis method for railway accidents based on HFACS-Railway Accidents (HFACS-RAs). *Saf. Sci.* **2017**, *91*, 232–250. [CrossRef]

12. Evans, A.W. Speed and rolling stock of trains in fatal accidents on Britain's main line railways: 1967–2000. In *Proceedings of the Institution of Mechanical Engineers, Part F: Journal of Rail and Rapid Transit*; SAGE Publications: London, UK, 2002; Volume 216, pp. 81–95. [CrossRef]

13. Dina, A.; Akanni, C.; Badejo, B. Evaluation of Railway Level Crossing Attributes on Accident Causation in Lagos, Nigeria. *Indones. J. Geogr.* **2017**, *48*, 108–117. [CrossRef]

14. Havarneanu, G.M.; Burkhardt, J.-M.; Paran, F. A systematic review of the literature on safety measures to prevent railway suicides and trespassing accidents. *Accid. Anal. Prev.* **2015**, *81*, 30. [CrossRef] [PubMed]

15. Ueda, M.; Sawada, Y.; Matsubayashi, T. The effectiveness of installing physical barriers for preventing railway suicides and accidents: Evidence from Japan. *J. Affect. Disord.* **2015**, *178*, 1–4. [CrossRef] [PubMed]

16. Kim, H.J.; Jeong, J.Y.; Kim, J.W.; Oh, J.K. A Factor Analysis of Urban Railway Casualty Accidents and Establishment of Preventive Response Systems. *Procedia Soc. Behav. Sci.* **2016**, *218*, 131–140. [CrossRef]

17. Edkins, G.D.; Pollock, C.M. The influence of sustained attention on Railway accidents. *Accid. Anal. Prev.* **1997**, *29*, 533–539. [CrossRef]

18. Evans, A.W. Fatal train accidents on Europe's railways: 1980–2009. *Accid. Anal. Prev.* **2011**, *43*, 391–401. [CrossRef] [PubMed]

19. Cheng, Y.-H.; Tsai, Y.-C. Railway-controller-perceived competence in incidents and accidents. *Ergonomics* **2011**, *54*, 1130–1146. Available online: http://www.tandfonline.com/doi/abs/10.1080/00140139.2011.622795 (accessed on 8 July 2020). [CrossRef] [PubMed]

20. Funk, T.; Hromádka, V.; Korytárová, J.; Vítková, E.; Seelmann, H. Statistic evaluation of occurrences on railways in the Czech Republic. In *Proceedings of the Economic and Social Development, 45th International Scientific Conference on Economic and Social Development–XIX International Social Congress (ISC 2019), Book of Proceedings, International Scientific Conference on Economic and Social Development, Varazdin, Croatia, Moscow, Russia, 17–18 October 2019*; Varazdin Development and Entrepreneurship Agency: Varazdin, Croatia, 2019; pp. 1–8.

21. Czech Railway Group. *Statistical Yearbook of the Czech*; Czech Railway Group: Prague, Czech Republic, 2018.

22. Hromádka, V.; Korytárová, J.; Vítková, E.; Seelmann, H.; Funk, T. Evaluation of Socio-economic Impacts of Incidents on the Railway Infrastructure. In Proceedings of the XV International Conference on Durability of Building Materials and Components (DBMC), Barcelona, Spain, 20–23 October 2020.

23. Hromádka, V.; Korytárová, J.; Vítková, E.; Seelmann, H.; Funk, T. Economic impact of occurrences on railways. In Proceedings of the CENTERIS—International Conference on Enterprise Information Systems/ProjMAN—International Conference on Project MANagement/HCist—International Conference on Health and Social Care Information Systems and Technologies, Vilamoura, Algarve, Portugal, 21–23 October 2020.

24. Czech Railway Infrastructure Administration (SŽDC). *Prohlášení o Dráze Celostátní a Regionální (National and Regional Railway Declaration)*; Czech Railway Infrastructure Administration (SŽDC): Prague, Czech Republic, 2015.

25. Czech Railway Infrastructure Administration (SŽDC). Jízdní řád (Timetable). Available online: https://www.spravazeleznic.cz/cestujici/jizdni-rad (accessed on 21 July 2020).

26. Czech Railway Infrastructure Administration (SŽDC). Abecední Seznam Stanic a Zastávek (Alphabetical List of Stations and Stops). Available online: https://www.spravazeleznic.cz/documents/50004227/50157161/seznam-stanic.pdf (accessed on 20 July 2020).

Publisher's Note: MDPI stays neutral with regard to jurisdictional claims in published maps and institutional affiliations.

applied sciences

MDPI

Article

Statistical Methods in Bidding Decision Support for Construction Companies

Agnieszka Leśniak

Faculty of Civil Engineering, Cracow University of Technology, 31-155 Krakow, Poland; alesniak@l7.pk.edu.pl

Abstract: On the border of two phases of a building life cycle (LC), the programming phase (conception and design) and the execution phase, a contractor is selected. A particularly appropriate method of selecting a contractor for the construction market is the tendering system. It is usually based on quality and price criteria. The latter may involve the price (namely, direct costs connected with works realization as well as mark-ups, mainly overhead costs and profit) or cost (based on the life cycle costing (LCC) method of cost efficiency). A contractor's decision to participate in a tender and to calculate a tender requires an investment of time and company resources. As this decision is often made in a limited time frame and based on the experience and subjective judgement of the contractor, a number of models have been proposed in the literature to support this process. The present paper proposes the use of statistical classification methods. The response obtained from the classification model is a recommendation to participate or not. A database consisting of historical data was used for the analyses. Two models were proposed: the LOG model—using logit regression and the LDA model—using linear discriminant analysis, which obtain better results. In the construction of the LDA model, the equation of the discriminant function was sought by indicating the statistically significant variables. For this purpose, the backward stepwise method was applied, where initially all input variables were introduced, namely, 15 identified bidding factors, and then in subsequent steps, the least statistically significant variables were removed. Finally, six variables (factors) were identified that significantly discriminate between groups: type of works, contractual conditions, project value, need for work, possible participation of subcontractors, and the degree of difficulty of the works. The model proposed in this paper using a discriminant analysis with six input variables achieved good performance. The results obtained prove that it can be used in practice. It should be emphasized, however, that mathematical models cannot replace the decision-maker's thought process, but they can increase the effectiveness of the bidding decision.

Keywords: bidding decision; LCC criterion; price criterion; construction; statistical method; classification; probability of winning

Citation: Leśniak, A. Statistical Methods in Bidding Decision Support for Construction Companies. *Appl. Sci.* **2021**, *11*, 5973. https://doi.org/10.3390/app11135973

Academic Editor: Igal M. Shohet

Received: 26 May 2021
Accepted: 24 June 2021
Published: 27 June 2021

1. Introduction

With the development of new technologies and advanced building materials, an increasing number of demands are placed on the construction industry. Modern buildings should have as little impact as possible on the environment [1–3] using sustainable materials (such as natural or recycled materials) [4–6] and environmentally friendly construction technologies [7–9]. They should have low energy consumption [10,11], demonstrate the ability to perform repairs resulting from wear and tear [12–14], as well as from possible breakdowns [15,16]. Preferably, they should allow the recycling or disposal [17,18] of the resulting construction waste. These aspects are considered by the participants in the investment process, both the investor, the contractor, and the user against the background of the different stages of the building life cycle. Phases identified in the literature include the following: the programming phase (study and conceptual analysis, as well as design), the execution phase (construction of the facility), the operation phase (operation, use, and maintenance of the facility) and the decommissioning phase (demolition of the facility). In

167

this paper, attention is paid to the programming phase, and in particular to the conclusion of the design phase, which must be followed here by the selection of the contractor for the construction work before the execution phase begins (Figure 1).

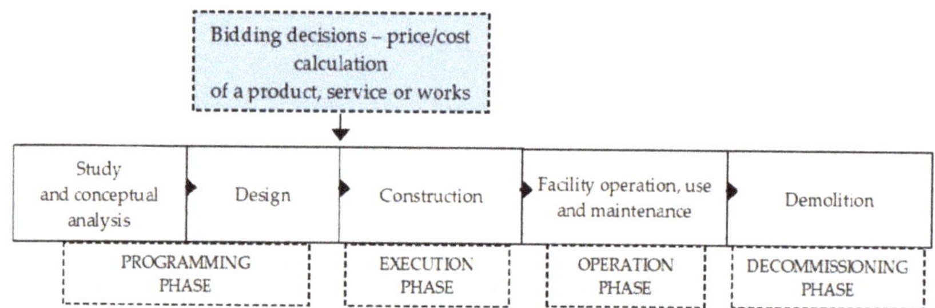

Figure 1. Bidding decisions of the building contractor within the building life cycle.

The methods of sourcing contractors in the construction market depend on the type of market (private or public sector) and the value of the project. Due to the individualized nature of construction production and the long production cycle, the tendering system is particularly suited to the operating conditions of the construction market [19]. The bidding procedure ensures that competition takes place properly and that its results are objective. It is also a factor conditioning the objectivity of prices in the construction industry. Bidding can be carried out by any investor looking for a contractor, but it is the potential contractor who must decide to tender and begin the laborious process of preparing a bid.

The selection of the most advantageous tender is normally based on quality and price criteria [20]. The price criterion may involve a price or cost and is based on a cost-effectiveness method, such as life-cycle costing (LCC). In the former case, the basis for determining a price are the direct costs connected with works realization as well as mark-ups, mainly overhead costs and profit [21–23]. In the latter case, life cycle costs (LCC) should be estimated, including the costs for planning, design, operation, maintenance, and decommissioning minus the residual value, if there is any [24]. In the literature, one can find many mathematical models prepared for the estimation of building life cycle costs [25–27], the description and comparison of which can be found, for example, in [28]. A contractor's decision to enter a tender requires action to prepare the tender and requires investment of time and commitment of staff, that is, the direct use of company resources. Irrespective of the outcome of the tender, the costs of preparing the tender will be incurred. Efficient bidding is certainly essential for every construction company. Choosing the right tender for a company has an impact on the creation of its image, its financial condition, and its aspiration to success [29].

The decision to participate in a tender often must be made by the contractor within a limited time frame and it is often based on his or her own experience. To improve the effectiveness of the decision, various models have been developed to support this process. In this case, a bidding decision model should be understood as a mathematical representation of reality, with a proposed technique to help the construction contractor decide to participate in the tender, avoiding errors and randomness. Efficient decision making is one of the greatest challenges of contemporary construction [30].

Different methods and tools are used to build models supporting construction contractors' decisions to bid. A summary of the selected existing (published after 2000) models is provided in Table 1.

Table 1. Examples of models supporting tender decisions presented in the literature after 2000.

Bidding Decision Model	Authors	Source	Year of Publication
Model based on Case-based reasoning approach	Chua D.K.H, Li D.Z, Chan W.T.	[31]	2001
Model based on the Analytical Hierarchy Process (AHP) method	Cagno E., Caron F., Perego A.	[32]	2001
Model based on artificial neural network	Wanous M., Boussabaine A. H., Lewis J.	[33]	2003
Model based on fuzzy linguistic approach	Lin Ch.-T., Chen Y.-T.	[34]	2004
Model based on logistic regression	Drew D., Lo H.P.	[35]	2007
Model based on a knowledge system	Egemen M., Mohamed A.	[36]	2008
Model based on data envelopment analysis (DEA)	El-Mashaleh M. S.	[37]	2010
Model based on a multi-criteria analysis and fuzzy set theory	Cheng M.-Y., Hsiang C. Ch., Tsai H.-Ch, Do H.-L.	[38]	2011
Model based on an ant colony optimisation algorithm and artificial neural network	Shi, H.	[39]	2012
Model based on fuzzy Analytical Hierarchy Process and regression-based simulation	Chou, J. S., Pham, A. D., Wang, H.	[40]	2013
Model based on a fuzzy set theory	Leśniak, A., Plebankiewicz, E.	[41]	2016
Model based on RBF neural networks	Leśniak, A.	[42]	2016
Model based on the simple additive weighted scoring	Chisala, M. L.	[43]	2017
Model based on the fuzzy AHP	Leśniak, A., Kubek, D., Plebankiewicz, E., Zima, K., Belniak, S.	[44]	2018
Model based on the game theory	Arya, A., Sisodia, S., Mehroliya, S., Rajeshwari, C. S.	[45]	2020
Model based on structural equation	Ojelabi, R. A., Oyeyipo, O. O., Afolabi, A. O., Omuh, I. O.	[46]	2020
Model Based on Projection Pursuit Learning Method	Zhang, X., Yu, Y., He, W., Chen, Y.	[47]	2021

It is worth noting that the indicated models differ in the methods used. Different methods, techniques, and approaches are sought and applied to obtain the most effective models. What is important, continuously for at least 20 years, modeling of a tender decision is still an object of research and interest of researchers.

The models proposed in the literature are generally based on factors, also called criteria, affecting the decision, and using them as input parameters. The number of publications on the identification of factors is considerable, as each country and region has a certain characteristic group of factors that will not be found in other markets [48–50]. It can therefore be concluded that the factors influencing tender decisions depend not only on the project to be tendered but also on the environment and market in which the contractor operates.

Bidding problems are also known in procurement auctions [51,52]. This paper [53] presents the analysis of the relation between the award price and the bidding price in the case of public procurement in Spain. An award price estimator was proposed as it is believed to be particularly useful for companies and public procurement agencies. Procurement auctions have long been employed in the logistics and transportation industry [54]. In combinatorial auctions, each carrier must determine the set of profitable contracts to bid on and the associated ask prices. This is known as the bid construction problem (BCP) [55]. Different approaches for the bid construction problem (BCP) in transportation procurement auctions are proposed in literature. One of them can be found in [56] where authors proposed solving the BCP problem for heterogeneous truckload using exact and heuristic methods.

The paper proposes the use of statistical methods to support the decision-making process of a construction contractor related to the preparation of a price offer and entering a tender. Two classification methods were used as decision support models. The response obtained from the classification model is a recommendation to participate in the tender (qualification into the W-winning class), or a recommendation to resign (allocation into the L-losing class). To perform the analyses, it was necessary to use a database consisting of historical data, that is, resolved tenders. The research framework diagram is presented in Figure 2.

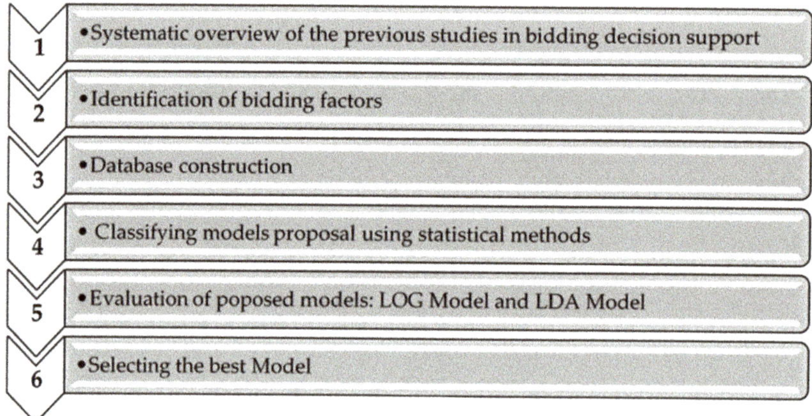

Figure 2. The research framework diagram.

2. Materials and Methods

2.1. Data Acquisition

In [57], a literature survey and research gap analysis of statistical methods used in the context of optimizing bids were presented. The paper attempts to build a decision-making model using two statistical methods: regression analysis and discriminant analysis. In the methods derived from regression analysis, the values of the Y variable (the explained variable) are given before determining the model and based on them and the adopted factors, the parameters of the model are determined. However, in the case of discriminant analysis, the values of the variable are obtained when the model is determined.

Factors influencing decision-making were proposed as input parameters of the models (explanatory variables). As a result of research (a questionnaire survey) conducted by the author in Poland, presented and described in previous works [29,44], 15 factors were identified: x_1—type of works, x_2—experience in similar projects, x_3—contractual conditions, x_4—investor reputation, x_5—project value, x_6—need for work, x_7—the size of the project, x_8—profits made in the past from similar undertakings, x_9—duration of the project, x_{10}—tender selection criteria, x_{11}—project location, x_{12}—time to prepare the offer, x_{13}—possible participation of subcontractors, x_{14}—the need for specialized equipment, and x_{15}—degree of difficulty of the works. The tender score was the model output variable (Y) representing the class:

- W—win—interpreted as a recommendation to take part in a tender,
- L—loss—interpreted as a recommendation to abandon the tender.

The starting point for the selected methods was the construction of a database. The research performed in Poland was of primary nature, based on information collected to solve a given decision problem. With regard to the type of research material, the study comprised quantitative research (evaluation of factors) and qualitative research: determination of the result obtained by the contractor in a given evaluated tender. The factors identified were used to evaluate the tenders entered into by the contractors participating in the research. Each factor, from x_1 to x_{15}, was rated on a scale from 1 to 7, where the numbers meant 1—very unfavorable, and 7—very favorable influence of the factor on the decision to participate in the tender. This scale has already been used successfully in previous works [44]. The result for each tender evaluated was then recorded (W—win, L—loss). In the end, the database contained 88 evaluated tenders, of which 64 were lost cases (L) and 24 won cases (W). Selected database records of evaluated tenders including factor evaluations with the corresponding result obtained in the tender (W—win, L—loss,) are presented in Table 2.

Table 2. Selected database records.

Record (Evaluated Tender)	Factors															Result
i	x_1	x_2	x_3	x_4	x_5	x_6	x_7	x_8	x_9	x_{10}	x_{11}	x_{12}	x_{13}	x_{14}	x_{15}	
4	6	7	5	5	3	6	3	4	5	3	6	5	2	6	7	L
12	4	6	5	6	1	6	1	4	6	5	6	6	7	3	5	W
62	7	7	4	4	4	5	3	4	5	3	5	5	4	6	6	L

2.2. Regression Analysis Model

The main task of the qualitative decision-making model will be to determine the probability of the contractor's success in the tender (winning) and to identify variables that significantly affect the outcome of the tender. A binomial (dichotomous) model is sought in which the explanatory variable Y is quantified by a zero-one variable. It takes two possible variants described by the codes "1"—W (win) and "0"—L (loss). If p_i is the probability of the event $Y_i = 1$, then $1 - p_i$ is the probability of the event $Y_i = 0$. The expected value of the variable Y_i is [58,59]:

$$E(Y_i) = 1 \cdot p_i + 0 \cdot (1 - p_i) = p_i \tag{1}$$

In binomial models, it is assumed that pi is a function of the vector of values of the explanatory variables x_i for the *i*-th object and the parameter vector β [58,59]:

$$P_i = P(y_i = 1) = F\left(x_i^T \beta\right) \tag{2}$$

Depending on the type of F-function, different types of models are distinguished [60]: a linear probability model, logit model, and probit model. Using the simplest of the binomial models—the linear probability model—has many negative consequences described in the literature [58,61]. Probit and logit models, on the other hand, as indicated by some authors [60], are similar to each other and in practice one of them is used. Therefore, the search for a binomial model for the phenomenon in question was limited to a logistic regression model. The general form of the logit model is as follows [58,59]:

$$Y_i^* = ln\frac{p_i}{1 - p_i} = \beta_0 \alpha_0 + \beta_1 X_{1i} + \beta_2 x X_{2i} + \cdots + \beta_k X_{ki} + u_i \tag{3}$$

where:

β_j—structural model parameters,

u_i—random component,

$ln\frac{p_i}{1-p_i}$—logit,

Y_i^*—unobservable qualitative variable,

X_{ji}—the values of the explanatory variables of the model,

p_i—the probability of taking the value "1" by the dependent variable Y_i calculated from the logistic distribution density function.

$$p_i = \frac{e^{x_i'\beta}}{1 + e^{X_i'\beta}} = \frac{1}{1 + e^{-X_i'\beta}} = \frac{1}{1 + e^{-(\beta_0 + \beta_1 X_{1i} + \beta_2 X_{2i} + \cdots + \beta_k X_{ki})}} \tag{4}$$

Unobservable variable Y_i^* is defined as a latent variable, as one can observe only the binary variable Y_i in the form:

$$Y_i = \begin{cases} 1; & Y_i^* > 0 \\ 0; & Y_i^* \leq 0 \end{cases} \tag{5}$$

Logit according to [53], denotes the odds ratio of accepting to not accepting the value "1" for the variable Y_i. It takes the value zero if $p_i = 0.5$. In the case when $p_i < 0.5$, the odds ratio takes a negative value, and when $p_i > 0.5$, a positive one.

2.3. Discriminant Analysis Model

Linear discriminant analysis (LDA), presented in 1936 [62], enables the classification of cases (objects) into one of the predetermined groups based on explanatory variables (case characteristics). The use of linear discriminant analysis to classify objects (cases) [63] or supporting decision-making processes [64] are commonly found in the literature. The aim of discriminant methods is to determine which of the explanatory variables differentiate groups the most. The discrimination problem can be solved by means of discriminant functions which are most often linear functions of input variables characterizing the cases [65]. If group sizes are not comparable, a modified form of the discriminant function should be used [65]:

$$K_r = c_{r0} + c_{r1}X_1 + c_{r2}X_2 + \ldots + c_{rm}X_m + ln\frac{n_r}{n}, \tag{6}$$

where:

K_r—classification function (for the r-th group of cases),
c_{rj}—the coefficient of the r-th classification function with j-th input variable of significant discriminatory power, $j = 0, 1, \ldots, m'$,
$c_{r0} = lnp_{ri}$—absolute term, probability p_i means the *a priori* probability of qualifying the i-th object to the r-th group,
n_r—denotes the size of a given group,
n—sample size.

Modeling takes place in several stages. In the first step of building the model, the discriminant function equation is sought by identifying variables that significantly discriminate groups. The next step is to check the statistical significance of the discriminant function and determine its coefficients. The next stage of the analysis is a classification procedure using classification functions.

2.4. Evaluation of the Proposed Models

To assess the quality and relevance of the performance of the proposed classification models [66], the following were proposed:

- A relevance matrix that indicates the number and often the proportion of correctly and incorrectly classified cases;
- Diagnostic test parameters: sensitivity (7), specificity (8), positive (9), and negative (10) predictive value, test reliability (11) based on the contingency matrix:

Sensitivity indicates the ratio of true positives to the sum of true positives and false negatives. In the problem under analysis, it describes the ability to detect the winning cases.

$$sensitivity = \frac{TP}{TP + FN} \tag{7}$$

Specificity means the ratio of true negatives to the sum of true negatives and false positives. In the problem examined, it describes the ability to detect the losing cases.

$$Specificity = \frac{TN}{TN + FP} \tag{8}$$

PPV (positive predictive value) denotes the probability that the case identified by the classifier as winning is indeed a winning case.

$$PPV = \frac{TP}{TP + FP} \tag{9}$$

NPV (negative predictive value) stands for the probability that the case identified by the classifier as loss is indeed a losing case.

$$NPV = \frac{TN}{FN + TN} \tag{10}$$

Effectiveness of the decision rule ACC (accuracy) implies the extent to which the results of the study reflect reality.

$$ACC = \frac{TP + TN}{TP + TN + FP + FN} \tag{11}$$

where:

TP—true positive results,
FP—false positive results,
TN—false negative results,
FN—true negative results.

3. Results and Discussion

3.1. LOG Model—The Model Using Regression Analysis

Using logit regression, an attempt was made to estimate the qualitative variable Y, also trying to explain which factors, with what strength and in what direction, affect the chance of a tender success (Y). The parameter estimates are summarized in Table 3.

By analyzing the obtained results with the assumed significance level $\alpha = 0.1$, only two variables significantly affect the model: x_3—contractual conditions and x_6—need for work. However, the p value for the variables x_{12}—time to prepare the offer and x_{15}—the degree of difficulty of the works, are slightly higher than 0.1, so it was decided to include these variables and recalculate the model. The parameter estimates for the logistic regression model (with four explanatory variables) are summarized in Table 4.

Finally, three variables were left (the non-significant variable x_{12}—time to prepare an offer, was discarded) and recalculations were made.

The parameter estimates for the logistic regression model (with three explanatory variables) are summarized in Table 5.

The form of the proposed logit model (LOG model) is as follows:

$$\hat{Y}_i = ln\frac{p_i}{1 - p_i} = -0.9532 \cdot x_3 - 2.2877 \cdot x_6 + 0.6012 \cdot x_{15} + 15.9217 \tag{12}$$

This means that the probability p_i (that is, situation $Y_i = 1$) is estimated as:

$$\hat{p}_i = \frac{\exp(-0.9532 \cdot x_3 - 2.2877 \cdot x_6 + 0.6012 \cdot x_{15} + 15.9217)}{1 + \exp(-0.9532 \cdot x_3 - 2.2877 \cdot x_6 + 0.6012 \cdot x_{15} + 15.9217)} \tag{13}$$

Statistical verification of the logit model consisting in determining the degree of the model fitting the data and testing the statistical significance of the parameters was successful. The odds quotient is 9.62 and is higher than 1 which means that the classification is nine times better than what would be expected by chance. Using the proposed logit model, it is possible to estimate the probability with which a given tender will be won.

Table 3. Parameter estimates for the logit model—15 explanatory variables.

Dependent Variable Y RESULT	Coefficient β	Standard Error	Walda Stat.	p Value	Statistically Significant *
Absolute term	181.2342	101.4030	3.1943	0.0739	*
x_1	−0.8114	2.1518	0.1422	0.7061	
x_2	−0.4787	1.2390	0.1493	0.6992	
x_3	−6.4249	2.8502	5.0814	0.0242	*
x_4	1.0285	1.2892	0.6364	0.4250	
x_5	0.6708	1.5074	0.1980	0.6563	
x_6	−6.6233	2.5664	6.6607	0.0099	*
x_7	−1.2769	1.5475	0.6809	0.4093	
x_8	−4.1032	6.4537	0.4042	0.5249	
x_9	0.8568	1.1536	0.5516	0.4577	
x_{10}	−3.3306	2.8215	1.3935	0.2378	
x_{11}	−2.9788	2.2852	1.6992	0.1924	
x_{12}	3.5272	2.3601	2.2335	0.1350	
x_{13}	−15.8525	11.8791	1.7808	0.1820	
x_{14}	−2.4435	2.2014	1.2320	0.2670	
x_{15}	4.9962	3.4396	2.1098	0.1464	

* Significance level $\alpha = 0.1$

Table 4. Parameter estimates for the logit model—four explanatory variables.

Dependent Variable Y RESULT	Coefficient β	Standard Error	Walda Stat.	p Value	Statistically Significant *
Absolute term	16.1289	4.6709	11.9235	0.0006	*
x_3	16.1289	4.6709	11.9235	0.0006	*
x_6	−0.9145	0.4582	3.9833	0.0460	*
x_{12}	−2.2736	0.5790	15.4179	0.0001	
x_{15}	−0.0840	0.5715	0.0216	0.8831	*

* Significance level $\alpha = 0.1$

Table 5. Parameter estimates for the logit model—three explanatory variables.

Dependent Variable Y RESULT	Coefficient β	Standard Error	Walda Stat.	p Value	Statistically Significant *
Absolute term	15.9217	4.4375	12.8739	0.0003	*
x_3	−0.9532	0.3769	6.3964	0.0114	*
x_6	−2.2877	0.5723	15.9812	0.0001	*
x_{15}	0.6012	0.3271	3.3777	0.0661	*

* Significance level $\alpha = 0.1$

3.2. LDA Model—The Model Using Discriminant Analysis

In the first step of building the model, the equation of the discriminant function was searched for, indicating variables that significantly discriminate groups. To achieve this, the backward stepwise method was applied. In this approach, all variables are entered into the model (step 0) and then, in subsequent steps, one variable that is the least statistically significant is removed. Results with all 15 input variables (step 0) indicated at the assumed significance level $\alpha = 0.1$, that only four variables significantly discriminate between groups (x_3, x_5, x_6, x_{13}).

The results for the model and the evaluation of all 15 input variables (step 0) are given in Table 6.

Table 6. Evaluation of the parameters of the discriminant function—15 explanatory variables.

Variables	Wilks' Lambda	Partial Lambda Wilks	The Value of the F Statistic	p Value	Tolerance	1-Tolerance	Statistically Significant *
x_1	0.430141	0.961162	2.90937	0.092377	0.377706	0.622294	
x_2	0.413447	0.999971	0.00207	0.963850	0.370875	0.629125	
x_3	0.449346	0.920082	6.25390	0.014669	0.376703	0.623297	*
x_4	0.421275	0.981391	1.36523	0.246487	0.522505	0.477495	
x_5	0.454559	0.909531	7.16170	0.009215	0.109267	0.890733	*
x_6	0.609746	0.678045	34.18763	0.000000	0.643071	0.356929	*
x_7	0.419668	0.985148	1.08548	0.300962	0.096009	0.903991	
x_8	0.434939	0.950558	3.74495	0.056895	0.577896	0.422104	
x_9	0.414878	0.996522	0.25127	0.617712	0.757040	0.242960	
x_{10}	0.417598	0.990032	0.72491	0.397361	0.672704	0.327296	
x_{11}	0.424018	0.975042	1.84297	0.178843	0.633470	0.366531	
x_{12}	0.430727	0.959854	3.01140	0.086960	0.512083	0.487917	
x_{13}	0.444633	0.929835	5.43308	0.022563	0.513878	0.486123	*
x_{14}	0.414026	0.998573	0.10292	0.749283	0.580513	0.419487	
x_{15}	0.434007	0.952600	3.58258	0.062408	0.391804	0.608196	

* Significance level $\alpha = 0.05$.

By analyzing the obtained results with the assumed significance level $\alpha = 0.1$, only four variables $(x_3; x_5; x_6; x_{13})$ discriminated significantly between groups. The model parameters are as follows: Wilks' Lambda = 0.41344, the corresponding F statistic (15.72) = 6.8100, and $p < 0.0000$.

During the first step of the analysis, the variable x_2 was removed—the least significantly discriminating group. Subsequent steps ($k = 2, \ldots, 15$) made it possible to select the most significant variables (Table 7).

Table 7. Evaluation of the discriminant function parameters—final model (six input variables).

Variables	Wilks' Lambda	Partial Lambda Wilks	The Value of the F Statistic	p Value	Tolerance	Wilks' Lambda
x_1	0.570048	0.825309	17.14504	0.000084	0.694765	*
x_3	0.495679	0.949133	4.341000	0.040358	0.771227	*
x_5	0.557289	0.844204	14.94840	0.000222	0.541454	*
x_6	0.740919	0.634977	46.56377	0.000000	0.776650	*
x_{13}	0.532386	0.883694	10.66070	0.001605	0.803942	*
x_{15}	0.500913	0.939217	5.242020	0.024650	0.789131	*

* Significance level $\alpha = 0.05$.

Finally, six input variables, x_1—type of works, x_3—contractual conditions, x_5—project value, x_6—need for work, x_{13}—possible participation of subcontractors, x_{15}—the degree of difficulty of the works, discriminate significantly between groups. The model parameters are as follows: Wilks' Lambda = 0.47047; the corresponding statistic F (6.81) = 15.195; $p < 0.0000$. It is worth noting that the smaller the value of Wilks' Lambda (from the range <0, 1>) the better the discriminating power the model has. In the analyzed example (0.47047), it is acceptable. Tolerance coefficient T_k determines the proportion of the variance of the variable x_k that is not explained by the variables in the model. If T_k coefficient takes a value smaller than the default 0.01, the variable is more than 99% redundant with other variables in the model. Entering variables with low tolerance coefficients into the model may cause its large inaccuracy. In the model under consideration, the T_k coefficients for the assumed variables exceed the value of 0.5.

The next step of the analysis is to check the statistical significance of the discriminant function (Table 8) and to determine its coefficients.

Table 8. Parameters for assessing the statistical significance of the discriminant function.

Eigenvalue	Canonical Correlation R	Wilks' Lambda	*Chi*-Square Statistics	*df* Number of Degrees of Freedom	*p* Value
1.125553	0.727691	0.470466	62.58464	6	0.000000

The eigenvalue of a discriminant function represents the ratio of the between-group variance to the within-group variance. Large eigenvalues characterize functions with high discriminatory power. Canonical correlation is a measure of the magnitude of the association between a grouping variable and the results of a discriminant function. It ranges from <0, 1>, where 0 means no relationship and 1 means maximum relationship. The value of 0.727691 means that the function is related to a grouping variable. The value of Wilks' Lambda is acceptable. The value of $p = 0.000000 < 0.05$. The proposed discriminant function is statistically significant and ultimately takes the following form:

$$D = -12.831 + 0.509x_1 + 0.437x_3 - 0.464x_5 + 1.502x_6 + 0.615x_{13} - 0.429x_{15} \quad (14)$$

The next stage of the analysis is the classification procedure using classification functions. In the problem under analysis, two classification functions were defined (two groups were assumed; W—win, L—loss), which take the following form:

- K_0 function, classifying to "L-loss" group:

$$K_0 = -181.383 + 7.139x_1 + 13.094x_3 + 0.148x_5 + 21.275x_6 + 15.141x_{13} + 9.958x_{15} + \ln\tfrac{64}{88} \quad (15)$$

- K_1 function, classifying to "W-win" group:

$$K_1 = -213.841 + 8.338x_1 + 14.123x_3 - 0.946x_5 + 24.813x_6 + 16.590x_{13} + 8.947x_{15} + \ln\tfrac{24}{88} \quad (16)$$

A given case is classified in the group for which the classification function takes the highest value.

3.3. Evaluation of Models—Discussion of Results

To evaluate the model, the classification efficiency expressed as the number of cases correctly classified into predefined classes was used. A summary of the performance of the proposed models is presented in Tables 9 and 10.

Table 9. Summary of classification for the LOG model.

Observed Numbers		Reality		Total
Answer from the Model		$y_i = 1$ (W—win)	$y_i = 0$ (L—loss)	
	$\hat{y}_i = 1$ (W—win)	13	7	20
	$\hat{y}_i = 0$ (L—loss)	11	57	68
LOG Model	Total	24	64	88
	Correct	54.17%	89.06%	79.55%
	Incorrect	45.83%	10.94%	20.45%
	Total	100.00%	100.00%	100.00%

Table 10. Summary of classification for the LDA model.

Observed Numbers		Reality		Total
Answer from the Model		$y_i = 1$ (W—win)	$y_i = 0$ (L—loss)	
LDA Model	$\hat{y}_i = 1$(W—win)	15	3	18
	$\hat{y}_i = 0$ (L—loss)	9	61	70
	Total	24	64	88
	Correct	62.50%	95.31%	86.36%
	Incorrect	37.50%	4.69%	13.64%
	Total	100.00%	100.00%	100.00%

The data in Tables 8 and 9 enable the basic parameters of the classification model to be determined. The results are given in Table 11.

Table 11. Basic parameters of LOG and LDA models as a classifier.

Model	Sensitivity	Specificity	Positive Predictive Value (PPV)	Negative Predictive Value (NPV)	Effectiveness (ACC)
LOG Model	54.17%	89.06%	65.00%	83.82%	79.55%
LDA Model	62.50%	95.31%	83.33%	87.14%	86.36%

From the values in Table 10, the LOG model correctly classified 79.55% of the cases, more correctly predicting tender failure (83.82%). The values obtained show a good fit of the model, but it is worrying that the model indicated only three tender factors as statistically significant: x_3—contractual conditions, x_6—need for work, and x_{15}—degree of difficulty of the works. In the case of the LDA model, classification into the set L—87.14% means that the model (analogous to the LOG model) more accurately predicts tender failure than winning (83.33%). The results obtained by the LDA model are better as it rendered 86% of correctly classified cases. The discriminant analysis, apart from the variables x_3, x_6, x_{15} indicated also x_1—type of works, x_5—value of the project, and x_{13}—possible participation of subcontractors, as significant variables for the model, where the greatest independent influence on the result of the discriminant function is exerted by the variable x_6—the need for work, while the least x_3—contractual conditions. The following is an extract from the LDA model results sheet with the values of the classification functions in relation to the observed (actual) values shown in Table 12.

Table 12. Classification function values for selected cases.

Case Number	Group Membership:	Group Membership:	Function Value	
	Observed	Model	K_0—win	K_1—loss
10	W	W	208.994	211.747
28	L	L	189.970	186.249
* 29	W	L	207.961	206.832
30	L	L	199.490	196.887
* 35	L	W	220.759	222.847

* Cases misclassified.

The analyses presented in this paper do not exhaust the issue of modeling contractors' decisions to participate in tenders for construction works. They can become a supplement of the models proposed so far in the literature. It should be noted that the construction of classification models requires having an appropriate database, which is built based on tender factors selected by the author of each model.

It is also worth emphasizing that in the face of fierce competition on the construction market, contractors are looking for solutions to maximize their chances of winning tenders. It is worth noting the observations of the authors of the study [67], who noted that the bid preparation process, which is time-consuming and requires a lot of effort, may create the need to have appropriate specialists. Typically, large companies are more able to employ such specialists, while small and medium-sized companies are definitely more likely to feel the need for tools to support the proper selection of orders and the decision to tender. It therefore appears that the proposal to build and use mathematical models is appropriate.

In further research, using the author's constructed database, the author of the paper intends to apply methods of artificial intelligence. The same database, model input and output parameters will allow to objectively compare the effectiveness of these two approaches.

4. Conclusions

The construction company at each stage of its activity has to make a number of important decisions related to the functioning of the company. One of them is the decision to enter a tender. Although it involves company finances and resources, the decision is usually taken quickly and based on subjectively perceived information. A number of models and mathematical methods have been proposed in the literature to assist the decision maker and to increase the effectiveness of the decisions taken. In this paper, two statistical classification methods are used for modeling: linear regression and linear discriminant analysis. The response obtained from the classification model is a recommendation to participate in the tender (qualification into class W—win), or a recommendation to resign (allocation into class L—loss). To perform the analyses, it was necessary to use a database consisting of historical data, that is, resolved tenders. The comparison of the classification models shows that the model using linear discriminant analysis performed well (86% correctly classified cases). The backward stepwise method was used to eliminate the least statistically significant variables. Finally, from a set of 15 identified factors, six input variables (factors) were identified that significantly discriminate between groups: x_1—type of works, x_3—contractual conditions, x_5—project value, x_6—need for work, x_{13}—possible participation of subcontractors, x_{15}—the degree of difficulty of the works. With these variables, the model achieved good performance. The paper by [44] presents the results of a survey in which the works contractors selected the following as the most important factors influencing the decision to enter a tender: type of works, contractual conditions, experience in similar projects, project value, need for work. As can be seen, they mostly coincide with the results obtained from statistical methods. The obtained results (effectiveness of classification and values of model evaluation parameters) testify to the possibility of using the LDA model in practice.

Funding: This research received no external funding.

Institutional Review Board Statement: Not applicable.

Informed Consent Statement: Not applicable.

Data Availability Statement: Data sharing not applicable.

Conflicts of Interest: The author declares no conflict of interest.

References

1. Bernardi, E.; Carlucci, S.; Cornaro, C.; Bohne, R.A. An Analysis of the Most Adopted Rating Systems for Assessing the Environmental Impact of Buildings. *Sustainability* **2017**, *9*, 1226. [CrossRef]
2. Sztubecka, M.; Skiba, M.; Mrówczyńska, M.; Mathias, M. Noise as a Factor of Green Areas Soundscape Creation. *Sustainability* **2020**, *12*, 999. [CrossRef]
3. Wałach, D. Analysis of Factors Affecting the Environmental Impact of Concrete Structures. *Sustainability* **2021**, *13*, 204. [CrossRef]
4. Antico, F.C.; Ibáñez, U.A.; Wiener, M.J.; Araya-Letelier, G.; Retamal, R.G. Eco-bricks: A sustainable substitute for construction materials. *Revista Construcción* **2017**, *16*, 518–526. [CrossRef]
5. Govindan, K.; Shankar, M.; Kannan, D. Sustainable material selection for construction industry—A hybrid multi criteria decision making approach. *Renew. Sustain. Energy Rev.* **2016**, *55*, 1274–1288. [CrossRef]

6.   Reddy, B.V. Sustainable materials for low carbon buildings. *Int. J. Low Carbon Technol.* **2009**, *4*, 175–181. [CrossRef]

7.   Leśniak, A.; Zima, K. Comparison of traditional and ecological wall systems using the AHP method. In Proceedings of the International Multidisciplinary Scientific GeoConference Surveying Geology and Mining Ecology Management, SGEM, Albena, Bulgaria, 18–24 June 2015; Volume 3, pp. 157–164.

8.   Zavadskas, E.K.; Antucheviciene, J.; Vilutiene, T.; Adeli, H. Sustainable Decision-Making in Civil Engineering, Construction and Building Technology. *Sustainability* **2018**, *10*, 14. [CrossRef]

9.   Apollo, M.; Siemaszko, A.; Miszewska, E. The selected roof covering technologies in the aspect of their life cycle costs. *Open Eng.* **2018**, *8*, 478–483. [CrossRef]

10. Sztubecka, M.; Skiba, M.; Mrówczyńska, M.; Bazan-Krzywoszańska, A. An Innovative Decision Support System to Improve the Energy Efficiency of Buildings in Urban Areas. *Remote Sens.* **2020**, *12*, 259. [CrossRef]

11. Chen, S.; Zhang, G.; Xia, X.; Setunge, S.; Shi, L. A review of internal and external influencing factors on energy efficiency design of buildings. *Energy Build.* **2020**, *216*, 109944. [CrossRef]

12. Nowogońska, B. A Methodology for Determining the Rehabilitation Needs of Buildings. *Appl. Sci.* **2020**, *10*, 3873. [CrossRef]

13. Słowik, M.; Skrzypczak, I.; Kotynia, R.; Kaszubska, M. The Application of a Probabilistic Method to the Reliability Analysis of Longitudinally Reinforced Concrete Beams. *Procedia Eng.* **2017**, *193*, 273–280. [CrossRef]

14. Konior, J.; Sawicki, M.; Szóstak, M. Damage and Technical Wear of Tenement Houses in Fuzzy Set Categories. *Appl. Sci.* **2021**, *11*, 1484. [CrossRef]

15. Ahn, S.; Kim, T.; Kim, J.-M. Sustainable Risk Assessment through the Analysis of Financial Losses from Third-Party Damage in Bridge Construction. *Sustainability* **2020**, *12*, 3435. [CrossRef]

16. Bednarz, L.; Bajno, D.; Matkowski, Z.; Skrzypczak, I.; Leśniak, A. Elements of Pathway for Quick and Reliable Health Monitoring of Concrete Behavior in Cable Post-Tensioned Concrete Girders. *Materials* **2021**, *14*, 1503. [CrossRef] [PubMed]

17. Spišáková, M.; Mésároš, P.; Mandičák, T. Construction Waste Audit in the Framework of Sustainable Waste Management in Construction Projects—Case Study. *Buildings* **2021**, *11*, 61. [CrossRef]

18. Tang, Z.; Li, W.; Tam, V.W.; Xue, C. Advanced progress in recycling municipal and construction solid wastes for manufacturing sustainable construction materials. *Resour. Conserv. Recycl.* **2020**, *6*, 100036. [CrossRef]

19. Smith, A.J. *Estimating, Tendering and Bidding for Construction Work*; Macmillan International Higher Education: London, UK, 2017.

20. Waara, F.; Bröchner, J. Price and Nonprice Criteria for Contractor Selection. *J. Constr. Eng. Manag.* **2006**, *132*, 797–804. [CrossRef]

21. Plebankiewicz, E.; Leśniak, A. Overhead costs and profit calculation by Polish contractors. *Technol. Econ. Dev. Econ.* **2013**, *19*, 141–161. [CrossRef]

22. Hegazy, T.; Moselhi, O. Elements of cost estimation: A survey in Canada and United States. *Cost Eng.* **1995**, *37*, 27–33.

23. Doloi, H.K. Understanding stakeholders' perspective of cost estimation in project management. *Int. J. Proj. Manag.* **2011**, *29*, 622–636. [CrossRef]

24. Boussabaine, A.; Kirkham, R. *Whole Life-Cycle Costing: Risk and Risk Responses*; Blackwell Publishing, Ltd.: Oxford, UK, 2008.

25. Plebankiewicz, E.; Meszek, W.; Zima, K.; Wieczorek, D. Probabilistic and Fuzzy Approaches for Estimating the Life Cycle Costs of Buildings under Conditions of Exposure to Risk. *Sustainability* **2020**, *12*, 226. [CrossRef]

26. Hromádka, V.; Korytárová, J.; Vítková, E.; Seelmann, H.; Funk, T. New Aspects of Socioeconomic Assessment of the Railway Infrastructure Project Life Cycle. *Appl. Sci.* **2020**, *10*, 7355. [CrossRef]

27. Biolek, V.; Hanák, T. LCC Estimation Model: A Construction Material Perspective. *Buildings* **2019**, *9*, 182. [CrossRef]

28. Plebankiewicz, E.; Zima, K.; Wieczorek, D. Review of methods of determining the life cycle cost of buildings. In Proceedings of the Creative Construction Conference, Kraków, Poland, 21–24 June 2015; Hajdu, M., Ed.; Elsevier B.V.: Amsterdam, The Netherlands, 2015; pp. 309–316.

29. Leśniak, A. Classification of the Bid/No Bid Criteria—Factor Analysis. *Arch. Civ. Eng.* **2015**, *61*, 79–90. [CrossRef]

30. Kapliński, O.; Tupenaite, L. Review of the multiple criteria decision making methods, intelligent and biometric systems applied in modern construction economics. *Transform. Bus. Econ.* **2011**, *10*, 166–181.

31. Chua, D.K.H.; Li, D.Z.; Chan, W.T. Case-Based Reasoning Approach in Bid Decision Making. *J. Constr. Eng. Manag.* **2001**, *127*, 35–45. [CrossRef]

32. Cagno, E.; Caron, F.; Perego, A. Multi-criteria assessment of the probability of winning in the competitive bidding process. *Int. J. Proj. Manag.* **2001**, *19*, 313–324. [CrossRef]

33. Wanous, M.; Boussabaine, A.H.; Lewis, J. A neural network bid/no bid model: The case for contractors in Syria. *Constr. Manag. Econ.* **2003**, *21*, 737–744. [CrossRef]

34. Lin, C.-T.; Chen, Y.-T. Bid/no bid decision-making—A fuzzy linguistic approach. *Int. J. Proj. Manag.* **2004**, *22*, 585–593. [CrossRef]

35. Oo, B.; Drew, D.S.; Lo, H.P. Applying a random coefficients logistic model to contractors' decision to bid. *Constr. Manag. Econ.* **2007**, *25*, 387–398. [CrossRef]

36. Egemen, M.; Mohamed, A. SCBMD: A knowledge-based system software for strategically correct bid/no bid and mark-up size decisions. *Autom. Constr.* **2008**, *17*, 864–872. [CrossRef]

37. El-Mashaleh, M.S. Decision to bid or not to bid: A data envelopment analysis approach. *Can. J. Civ. Eng.* **2010**, *37*, 37–44. [CrossRef]

38. Cheng, M.-Y.; Hsiang, C.-C.; Tsai, H.-C.; Do, H.-L. Bidding decision making for construction company using a multi-criteria prospect model. *J. Civ. Eng. Manag.* **2011**, *17*, 424–436. [CrossRef]

39. Shi, H. ACO trained ANN-based bid/no-bid decision-making. *Int. J. Model. Identif. Control* **2012**, *15*, 290–296. [CrossRef]
40. Chou, J.-S.; Pham, A.-D.; Wang, H. Bidding strategy to support decision-making by integrating fuzzy AHP and regression-based simulation. *Autom. Constr.* **2013**, *35*, 517–527. [CrossRef]
41. Leśniak, A.; Plebankiewicz, E. Modeling the Decision-Making Process Concerning Participation in Construction Bidding. *J. Manag. Eng.* **2015**, *31*, 04014032. [CrossRef]
42. Leśniak, A. Supporting contractors' bidding decision: RBF neural networks application. *AIP Conf. Proc.* **2016**, *1738*, 200002. [CrossRef]
43. Chisala, M.L. Quantitative Bid or No-Bid Decision-Support Model for Contractors. *J. Constr. Eng. Manag.* **2017**, *143*, 04017088. [CrossRef]
44. Leśniak, A.; Kubek, D.; Plebankiewicz, E.; Zima, K.; Belniak, S. Fuzzy AHP Application for Supporting Contractors' Bidding Decision. *Symmetry* **2018**, *10*, 642. [CrossRef]
45. Arya, A.; Sisodia, S.; Mehroliya, S.; Rajeshwari, C. A novel approach to Optimize Bidding Strategy for Restructured Power Market using Game Theory. In Proceedings of the 2020 IEEE International Symposium on Sustainable Energy, Signal Processing and Cyber Security (iSSSC), Gunupur Odisha, India, 16–17 December 2020; pp. 1–6.
46. Ojelabi, R.A.; Oyeyipo, O.O.; Afolabi, A.O.; Omuh, I.O. Evaluating barriers inhibiting investors participation in Public-Private Partnership project bidding process using structural equation model. *Int. J. Constr. Manag.* **2020**. [CrossRef]
47. Zhang, X.; Yu, Y.; He, W.; Chen, Y. Bidding Decision-Support Model for Construction Projects Based on Projection Pursuit Learning Method. *J. Constr. Eng. Manag.* **2021**, *147*, 04021054. [CrossRef]
48. Shokri-Ghasabeh, M.; Chileshe, N. Critical factors influencing the bid/no bid decision in the Australian construction industry. *Constr. Innov.* **2016**, *16*, 127–157. [CrossRef]
49. Jarkas, A.M.; Mubarak, S.A.; Kadri, C.Y. Critical Factors Determining Bid/No Bid Decisions of Contractors in Qatar. *J. Manag. Eng.* **2014**, *30*, 05014007. [CrossRef]
50. Li, G.; Chen, C.; Zhang, G.K.; Martek, I. Bid/no-bid decision factors for Chinese international contractors in international construction projects. *Eng. Constr. Arch. Manag.* **2020**, *27*, 1619–1643. [CrossRef]
51. Song, J.; Regan, A. Approximation algorithms for the bid construction problem in combinatorial auctions for the procurement of freight transportation contracts. *Transp. Res. Methodol.* **2005**, *39*, 914–933. [CrossRef]
52. Hanák, T.; Serrat, C. Analysis of construction auctions data in Slovak public procurement. *Adv. Civ. Eng.* **2018**, *2018*, 9036340. [CrossRef]
53. Rodríguez, M.J.G.; Montequín, V.R.; Fernández, F.O.; Balsera, J.M.V. Public Procurement Announcements in Spain: Regulations, Data Analysis, and Award Price Estimator Using Machine Learning. *Complexity* **2019**, *2019*, 2360610. [CrossRef]
54. Kuyzu, G.; Akyol, Ç.G.; Ergun, Ö.; Savelsbergh, M. Bid price optimization for truckload carriers in simultaneous transportation procurement auctions. *Transp. Res. Methodol.* **2015**, *73*, 34–58. [CrossRef]
55. Hammami, F.; Rekik, M.; Coelho, L.C. Exact and hybrid heuristic methods to solve the combinatorial bid construction problem with stochastic prices in truckload transportation services procurement auctions. *Transp. Res. Methodol.* **2021**, *149*, 204–229. [CrossRef]
56. Hammami, F.; Rekik, M.; Coelho, L.C. Exact and heuristic solution approaches for the bid construction problem in transportation procurement auctions with a heterogeneous fleet. *Transp. Res. Logist. Transp. Rev.* **2019**, *127*, 150–177. [CrossRef]
57. Aasgård, E.K.; Fleten, S.-E.; Kaut, M.; Midthun, K.; Perez-Valdes, G.A. Hydropower bidding in a multi-market setting. *Energy Syst.* **2019**, *10*, 543–565. [CrossRef]
58. Maddala, G.S.; Lahiri, K. *Introduction to Econometrics*; Macmillan: New York, NY, USA, 1992.
59. Larose, D.T. *Data Mining Methods and Models*; Wiley-Interscience: Hoboken, NJ, USA, 2006.
60. Judge, G.G.; Hill, C.; Griffiths, W.E.; Lütkepohl, H.; Lee, T. *The Theory and Practice of Econometrics*; John Wiley & Sons: New York, NY, USA, 1985.
61. Gatnar, E. *Symbolic Methods of Data Classification*; Polish Scientific Publishers PWN: Warsaw, Poland, 1998. (In Polish)
62. Fisher, R. Has Mendel's work been rediscovered? *Ann. Sci.* **1936**, *1*, 115–137. [CrossRef]
63. Steiniger, S.; Langenegger, T.; Burghardt, D.; Weibel, R. An Approach for the Classification of Urban Building Structures Based on Discriminant Analysis Techniques. *Trans. GIS* **2008**, *12*, 31–59. [CrossRef]
64. Fernandez, G.C. Discriminant analysis: A powerful classification technique in data mining. In Proceedings of the SAS Users International Conference, Orlando, FL, USA, 14–17 April 2002; pp. 247–256.
65. Huberty, C.J.; Barton, R.M. An Introduction to Discriminant Analysis. *Meas. Eval. Couns. Dev.* **1989**, *22*, 158–168. [CrossRef]
66. Panek, T.; Zwierzchowski, J. *Statistical Methods of Multivariate Comparative Analysis. Theory and Applications*; SGH Publishing House: Warsaw, Poland, 2013. (In Polish)
67. Bageis, A.S.; Fortune, C. Factors affecting the bid/no bid decision in the Saudi Arabian construction contractors. *Constr. Manag. Econ.* **2009**, *27*, 53–71. [CrossRef]

applied sciences

Article

Improvement of Performance Level of Steel Moment-Resisting Frames Using Tuned Mass Damper System

Masoud Dadkhah [1], Reza Kamgar [1], Heisam Heidarzadeh [1], Anna Jakubczyk-Gałczyńska [2,*] and Robert Jankowski [2]

[1] Department of Civil Engineering, Shahrekord University, Shahrekord 88186-34141, Iran; masouddadkhah1354@gmail.com (M.D.); kamgar@sku.ac.ir (R.K.); heidarzadeh@sku.ac.ir (H.H.)

[2] Faculty of Civil and Environmental Engineering, Gdansk University of Technology, 80-233 Gdansk, Poland; jankowr@pg.edu.pl

* Correspondence: annjakub@pg.edu.pl

Received: 2 April 2020; Accepted: 12 May 2020; Published: 14 May 2020

Abstract: In this paper, parameters of the tuned mass dampers are optimized to improve the performance level of steel structures during earthquakes. In this regard, a six-story steel frame is modeled using a concentrated plasticity method. Then, the optimum parameters of the Tuned Mass Damper (TMD) are determined by minimizing the maximum drift ratio of the stories. The performance level of the structure is also forced to be located in a safety zone. The incremental dynamic analysis is used to analyze the structural behavior under the influence of the artificial, near- and far-field earthquakes. The results of the investigation clearly show that the optimization of the TMD parameters, based on minimizing the drift ratio, reduces the structural displacement, and improves the seismic behavior of the structure based on Federal Emergency Management Agency (FEMA-356). Moreover, the values of base shear have been decreased for all studied records with peak ground acceleration smaller or equal to 0.5 g.

Keywords: dynamic analysis; steel frames; Tuned Mass Damper; optimization; drift ratio

---

1. Introduction

Structural disruption resulting in dangerous vibrations might be inevitable due to dynamic loads, such as wind and earthquake (see, for example, [1–4]). The use of different methods to minimize these vibrations and make structures more resistant to dynamic loads is growing day by day [5–8]. Accordingly, many researchers have focused their investigations on studying the effectiveness of different control systems, including dampers, base isolation, etc. [9–15]. Some of the most popular types of dampers include the Tuned Mass Damper (TMD; see [16–19]) and the Tuned Liquid Damper (TLD) [20]. Both are known as passive control systems. These vibrations control systems can absorb some of the input energy due to the earthquake motion [21]. There are also other passive control vibration systems, such as friction tuned mass damper (see [22–24]), viscous damper (see [25–27]), magnetorheological damper (see [28,29]), tuned mass-damper–inerter (see [30,31]), and pendulum tuned mass damper (see [32]).

Numerous researchers used various meta-heuristic algorithms to optimize the parameters of the TMD systems. Among all existing meta-heuristic algorithms, some lead to more suitable results for nonlinear optimization models (see [33], for example). Some researchers used the algorithm of the Genetic Algorithm (GA) and Charged System Search (CSS) to optimize the parameters of the TMD subjected to the critical earthquake [16–18,34–38]. Kamgar et al. [17] optimized the parameters of the TMD system using the Grey Wolf Optimization (GWO) method. The results of their research show that the maximum structural responses (i.e., roof displacement and stroke ratio) may be reduced by optimizing parameters of the TMD system and the dynamic behavior of the structure in terms of minimizing the input, and kinematic energies can be improved. Khatibinia et al. [19] investigated the optimum parameters of the TMD system intending to minimize the sum of the root-mean-square of drifts in the frequency domain under the critical earthquake for a 10-story shear building. Wong [39] evaluated the problem of the input energy dissipation for non-elastic structures equipped with the TMD. The author demonstrated the effectiveness of the TMD system in reducing the dynamic responses of the structures and concluded that the control of the structure using the TMD system could absorb more input energy, and subsequently, could damp the absorbed energy. This damping energy improves the dynamic performance of the structure. Nigdeli et al. [40] optimized the parameters of the TMD system based on the minimization of the acceleration transfer function. The results of their research indicate that the use of meta-heuristic algorithms to optimize the TMD parameters is more effective than classical methods. Kamgar et al. [16] calculated the optimum parameters of the TMD system, taking into account the soil-structure interaction effect adapting the whale optimization algorithm. They showed that the soil type and the objective function were very effective in the optimal parameters obtained for the TMD system.

An accurate estimation of the dynamic capacity of structures is one of the most critical challenges for researchers. The use of the Incremental Dynamic Analysis (IDA) is essential to evaluate accurately the dynamic performance of structures subjected to earthquake loads. For the first time, Bertero in [41] presented the IDA's time history analysis by scaling the earthquake records step by step and incrementally. Vamvatsikos and Cornell [42] introduced the IDA analysis as a method often used today. The most important advantage of the IDA analysis, in comparison with other methods, is the high ability of this type of analysis to show the actual attitude of the structure from the elastic state to the inelastic one. Additionally, this method can consider structural instability due to entering the structure from the elastic state into the inelastic one [43]. The optimization algorithm has also been utilized to design semi-rigid steel frames and reinforced concrete sections (see [44,45]).

In the present paper, a moment-resisting steel frame equipped with the TMD system is selected (see [39,46]). The nonlinear behavior of the joints is modeled using zero-length spring elements and rotational springs at the end of the beam-column elements [47]. Additionally, OpenSees software is used to simulate the dynamic structural behavior. Herein, the parameters of the TMD system are optimized to minimize the maximum drift ratio of structures subjected to earthquakes. Then, the effect of an optimal TMD on the development of the nonlinear seismic efficiency of steel moment-resisting frames is investigated. The main aim of the paper is to investigate a system that, in addition to lower cost, can improve the performance of the structure in comparison with the other vibration control systems.

The authors of previously published papers focused on the optimization of the parameters of the TMD system by calculating the optimal values of the TMD system numerically using the optimization algorithms [16–19,34–40]. The most important question is whether these optimal values can still control the dynamic responses of structures by changing the characteristics of the earthquake. Therefore, in this paper, the controlled structure is subjected to several incremental dynamics analyses using the optimal values for the TMD system to answer this question. The answer to the question that the controlled structure can withstand large earthquakes is vital in this regard. Additionally, the other question is whether the controlled structure can minimize the structural responses in other ranges of the peak ground acceleration (PGA) rather than the studied earthquakes? In fact, the structure is first designed to an earthquake with the PGA equal to 1 g (g is the acceleration of gravity) using a TMD system. Then the responses of the controlled structure are examined using the incremental dynamic analyses by changing the PGA from 0.1 g to 1 g. Additionally, in this paper, for the first time, the Life Safety constraint is introduced to keep the structure in a safe zone based on FEMA-356 [48].

2. Optimal Design of the TMD System

The paper is focused on an optimization method of the parameters of the TMD system (including damping, stiffness, and mass). The optimization is conducted according to a reduction in the maximum drift ratio of structures exposed to earthquakes. This criterion is upon the limitation proposed by FEMA-356 [48] for the maximum allowable drift ratio of the steel moment-resisting frame. Therefore, the optimal design of the TMD system for a steel moment-resisting frame can be formulated as:

$$
\begin{aligned}
&Find: && M_d,\ K_d,\ C_d \\
&Minimize: && \max\left(\frac{\max|drift_i|_{with\ TMD}}{\max|drift_i|_{without\ TMD}}\right)\times 100, \quad i=1,2,3,\ldots,n \\
&Subjected\ to: && M_d^{min} \le M_d \le M_d^{max} \\
& && K_d^{min} \le K_d \le K_d^{max} \\
& && C_d^{min} \le C_d \le C_d^{max} \\
& && \max(|u_d(t)-x_{roof}(t)|) \le 1000\ (mm) \\
& && \max|drift_i|_{with\ TMD} \le 0.025, \quad i=1,2,3,\ldots,n
\end{aligned}
\tag{1}
$$

where M_d, K_d and C_d indicate the mass, stiffness, and damping coefficient for the TMD system, respectively. M_d^{min}, M_d^{max}, K_d^{min}, K_d^{max}, C_d^{min} and C_d^{max} are the lower and upper bounds of the TMD mass, stiffness and damping constants, respectively. These lower and upper bounds have been selected based on the work [46].

3. Passive Control Systems

Currently, the control of the seismic response of the structures subjected to dynamic loads is a method that can help engineers to design structures. Among various methods of seismic control, the passive control method is one of the most popular, due to the lower cost of construction and maintenance. One of the passive control systems is the application of TMD. This system, consisting of a mass, damping, and linear spring, is typically installed on the roof of the structure. The system reduces the dynamic response of the structure by affecting its dominant mode. The mass of the TMD system moves with a different phase relative to the structure, and it improves the seismic structural response by the dissipation of energy [49–51]. The efficiency of the TMD system is highly dependent on its parameters. Therefore, the optimization of these parameters for the seismic control of tall structures against dynamic loads is one of the crucial issues.

A schematic view of the TMD system is exhibited in Figure 1. In this study, the Water Cycle Algorithm (WCA) has been used to reduce the relative displacement of the considered frame by the LS performance level presented in FEMA-356 [48] (i.e., the maximum allowable drift ratio for the steel moment-resisting frame is 2.5%). Therefore, the dynamic capacity and the performance of the controlled structure have been investigated using the IDA analysis.

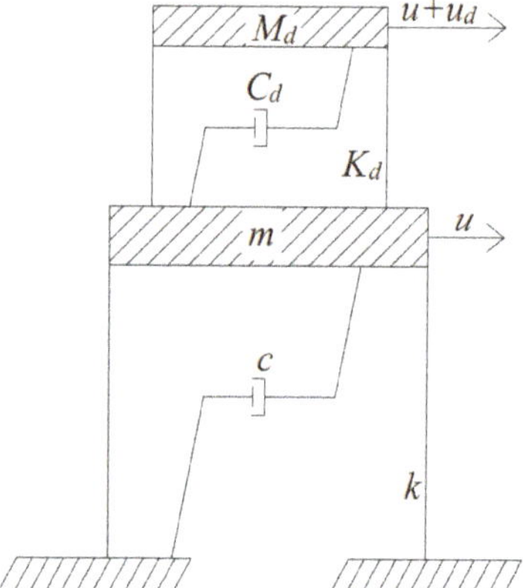

Figure 1. Schema of a single-degree-of-freedom system controlled by a TMD system.

4. Water Cycle Algorithm (WCA)

The optimization algorithms are generally based on natural phenomena, social events, and physical laws to solve different problems. The clever way used by these algorithms for computation is based on iterations to improve the performance of a system. Eskandar et al. [52] proposed WCA based on the rotational cycle of water in nature (i.e., the flow of streams toward the sea). The flow of water in the environment is like a tree or root of a tree. The small branches of this stream are small rivers that form the rivers by joining together. A sea is a place with the lowest elevation, and eventually, the rivers flow into it. The initial population is computed to formulate the equations of the water cycle algorithm, as in other population-based algorithms. Detailed information has been illustrated in [52]. The streams in the rivers arise from existing differences in the levels of two points, i.e., water flows from higher altitudes to lower altitudes. After rain comes down, the streams and rivers are formed and move to the lowest area, usually to the sea. The water cycle in nature consists of three processes: (1) the precipitation that creates the initial population, (2) the surface movement of the rivers and streams to the sea, and (3) the procedure of the evaporation and condensation.

Therefore, to formulate the first step, Equations (2) and (3) are used.

$$Raindrop = [x_1, x_2, x_3, \ldots, x_N] \tag{2}$$

$$Population\ of\ raindrops = \begin{bmatrix} Raindrops_1 \\ Raindrops_2 \\ Raindrops_3 \\ \vdots \\ Raindrops_{N_{pop}} \end{bmatrix} = \tag{3}$$

$$\begin{bmatrix} x_1^1 & x_2^1 & x_3^1 & \cdots & x_{N_{var}}^1 \\ x_1^2 & x_2^2 & x_3^2 & \cdots & x_{N_{var}}^2 \\ \vdots & \vdots & \vdots & \vdots & \vdots \\ x_1^{N_{pop}} & x_2^{N_{pop}} & x_3^{N_{pop}} & \cdots & x_{N_{var}}^{N_{pop}} \end{bmatrix}$$

where the values of the decision variables $(x_1, x_2, x_3, \ldots, x_N)$ can be expressed by several floating points of the problem. The cost of *Raindrop* is also obtained by Equation (4) as follows:

$$C_i = Cost_i = f\left(x_1^i, x_2^i, \ldots, x_{N_{var}}^i\right) i = 1, 2, 3, \ldots, N_{pop} \tag{4}$$

in which N_{pop} and N_{var} represent the stream population as the initial population and some design variables, respectively. $Cost_i$ shows the cost estimation function of the variables, see for example [53]. Firstly, the parameter N_{pop} is generated, and a number is selected for the N_{sr} parameter as the best value (minimum values) for the rivers and sea in the first step. The number of rivers and the sea is calculated as a variable N_{sr} as follows:

$$N_{sr} = Number\ of\ Rivers + \underbrace{1}_{Sea} \tag{5}$$

$$N_{Raindrop} = N_{pop} - N_{sr} \tag{6}$$

The parameter $N_{Raindrop}$ is the rest of the population that make up possible routes to the sea or the rivers. Equation (7) is used depending on the intensity of the flow to determine and assign raindrops in rivers and sea:

$$NS_n = Round\left\{ \left| \frac{Cost_n}{\sum_{i=1}^{N_{sr}} Cost_i} \right| \times N_{Raindrops} \right\}, n = 1, 2, \ldots, N_{SR} \tag{7}$$

in which NS_n is a number of streams that flow into some specific rivers or the sea. Besides, the new positions of the streams and rivers are also expressed as follows:

$$\vec{X}_{Stream}^{t+1} = \vec{X}_{Stream}^{t} + rand \times C \times \left(\vec{X}_{River}^{t} - \vec{X}_{Stream}^{t} \right) \tag{8}$$

$$\vec{X}_{River}^{t+1} = \vec{X}_{River}^{t} + rand \times C \times \left(\vec{X}_{Sea}^{t} - \vec{X}_{River}^{t} \right) \tag{9}$$

in which *rand* is a uniformly distributed random number from the interval $\langle 0, 1 \rangle$. Additionally, the parameter C has a number between 1 and 2, and it is usually close to 2. If the evaporation conditions are taken into account, it can help the algorithm to prevent premature convergence:

$$If \|\vec{X}_{Sea}(t) - \vec{X}_{River}^{i}(t)\| < d_{max}(t) or rand \in \langle 0, 1 \rangle, \tag{10}$$
$$n = 1, 2, 3, \ldots, N_{sr} - 1$$

where d_{max} is a small number (close to zero). It indicates that the river is connected to the sea when the river is away from the sea less than the value of d_{max}. The parameter d_{max} can control the optimal solution, and its value is updated as follows:

$$d_{max}(t + 1) = d_{max}(t) - \frac{d_{max}(t)}{MaxIteration} \tag{11}$$

Additionally, after considering the evaporation conditions for the algorithm, the new streams are randomly generated in the search space as follows:

$$\vec{X}_{Stream}^{New} = LB + rand \cdot (UB - LB) \tag{12}$$

in which the parameters UB, LB are the minimum and maximum boundary conditions, respectively. Then, the river flowing to the sea is selected from the best new raindrops, and the other remaining new raindrops create streams that can flow to the river or sea. Equation (13) is utilized for the streams that have flowed to the sea to check the computational performance and convergence rate of the optimization problem:

$$\vec{X}_{Stream}^{New} = X_{sea} + \sqrt{\mu} \times randn(1, N_{var}) \tag{13}$$

where μ is a coefficient, and the value of 0.1 has been proposed for it in [52].

5. Incremental Dynamic Analysis (IDA)

IDA is an accurate analysis method that is capable of estimating the seismic behavior of structures subjected to different earthquakes. One of the principles of the functional design of structures is the use of the nonlinear dynamic capacity of load-bearing members in the design of structures.

The stiffness values of the structural elements decrease when a severe dynamic load is applied to the structure. As a result, the stiffness matrix of the structure will decrease when the plastic joints are formed in the members. Finally, it results in the redistribution of the forces among the members of the structure. Finally, the structure experiences more deformation because of yielding in some elements, and it leads to more significant energy absorption and damping.

In IDA, a structure is subjected to a mapping acceleration. The mapping acceleration is selected from an earthquake. Then, the earthquake is scaled so that its maximum acceleration is equal to 1 g. Hence, the structure is subjected to the scaled earthquake record. Next, the maximum acceleration is added to the value of 0.2 g, and the structure is re-analyzed. This process continues until the maximum acceleration of the earthquake is equal to 1 g. Finally, the IDA curve is plotted for the frame and earthquake. The IDA shows the nonlinear dynamic response of the structure against the intensity of the seismic excitation.

6. Modeling and Verification

A six-story steel frame is selected in this study (see Figure 2). The frame has already been studied by several researchers (see [39,46]). A finite element software named OpenSees [47] is used to model the frame. Additionally, the modified Ibarra–Krawinkler (IMK) deterioration model (see [54–56]) is utilized to simulate the nonlinear behavior of the plastic joints. According to this constitutive model, the zero-length elements are considered as concentrated plasticity at the beam-column connections. The IMK model takes into account the cyclic response of the springs, which shows the nonlinear behavior of the frame. Due to the presence of the gravitational loads, as well as the TMD mass, the P-delta effect must be taken into account to consider the enhanced structural responses. For this purpose, a virtual column with truss elements attached to the base of the structure is analyzed. Additionally, all columns are considered to be fixed at the base of the structure. Since the rotational springs and frame elements are connected in series, the stiffness of the rotational elements must be modified in such a way that the stiffness value of elements and the actual stiffness of the frame are the same. For this purpose, the stiffness of the rotational springs is considered to be $n = 10$ times higher than the rotational stiffness of the elastic elements. The stiffness and also the moment of inertia of the elastic element are multiplied into the $(n + 1)/n$. Finally, to match the nonlinear behavior of the elements and the actual behavior of the frame, the strain coefficient of the plastic joint should be modified according to Equation (14)—see [54–57].

$$\alpha_{s,spring} = \alpha_{s,mem} / (1 + n \cdot (1 - \alpha_{s,mem})) \tag{14}$$

where $\alpha_{s,mem}$ shows the real strain coefficient of the frame, and $\alpha_{s,spring}$ is the strain hardening coefficient of the rotational spring. All beam elements are subjected to the uniformly distributed load equal to 21.89 kN/m. Additionally, the following relations (Equations (15)–(22)) are used to determine the parameters required in the definition of the behavioral curve of the rotational springs [54–57]:

$$\theta_p = 0.318\left(\frac{h}{t_w}\right)^{-0.55}\left(\frac{b_f}{2t_w}\right)^{-0.345}\left(\frac{L_b}{r_y}\right)^{-0.023}\left(\frac{L}{d}\right)^{0.09}\left(\frac{c^1_{unit}\cdot d}{533}\right)^{-0.33}\left(\frac{c^2_{unit}\cdot F_y}{355}\right)^{-0.13}$$
$$d \geq 21in \tag{15}$$

$$\theta_{pc} = 5.63\left(\frac{h}{t_w}\right)^{-0.565}\left(\frac{b_f}{2t_w}\right)^{-0.8}\left(\frac{c^1_{unit}\cdot d}{533}\right)^{-0.28}\left(\frac{c^2_{unit}\cdot F_y}{355}\right)^{-0.43}$$
$$d \geq 21in \tag{16}$$

$$\Lambda = 536\left(\frac{h}{t_w}\right)^{-1.26}\left(\frac{b_f}{2t_w}\right)^{-0.525}\left(\frac{L_b}{r_y}\right)^{-0.13}\left(\frac{c^2_{unit}\cdot F_y}{355}\right)^{-0.291} \qquad (17)$$
$$d \geq 21in$$

$$\theta_{pc} = 7.5\left(\frac{h}{t_w}\right)^{-0.61}\left(\frac{b_f}{2t_w}\right)^{-0.71}\left(\frac{L_b}{r_y}\right)^{-0.11}\left(\frac{c^1_{unit}\cdot d}{533}\right)^{-0.161}\left(\frac{c^2_{unit}\cdot F_y}{355}\right)^{-0.32} \qquad (18)$$
$$d < 21in$$

$$\theta_p = 0.0865\left(\frac{h}{t_w}\right)^{-0.365}\left(\frac{b_f}{2t_w}\right)^{-0.14}\left(\frac{L}{d}\right)^{0.34}\left(\frac{c^1_{unit}\cdot d}{533}\right)^{-0.721}\left(\frac{c^2_{unit}\cdot F_y}{355}\right)^{-0.23} \qquad (19)$$
$$d < 21in$$

$$\Lambda = 495\left(\frac{h}{t_w}\right)^{-1.34}\left(\frac{b_f}{2t_w}\right)^{-0.595}\left(\frac{c^2_{unit}\cdot F_y}{355}\right)^{-0.36} \qquad (20)$$
$$d < 21in$$

$$M_y = 1.17\cdot Z\cdot F_y \qquad (21)$$

$$\theta_y = \left(1.17\cdot Z\cdot F_y/6EI\right)/L \qquad (22)$$

where h is the web depth; b_f is the width flange of a beam; L_b is the distance from the column face to the nearest lateral brace; d is the beam depth. Additionally, r_y shows the radius of gyration about the y-axis of the beam, and t_w is the web thickness of the beam. The parameter Z is the plastic section modulus, and F_y shows the expected yield strength. Moreover, the parameters c^1_{unit} and c^2_{unit} are the two coefficients for unit conversion, while E depicts Young's modulus, and I shows the moment of inertia. The parameters θ_p, θ_{pc}, θ_y, exhibit pre-capping plastic rotation in monotonic loading, post-capping plastic rotation, and yield rotation, respectively. The parameter L is the beam shear span (distance from plastic hinge location to the point of inflection); Λ shows the capacity of the reference cumulative rotation, and M_y is the effective yield moment [56].

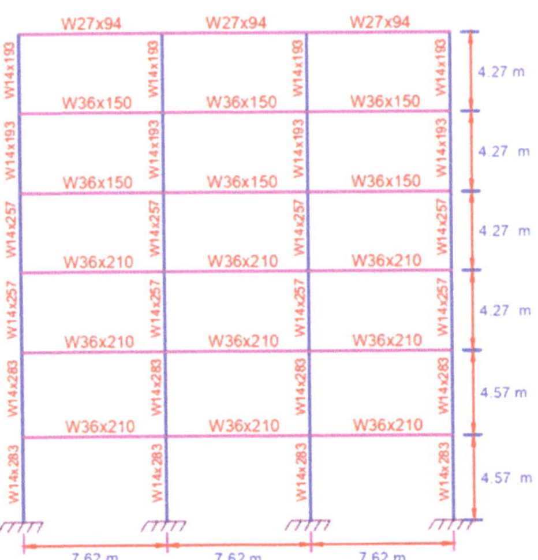

Figure 2. Six-story steel moment-resisting frame.

The mass of all stories and also the values of the damping ratio for all modes are assumed to be equal to m = 300,000 kg, and 3%, respectively. The Young's modulus is 200 MPa. The maximum allowable stroke for TMD is considered to be equal to 1000 mm. The values for the first six natural frequencies of the six-story steel frame are calculated numerically and compared with the results presented by Wong in [39] and Bilondi et al. in [46] to validate the model (see Table 1). It can be seen from Table 1 that the differences are negligible. It confirms the accuracy of the modeling approach.

Table 1. The benchmark Special Moment Resisting Frame's (SMRF's) natural frequencies of vibration.

Mode Number	Natural Periods of Vibration (rad/s)		
	Wong [39]	Bilondi et al. [46]	Present Study
1	5.15	5.07	5.07
2	14.28	13.96	14.22
3	25.13	25.13	25.13
4	34.91	34.91	34.91
5	44.88	44.88	44.88
6	57.12	57.12	57.12

Additionally, the time history of cumulative hysteresis energy, E_h, (see Equation (23)) for the structure with (W) and without (W/O) TMD are compared with [46] (see Figure 3) subjected to the Northridge earthquake (see Figure 4). The values of mass, stiffness, and damping ratio of the TMD system are considered to be equal to 180,000 N·s²/m), 5,264,000 N/m, and 0.05 based on [46], respectively. Figure 3 also shows the verification of the result obtained by this study and described in [46]:

$$E_h = \sum_{k=1}^{n_e} E_{h,k} \tag{23}$$

where $E_{h,k}$, k, and n_e present the summation of work done by internal forces (e.g., moment, axial, shear), an integer counter, and the number of elements, respectively.

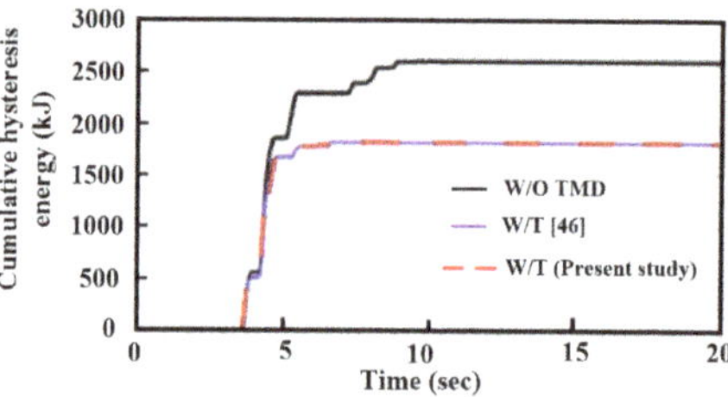

Figure 3. The cumulative hysteresis energy of the controlled and uncontrolled six-story steel frame building.

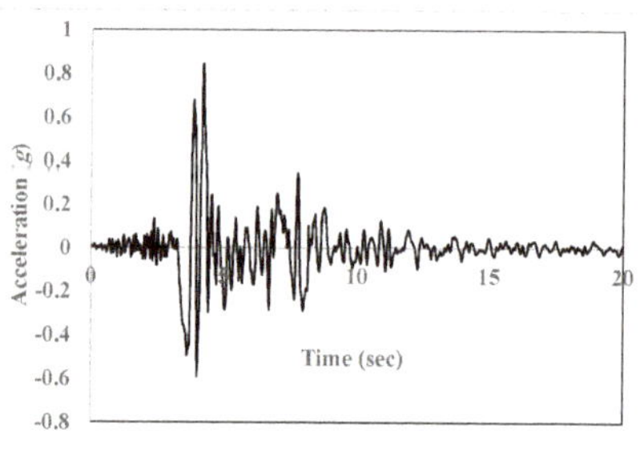

Figure 4. Acceleration time history of the Northridge 1994 earthquake.

In this paper, to optimize and analyze the structure dynamically, several real earthquakes and one artificial earthquake are selected (see Table 2). Figure 5 shows the acceleration time history of the artificial earthquake. The artificial earthquake is calculated using the Gaussian White Noise process and based on the Kanai–Tajimi filter and power spectral density function (PSDF) [58–60]:

$$S_{Kanai-Tajimi}(\omega) = S_0\left[\frac{\omega_g^4+(2\times\omega_g\times\xi_g\times\omega)^2}{(\omega^2-\omega_g^2)^2+(2\times\omega_g\times\xi_g\times\omega)^2}\right]$$
$$S_0 = \frac{0.03\times\xi_g}{\pi\times\omega_g\times(4\times\xi_g^2+1)}$$

(24)

where S_0, ω_g, and ξ_g are the intensity of the PSDF, frequency, and damping of the soil, respectively. In this paper, ω_g and ξ_g are considered to be equal to 25.13 rad/s and 0.8 rad/s, respectively, based on [61], which indicates that the structure has been located on the stiff soil. The strong ground motion of the artificial earthquake is big enough, and it can be used in the dynamic analysis of the structures based on Uniform Building Code 97 [62].

It should be noted that Chandler's classification has been considered in the selection of the earthquake to cover all existing categories for the earthquakes (see [8,63]).

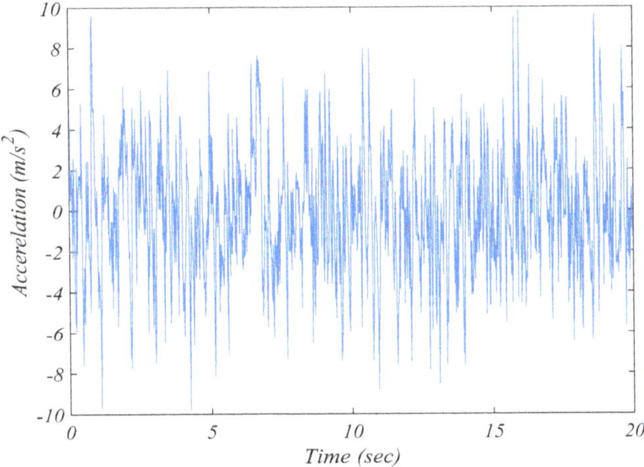

Figure 5. Acceleration time history of an artificial earthquake.

Table 2. Characteristics of the selected scaled earthquakes.

Abbreviation	AE	NE3	NE2	NE1	FE3	FE2	FE1
Earthquake	An artificial earthquake	Kocaeli Turkey	Gazli USSR	Chi-Chi Taiwan	Superstition Hills-02	San Fernando	Duzce Turkey
Station	—	Arcelik	Karakyr	CHY101	El Centro Imp. Co. Cent	LA—Hollywood Stor FF	Bolu
PGA (g)	1	1	1	1	1	1	1
PGA/PGV (g·s/m)	0.206	0.601	1.277	0.630	0.743	1.035	1.32
Strong ground motion duration (s)	18	13.265	6.956	28.55	27.99	13.15	8.55
Predominant Period (s)	0.2	0.28	0.14	1.08	0.22	0.24	0.32
Total time duration (s)	19.98	29.995	13.0878	88.995	59.99	79.44	55.89
Arias Intensity (m/s)	33.311	15.370	7.599	12.291	8.646	13.396	6.806

PGA—Peak Ground Acceleration; PGV—Peak Ground Velocity.

7. Results and Discussion

7.1. Optimization of the Parameters of the TMD System

In this paper, a six-story steel frame equipped with the TMD system is studied. One of the most critical issues with the TMD systems is to determine their optimal parameters. Therefore, different methods are used to optimize these parameters. Among them, the meta-heuristic algorithms are commonly used due to the current uncertainty of the mathematical problem as well as different scenarios used for the objective function. In general, all meta-heuristic algorithms start from a local search and eventually reach the desired values. In this paper, a WCA (see Section 4) meta-heuristic algorithm has been selected because of its ability to solve constraint problems. Therefore, the optimum parameters for the TMD system are calculated by WCA to minimize the relative displacement of the stories adapted to the LS performance level presented by FEMA-356 [48]. The optimum parameters of the TMD system exposed to the near- and far-field earthquakes, as well as to the artificial earthquake, have been presented in Table 3.

In fact, the water cycle algorithm has been used to optimize the parameters of the TMD system subjected to the far-, near-field, and artificial earthquakes. Equations (2)–(13) are utilized to perform an optimization problem using Matlab software. Additionally, Equation (1) is used to make an optimization problem with an upper and lower boundary for the optimization variables. Table 3 shows the optimum parameters of the TMD system subjected to different studied earthquakes.

Table 3. Optimum parameters achieved for TMD subjected to the drift ratio constraint.

Earthquake	K_{tmd} (N/mm)	C_{tmd} (N·s/mm)	M_{tmd} (N·s^2/mm)
FE1	404.94	770.18	180
FE2	543.76	304.28	180
FE3	490.1	942.89	180
NE1	421.67	1753.1	180
NE2	315.87	972.87	141
NE3	405.02	500.0	180
AE	405.47	935.66	180

7.2. The Structural Seismic Performance

The results for the structure controlled with the TMD system are shown in Table 4. As it is observed from the table, the reduction in the base shear is less than 3.49% under different earthquakes, even though the base shear has increased slightly during the earthquakes NE1 and AE. Hence, due to the insignificant reduction in the base shear under different earthquakes, it might be concluded that the optimization of the TMD parameters focused on reducing the drift will not have a significant effect on the base shear of structures. Table 4 shows that the average acceleration of the structure equipped with the TMD system has increased under the FE2 and AE earthquakes by about 5% but, for the rest of earthquakes, the parameters of TMD optimized based on drift can reduce the acceleration of the structure. Similar results have been obtained for the average displacement of the structure controlled with the TMD system. In fact, for the FE2, NE3 earthquakes, the average displacement of the controlled structure has increased while this parameter has decreased for other earthquakes. It should be noted that the negative and positive signs in Table 4 show the decrease and increase in the average reduction, respectively. In other words, the positive value shows that the parameter is reduced, and the negative value shows that it is increased.

The maximum drift ratio of all stories and the roof displacement time history of the structure are depicted in Figures 6 and 7 with and without the TMD system subjected to the near- and far-field earthquakes. It is clear from the table that the maximum drift ratio of the stories has been reduced for all earthquakes. It is also observed that the roof displacement has increased for some earthquakes (e.g., FE2, and NE3). The controlled structure has been vibrated about a new plastic axis subjected to these earthquakes, and the non-elastic deformation of the structure has increased. This indicates that the drift-optimized TMD system is unable to improve the seismic performance of the structure for some earthquakes.

Table 4. The average reduction in the structural responses.

Earthquake	Average Reduction (%)			
	Displacement	**Drift**	**Acceleration**	**Base Shear**
FE1	5.78	5.02	8.10	2.61
FE2	−20.56	25.08	−5.25	0.81
FE3	11.14	6.53	10.06	2.75
NE1	14.81	10.96	3.15	−2.49
NE2	12.26	4.18	9.03	0.65
NE3	−31.34	42.92	19.69	3.49
AE	31.55	19.42	−5.23	−1.12

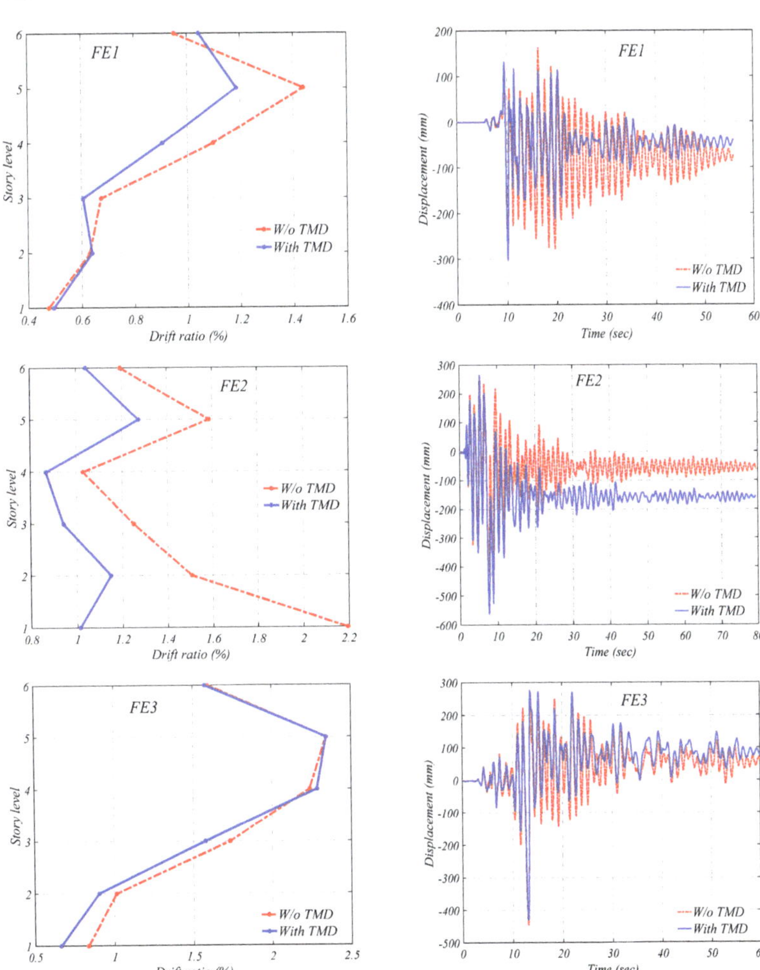

Figure 6. Drift ratios of the stories and the time history of the displacement of the roof in the controlled and uncontrolled structure subjected to the far-field earthquakes.

7.3. The Results of the Incremental Dynamic Analysis

In this section, the parameters of the TMD system subjected to the different earthquakes are optimized based on minimizing the relative displacement criterion presented by FEMA-356 [48]. Next, the IDA analysis is used to investigate the dynamic performance of the six-story steel frame equipped with the TMD system. The IDA analysis is calculated for the maximum base shear, maximum acceleration, maximum displacement, and maximum drift ratio of the desired structure, and its curves for near- and far-field earthquakes are plotted in Figures 8–11.

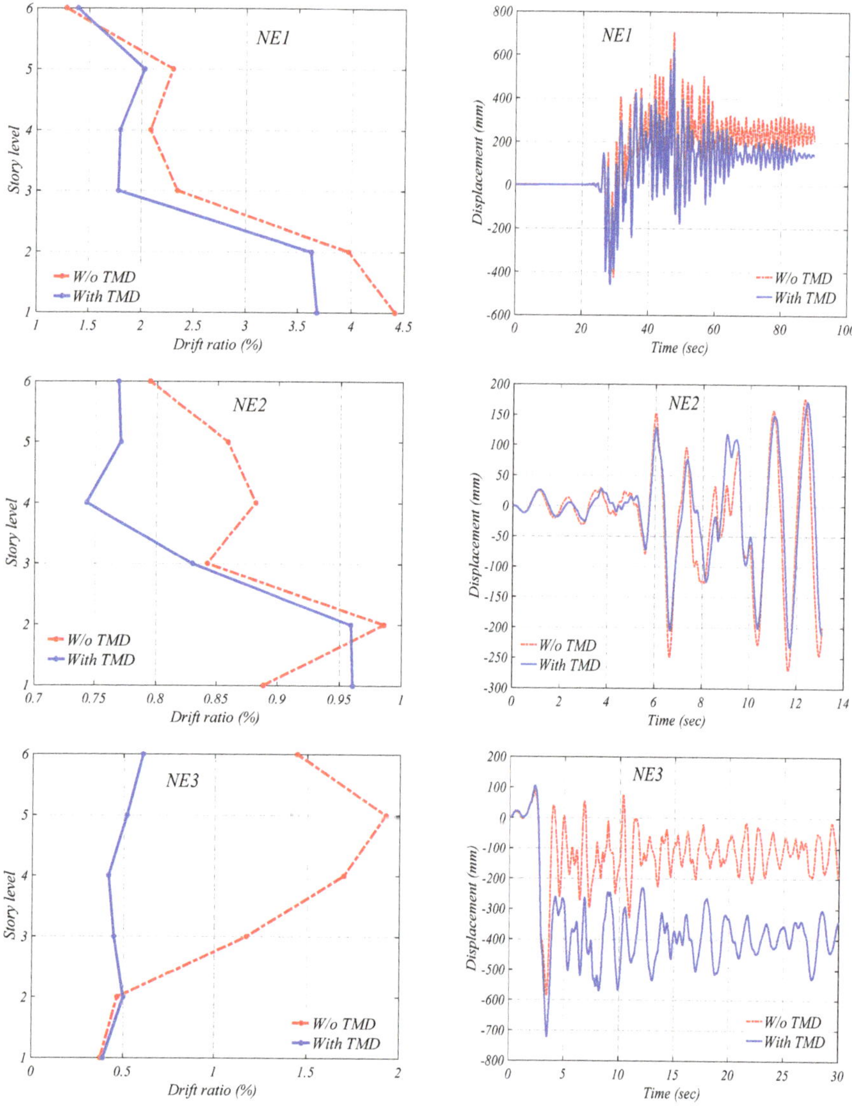

Figure 7. Drift ratios of the stories and the time history of the displacement of the roof in the controlled and uncontrolled structure subjected to the near-field earthquakes.

7.3.1. The Results of IDA for the Drift Ratio

The drift ratio of the structures with and without the TMD system is shown in Figure 8. It can be concluded that the maximum drift ratio of all stories has been decreased for the structure controlled with the TMD system subjected to the FE1, NE1, and NE2 earthquakes. The drift ratio has increased for the FE2 earthquake with PGA equal to 0.9 g and 1 g. Additionally, the maximum drift ratio of the structure has increased under the FE3 earthquake in the range of $0.6g \leq PGA \leq 0.9g$. A disturbance has taken place for the NE3 earthquake in the range of $0.7g \leq PGA \leq 0.9g$. Therefore, it might be concluded that the TMD parameters should be evaluated for PGA values larger than 0.5 g in controlling the drift ratio.

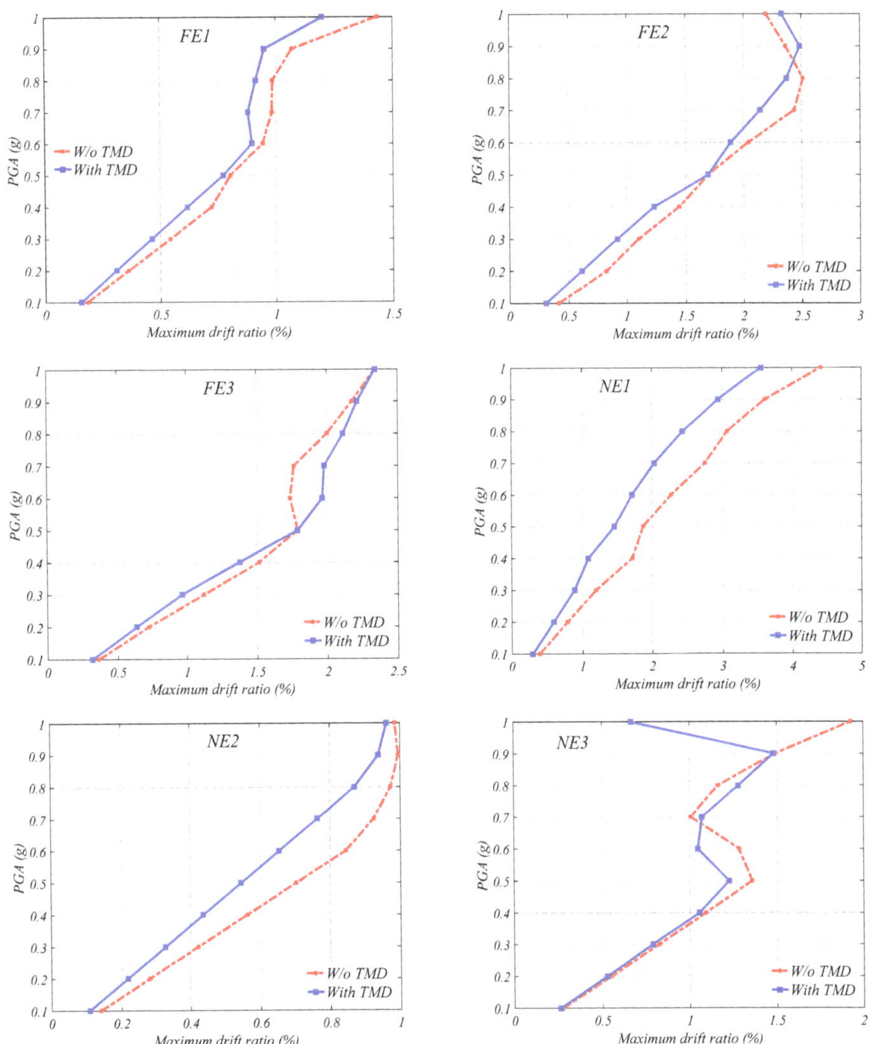

Figure 8. Maximum drift ratio of the frame subjected to different far- and near-field earthquakes.

7.3.2. The Results of IDA for the Maximum Displacement

The maximum displacements of the structures subjected to different earthquakes are presented in Figure 9. This figure shows that the maximum displacement of the structure has increased subjected to the FE3 earthquake in the range of 0.6g ≤ PGA ≤ 0.9g. Therefore, it could be concluded that the optimized TMD based on the reduction of the drift ratio can reduce the maximum displacement of the structure. Additionally, it has to be noticed that the TMD parameters should be evaluated for PGA values larger than 0.5 g in controlling the maximum displacement.

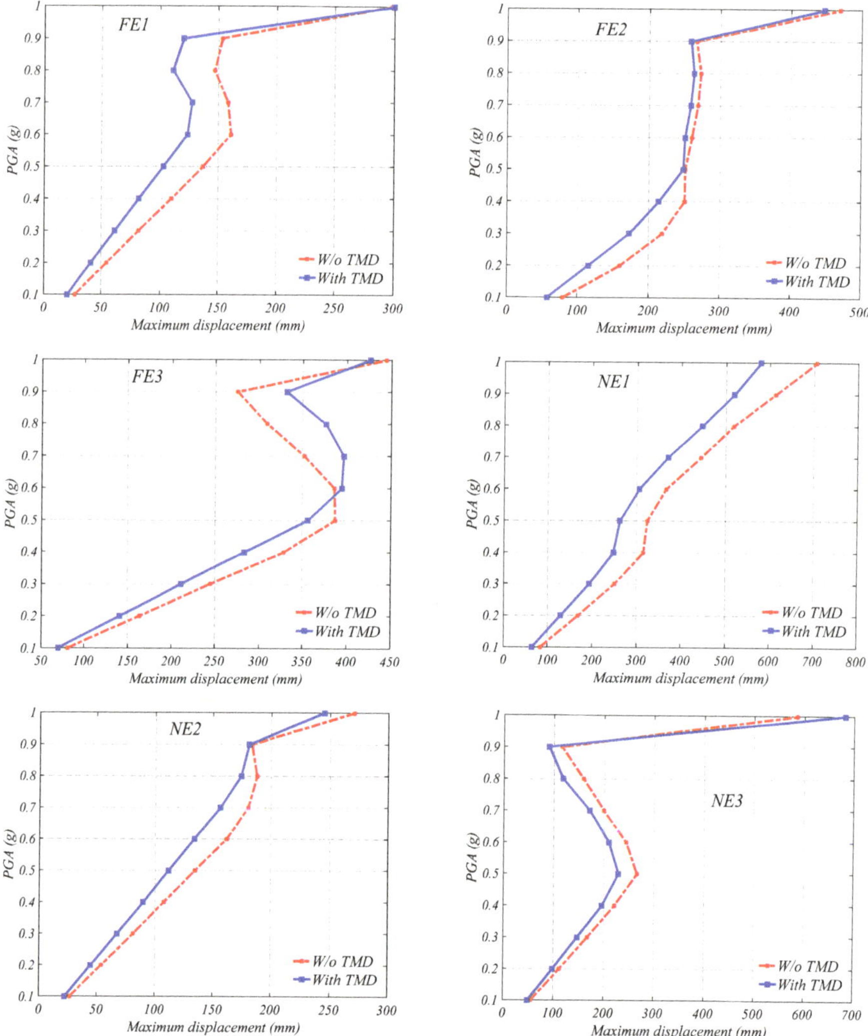

Figure 9. The maximum displacement of the frame subjected to the different far- and near-field earthquakes.

7.3.3. The Results of IDA for the Maximum Base Shear

Figure 10 shows the performance of the controlled and uncontrolled structures based on the base shear. As it can be seen from the figure, for all the selected earthquakes and PGA values equal or less than 0.5 g, the maximum base shear of the controlled structure has decreased. Additionally, the base shear has not changed for PGA larger than 0.5 g. Therefore, it could be concluded that the optimized TMD system based on minimizing the drift ratio does not have any particular effect on the base shear for PGA values larger than 0.5 g.

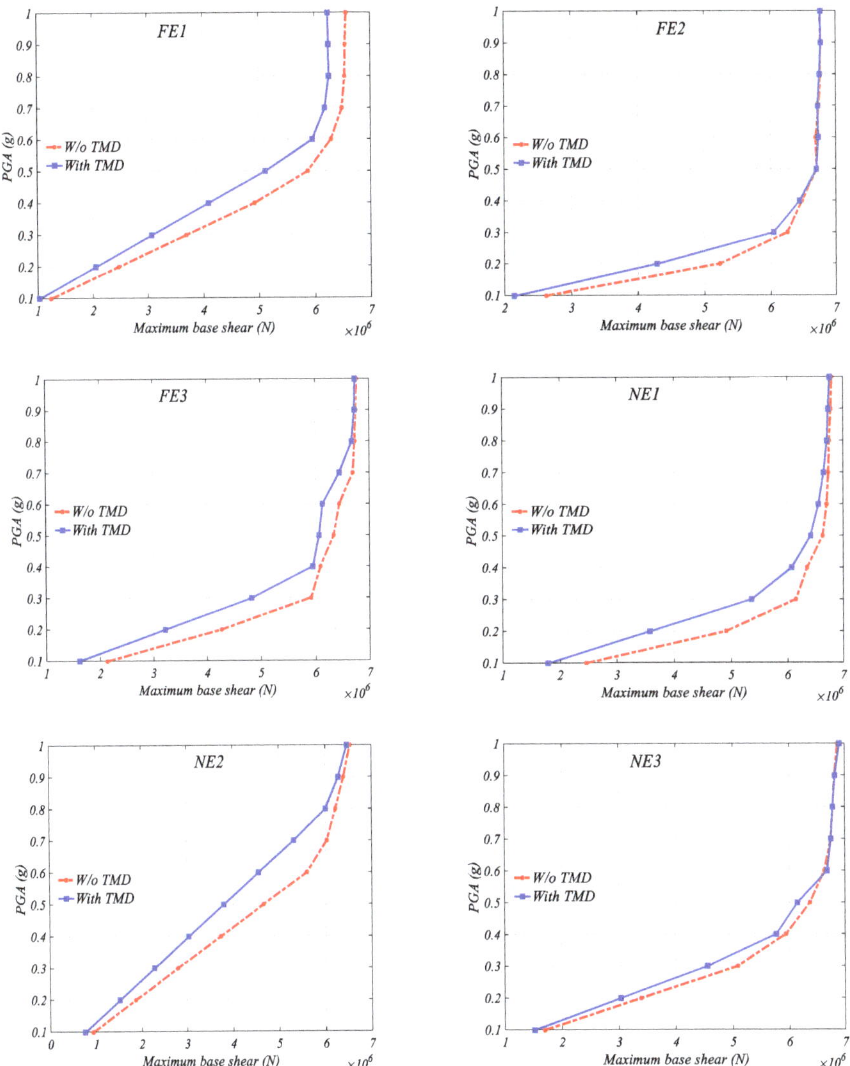

Figure 10. Maximum base shear of the frame subjected to different far- and near-field earthquakes.

7.3.4. The Results of IDA for the Maximum Acceleration of the Structure

A comparison between the results for the maximum acceleration is shown in Figure 11. It can be concluded that the performance of the structure equipped with the optimized TMD system has been improved in minimizing the maximum acceleration of the structure subjected to the FE1, NE1, and NE2 earthquakes. The maximum acceleration of the structure has increased in the range of $0.8g \leq PGA \leq 1.0g$ under the FE2 earthquake. Besides, an increase in the acceleration response has occurred for the FE3 earthquake when PGA is smaller than 0.5 g. Therefore, the performance of the controlled structure subjected to the NE3 earthquake is different in comparison to the structure subjected to other earthquakes. An increase in the maximum acceleration can be seen for the PGA equal to 0.9 g, while a sharp decrease has occurred for the PGA equal to 1.0 g. Therefore, it could be concluded again that the TMD parameters should be evaluated for the PGA values larger than 0.5 g in controlling the maximum acceleration.

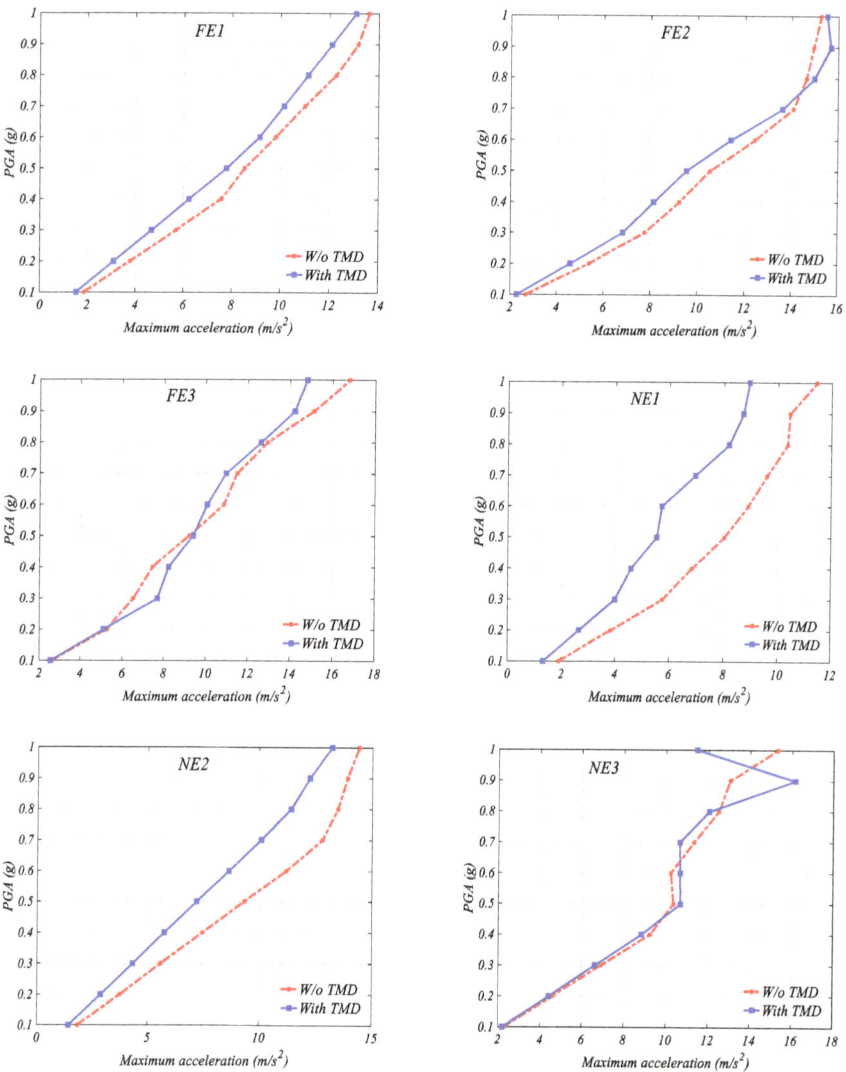

Figure 11. Maximum acceleration of the frame subjected to the different far- and near-field earthquakes.

7.4. The Results for the Artificial Earthquake

7.4.1. Seismic Performance of the Controlled and Uncontrolled Structures Exposed to the Artificial Earthquake

Figure 12 presents the responses of the controlled and uncontrolled structures (i.e., drift ratio, maximum displacement, maximum base shear, and the maximum acceleration) exposed to the artificial earthquake. The figure indicates that the TMD system optimized based on the drift ratio can reduce all structural responses, except for the maximum acceleration of the structure. The optimized TMD system has also controlled the maximum acceleration of the fifth and sixth stories, but it has increased this value for the lower stories.

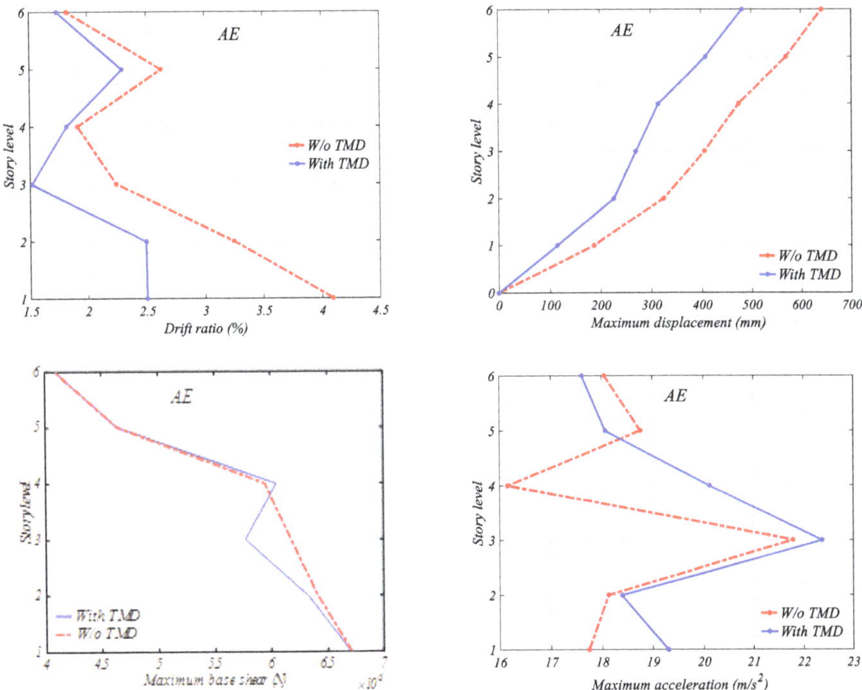

Figure 12. Maximum drift ratio, displacement, base shear, and acceleration of the frame subjected to the artificial earthquake.

7.4.2. IDA for the Dynamic Performance of the Controlled and Uncontrolled Structures Exposed to the Artificial Earthquake

The results of IDA for the structure subjected to artificial earthquakes are shown in Figure 13. The performance of the controlled structure shows an increase in the maximum acceleration values in the range of 0.7g ≤ PGA ≤ 0.8g. The maximum acceleration of the structure has decreased for other PGA values. Moreover, the value of base shear has decreased for PGA smaller than 0.5 g and remained constant for PGA larger than 0.5 g. Additionally, the results show that the performance of the structure has improved for the maximum displacement and drift ratio.

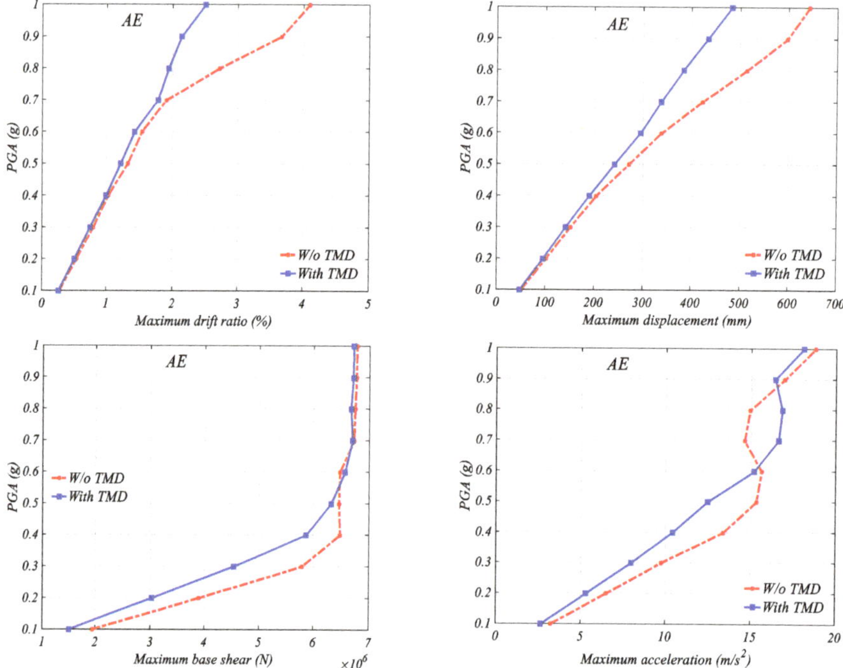

Figure 13. Results of Incremental Dynamic Analysis (IDA) analysis for the frame subjected to the artificial earthquake.

7.4.3. Investigation of the Performance Level of the Structure Subjected to the Artificial Earthquake

The performance levels of the structure subjected to the artificial earthquake, based on the FEMA-356 [48], are presented in Figure 14. Based on FEMA-356, the maximum drift ratios of the controlled and uncontrolled structures have been categorized in the figure as:

1. Immediate Occupancy ≤ 0.7%;
2. 0.7% ≤ Life Safety ≤ 2.5%;
3. 2.5% ≤ Collapse Prevention ≤ 5%.

As it can be observed from Figure 14, the six-story steel moment-resisting frame without TMD system has entered into the collapse prevention level for PGA larger than 0.8 g, but the performance level of the structure controlled with the TMD system has remained in the life safety range. Therefore, results indicate that the optimization of the TMD parameters to minimize the maximum drift ratio smaller than 2.5% substantially improve the seismic performance of the structure subjected to the artificial earthquake. It leads to a delay in the fracture of the bending connections (see FEMA-356 [48]), and it can help the beams and columns to sustain distortion.

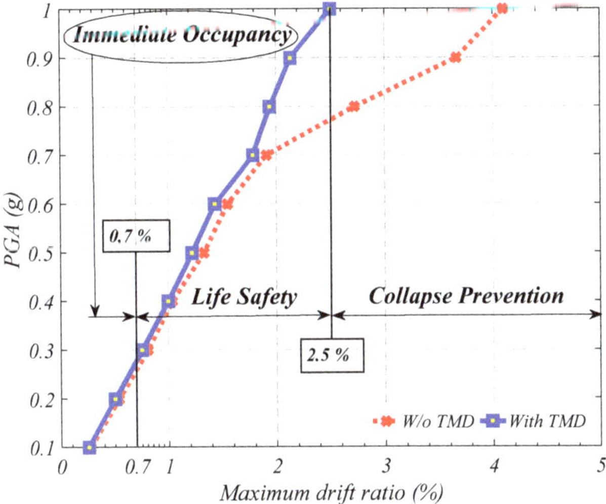

Figure 14. The performance level of the frame subjected to the artificial earthquake.

Besides, to evaluate the capability of the method presented in this paper, the six-story steel moment-resisting frame structure is considered. For this purpose, two other objective functions (see Equation (25)) are considered, and the optimal values of the TMD system are computed under the artificial earthquake. Table 5 shows the optimal values of the TMD system for different objective functions. It should be noted that the constraints and upper and lower bounds are considered as in Equation (1) and Ref. [46].

$$Find: \quad M_d, \; K_d, \; C_d$$

$$Minimize: \quad \max\left(\frac{\max|drift_i|_{with\ TMD}}{\max|drift_i|_{without\ TMD}}\right) \times 100, \quad i = 1, 2, 3, \ldots, n \quad (O.F\ 1)$$

$$Minimize: \quad \max\left(\frac{\max|Acc_{roof}|_{with\ TMD}}{\max|Acc_{roof}|_{without\ TMD}}\right) \times 100, \quad i = 1, 2, 3, \ldots, n \quad (O.F\ 2) \qquad (25)$$

$$Minimize: \quad \max\left(\frac{\max|Disp_{roof}|_{with\ TMD}}{\max|Disp_{roof}|_{without\ TMD}}\right) \times 100, \quad i = 1, 2, 3, \ldots, n \quad (O.F\ 3)$$

where O.F., Acc_{roof}, and $Disp_{roof}$ show the objective function, roof acceleration, and displacement, respectively. The last-second objective functions are usually used by researchers to optimize the parameters of vibration control systems [18,34,37,40]. Additionally, the computational workload, in the form of required running time, for each objective function is presented in Table 5. The calculations have been run on a computer with a 64-bit operating system and CPU Intel(R) Core(TM) i3-4170 CPU 3.70 GHz with 4 GB RAM.

Table 5. Optimum parameters and required running time achieved for the TMD subjected the artificial earthquake for different objective functions.

O.F	K_{tmd} (N/mm)	C_{tmd} (N·s/mm)	M_{tmd} (N·s²/mm)	Required Running Time (s)
O.F 1	405.47	935.66	180	997.854
O.F. 2	211.021	1130.6	122.87	865.528
O.F. 3	250.025	749.317	163.156	1114.548

It can be seen from Table 5 that the required running time for O.F. 2 is shorter than for two other objective functions. Moreover, the computational workload of the proposed objective function (O.F 1) is between those of the two traditional objective functions (O.F. 2 and O.F. 3).

Table 6 presents the maximum responses of the controlled and uncontrolled structures subjected to the artificial earthquake for different objective functions. The maximum roof displacement, roof acceleration, and drift ratio for the uncontrolled structures are 0.642 m, 18.06 m/s^2, and 4.09%, respectively. Table 6 shows that O.F 1 has the best performance between all studied objective functions in reducing the maximum responses of the structure and therefore improving the seismic behavior of the structures.

Table 6. Maximum responses of the controlled structure subjected to the artificial earthquake for different objective functions.

Objective Function	Max. Roof Disp. (m)	Max. Roof Acc. (m/s^2)	Max. Drift Ratio (%)
O.F 1	0.504	17.74	2.45
O.F. 2	0.509	17.52	2.82
O.F. 3	0.515	17.75	2.61

Additionally, to evaluate the capability of the method presented in this paper, the 10-story steel moment-resisting frame structure is considered here that has been studied by Wong and Johnson [64]. The mass of all stories and the damping ratio for all 10 modes are assumed to be equal to $m = 218,900$ kg, and 3%, respectively [64]. The cross-sections and lengths for the beams and columns of the structure are shown in Figure 15. The material has the yield stress equal to 248.2 MPa. Additionally, a gravity uniformly distributed load, equal to 21.89 kN/m, is applied to all beams. The modulus of elasticity is considered to be equal to 200 GPa. All beam-to-column connections are considered to be rigid [64]. The first natural frequency of the 10-story steel frame is 4.19 rad/s, which is equal to the results presented by Wong and Johnson in [64]. It can be seen that the differences are negligible, therefore, it confirms the accuracy of the modeling approach. Then, the optimum parameters of the TMD system are computed using WCA for the first objective function (O.F 1) subjected to the artificial earthquake, a far- and near-field earthquake. The optimal values of the TMD system are shown in Table 7.

Table 7. Optimum parameters achieved for the TMD system subjected the far-, near- and artificial earthquakes.

Earthquake	K_{tmd} (N/mm)	C_{tmd} (N·s/mm)	M_{tmd} (N·s^2/mm)
FFE3	263.34	130.87	91.08
NE2	170.78	89.06	59.28
AE	84.21	2000	131.59

Finally, for the controlled structure using the optimum parameters presented in Table 7, the IDA analysis is run. Figures 16–18 show the IDA curves for the responses of the controlled structure in comparison with the uncontrolled one.

Figure 16 shows that the performance of the controlled structure shows a decrease in the maximum acceleration, maximum displacement, and drift ratio for all regions of PGA; but, the maximum base shear has been increased in the range of 0.5g $\leq$ PGA $\leq$ 0.6g. It can also be seen from Figure 17 that the performance of the controlled structure shows a decrease in the maximum displacement in the range of PGA $\leq$ 0.6g. The maximum acceleration of the structure has decreased for all values of PGA, except for 0.8 g, and 1 g. Moreover, the value of base shear has decreased for PGA smaller than 0.9 g. Additionally, the results show that the performance of the structure has improved for the maximum drift ratio for the PGA smaller than 0.5 g. Figure 18 indicates that the performance of the controlled structure shows a decrease in all regions of PGA for the maximum drift ratio, acceleration, and base shear. The maximum displacement of the structure has decreased for all values of PGA in the range of PGA $\leq$ 0.8g. Therefore, the results show that the performance of the structure has improved significantly.

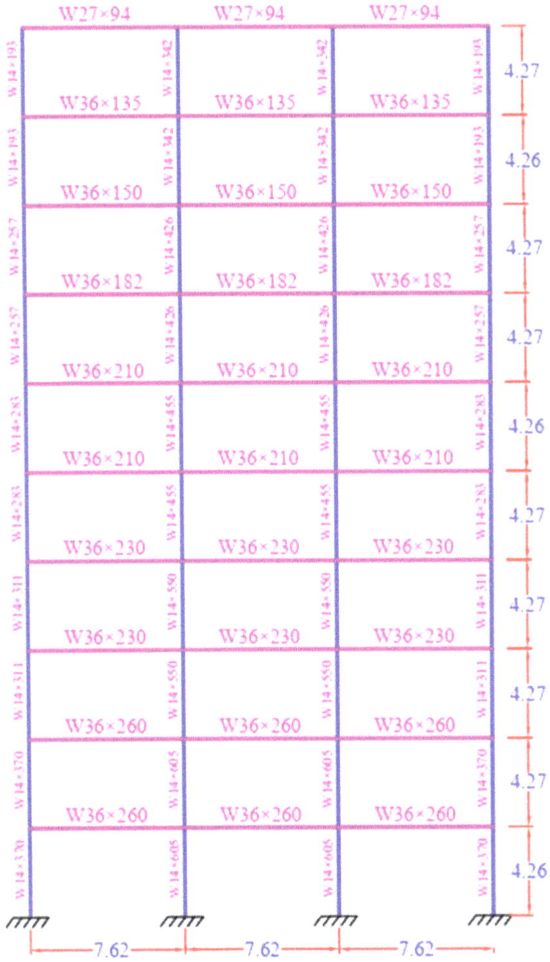

Figure 15. Ten-story steel moment-resisting frame.

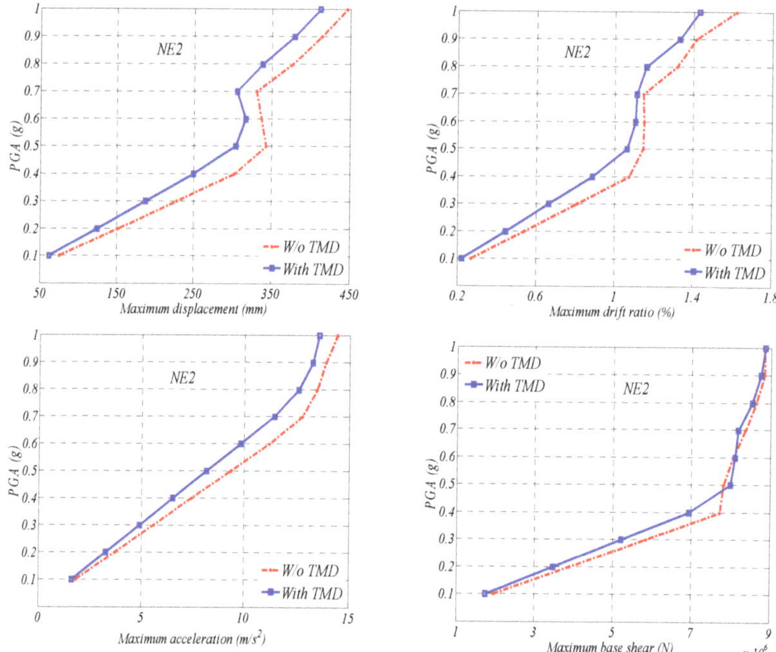

Figure 16. Results of IDA analysis for the frame subjected to a near-field earthquake (Gazli earthquake).

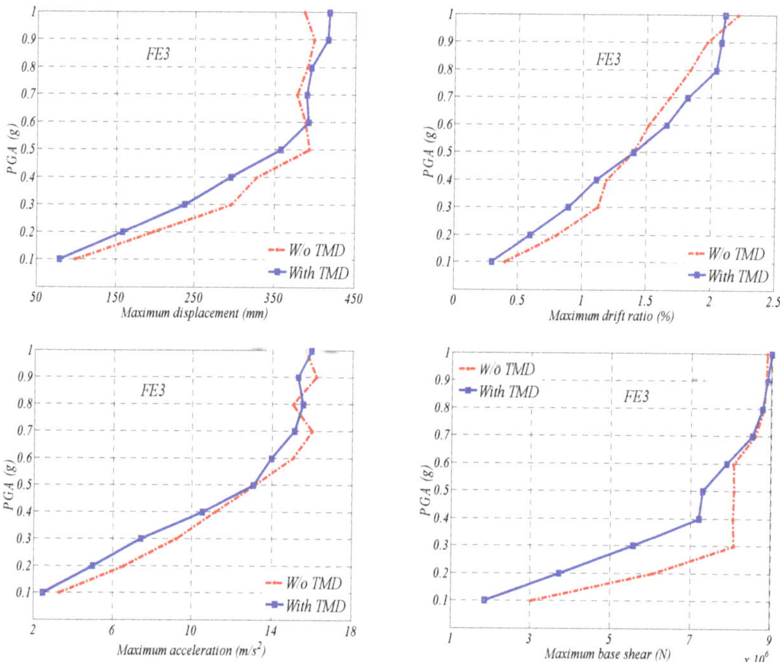

Figure 17. Results of IDA analysis for the frame subjected to a far-field earthquake (Superstition earthquake).

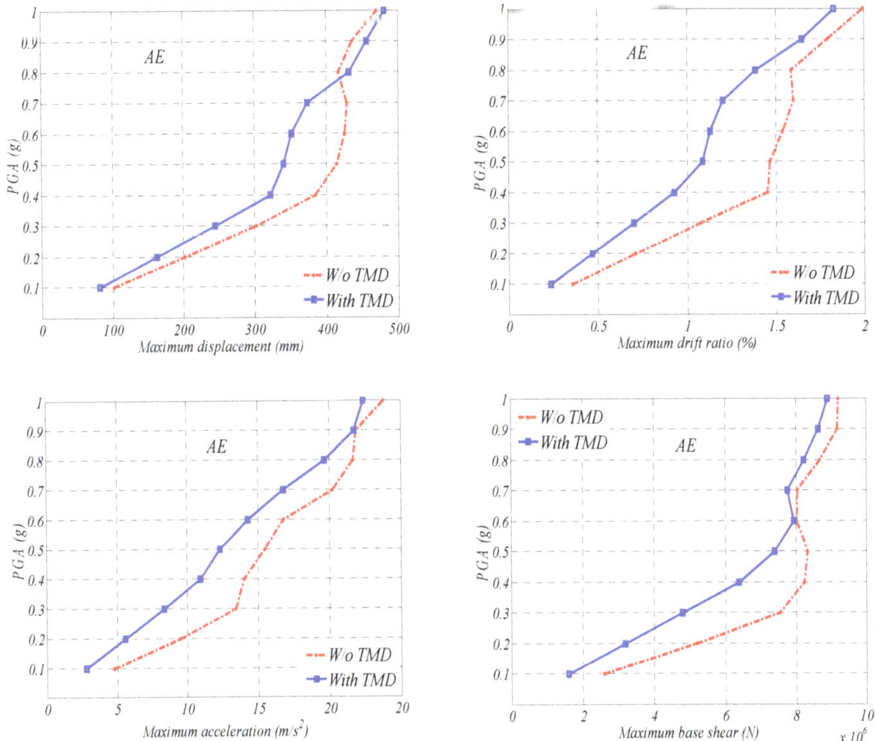

Figure 18. Results of IDA analysis for the frame subjected to the artificial earthquake.

8. Conclusions

The performance-based design of the steel structure controlled with the TMD system has been investigated in this paper. All studied earthquakes have been scaled in such a way that all have had the PGA value equal to 1 g. Then, the optimum parameters of the TMD system have been calculated using a meta-heuristic algorithm (i.e., WCA) focused on minimizing the maximum drift ratio of the stories based on the FEMA-356 subjected to the scaled earthquakes. For this purpose, two different frame buildings (i.e., a six-story and a 10-story moment-resisting frame) have been considered and modeled with the OpenSees software. The optimum parameters of the TMD system have been computed subjected to the scaled earthquakes. Then, PGA has been changed in the range of 0.1 g to 1 g in IDA, and the responses of the controlled structure have been examined. The results of the study show that the base shear decreases for the PGA value smaller than 0.5 g under all earthquakes studied. At the same time, for all records with PGA larger than 0.5 g, the TMD system does not make any considerable reduction in the base shear value of the controlled structure. This sentence is correct for the maximum drift ratio. Finally, the responses of controlled structure (i.e., the maximum acceleration and displacement) have almost decreased for all regions of PGA.

Moreover, the results of the investigation show that optimizing the TMD parameters, based on minimizing the drift ratio, decreases the structural displacement, and improves the seismic behavior of the structure based on FEMA-356. The results also indicate that the response of the controlled and uncontrolled structure (e.g., drift ratio, maximum displacement, maximum base shear) is reduced during the artificial earthquake. Additionally, the optimization of the TMD parameters, to keep the maximum drift ratio smaller than 2.5%, improves the seismic performance of the structure subjected to the artificial earthquake. It leads to a delay in the fracture of the bending connections, and it can help the beams and columns to sustain distortion.

Finally, a comparison between the traditional objective functions and the proposed objective function (i.e., the maximum drift ratio of less than 2.5%) has been presented. The results show that the optimum parameters of the TMD system based on the proposed objective function have a better performance in reducing the structural responses in comparison with the other previously used objective functions.

Author Contributions: Conceptualization, M.D., R.K., and H.H.; methodology, M.D., R.K., H.H., A.J.-G., and R.J.; software, M.D., R.K., and H.H.; validation, M.D., R.K., H.H., A.J.-G., and R.J.; formal analysis, M.D., R.K., and H.H.; investigation, M.D. and R.K.; writing—original draft preparation, M.D., R.K., and H.H.; writing—review and editing, A.J.-G. and R.J. All authors have read and agreed to the published version of the manuscript.

Funding: This research received no external funding.

Conflicts of Interest: The authors declare no conflicts of interest.

References

1. Elwardany, H.; Seleemah, A.; Jankowski, R. Seismic pounding behavior of multi-story buildings in series considering the effect of infill panels. *Eng. Struct.* **2017**, *144*, 139–150. [CrossRef]
2. Elwardany, H.; Seleemah, A.; Jankowski, R.; El-Khoriby, S. Influence of soil-structure interaction on seismic pounding between steel frame buildings considering the effect of infill panels. *Bull. Earthq. Eng.* **2019**, *17*, 6165–6202. [CrossRef]
3. Jankowski, R. Pounding between superstructure segments in multi-supported elevated bridge with three-span continuous deck under 3D non-uniform earthquake excitation. *J. Earthq. Tsunami* **2015**, *9*, 1550012. [CrossRef]
4. Sołtysik, B.; Jankowski, R. Non-linear strain rate analysis of earthquake-induced pounding between steel buildings. *Int. J. Earth Sci. Eng.* **2013**, *6*, 429–433.
5. Kamgar, R.; Rahgozar, R. Determination of optimum location for flexible outrigger systems in non-uniform tall buildings using energy method. *Int. J. Optim. Civ. Eng.* **2015**, *5*, 433–444.
6. Tavakoli, R.; Kamgar, R.; Rahgozar, R. The best location of belt truss system in tall buildings using multiple criteria subjected to blast loading. *Civ. Eng. J.* **2018**, *5*, 1338–1353. [CrossRef]
7. Tavakoli, R.; Kamgar, R.; Rahgozar, R. Seismic performance of outrigger–belt truss system considering soil–structure interaction. *Int. J. Adv. Struc. Eng.* **2019**, *11*, 45–54. [CrossRef]
8. Kamgar, R.; Shojaee, S.; Rahgozar, R. Rehabilitation of tall buildings by active control system subjected to critical seismic excitation. *Asi. J. Civ. Eng.* **2015**, *16*, 819–833.
9. Chung, L.L.; Wu, L.Y.; Huang, H.H.; Chang, C.H.; Lien, K.H. Optimal design theories of tuned mass dampers with nonlinear viscous damping. *Earthq. Eng. Eng. Vib.* **2009**, *8*, 547–560. [CrossRef]
10. Fei, Z.; Jinting, W.; Feng, J.; Liqiao, L. Control performance comparison between tuned liquid damper and tuned liquid column damper using real-time hybrid simulation. *Earthq. Eng. Eng. Vib.* **2019**, *18*, 695–701. [CrossRef]
11. Housner, G.W.; Bergman, L.A.; Caughey, T.K.; Chassiakos, A.G.; Claus, R.O.; Masri, S.F.; Skelton, R.E.; Soong, T.T.; Spencer, B.F.; Yao, J.T.P. Structural control: Past, present, and future. *J. Eng. Mech.* **1997**, *123*, 897–971. [CrossRef]
12. Kamgar, R.; Askari Dolatabad, Y.; Babadaei Samani, M.R. Seismic optimization of steel shear wall using shape memory alloy. *Int. J. Optim. Civ. Eng.* **2019**, *9*, 671–687.
13. Lasowicz, N.; Kwiecień, A.; Jankowski, R. Experimental study on the effectiveness of polymer damper in damage reduction of temporary steel grandstand. *J. Phys. Conf. Ser.* **2015**, *628*, 012051. [CrossRef]
14. Falborski, T.; Jankowski, R. Experimental study on effectiveness of a prototype seismic isolation system made of polymeric bearings. *Appl. Sci.* **2017**, *7*, 808. [CrossRef]
15. Miari, M.; Choong, K.K.; Jankowski, R. Seismic pounding between adjacent buildings: Identification of parameters, soil interaction issues and mitigation measures. *Soil Dyn. Earthq. Eng.* **2019**, *121*, 135–150. [CrossRef]
16. Kamgar, R.; Khatibinia, M.; Khatibinia, M. Optimization criteria for design of tuned mass dampers including soil–structure interaction effect. *Int. J. Optim. Civ. Eng.* **2019**, *9*, 213–232.

17. Kamgar, R.; Samea, P.; Khatibinia, M. Optimizing parameters of tuned mass damper subjected to critical earthquake. *Struct. Des. Tall Spec. Build.* **2018**, *27*, 1460. [CrossRef]

18. Kaveh, A.; Mohammadi, S.; Hosseini, O.K.; Keyhani, A.; Kalatjari, V.R. Optimum parameters of tuned mass dampers for seismic applications using charged system search. *Iran. J. Sci. Technol. Trans. Civil Eng.* **2015**, *39*, 21.

19. Khatibinia, M.; Gholami, H.; Kamgar, R. Optimal design of tuned mass dampers subjected to continuous stationary critical excitation. *Int. J. Dyn. Control* **2018**, *6*, 1094–1104. [CrossRef]

20. Kamgar, R.; Gholami, F.; Zarif Sanayei, H.R.; Heidarzadeh, H. Modified tuned liquid dampers for seismic protection of buildings considering soil-structure interaction effects. *Iran. J. Sci. Technol. Trans. Civ. Eng.* **2020**, *44*, 339–354. [CrossRef]

21. Soong, T.T.; Dargush, G.F. *Passive Energy Dissipation Systems in Structural Engineering*; Wiley: Hoboken, NJ, USA, 1997.

22. Mualla, I.H.; Belev, B. Performance of steel frames with a new friction damper device under earthquake excitation. *Eng. Struc.* **2002**, *24*, 365–371. [CrossRef]

23. Jarrahi, H.; Asadi, A.; Khatibinia, M.; Etedali, S. Optimal design of rotational friction dampers for improving seismic performance of inelastic structures. *J. Buil. Eng.* **2020**, *27*, 100960. [CrossRef]

24. Miguel, L.F.F.; Miguel, L.F.F.; Lopea, R.H. Simultaneous optimization of force and placement of friction dampers under seismic loading. *Eng. Opt.* **2016**, *48*, 1–21. [CrossRef]

25. Shen, H.; Zhang, R.; Weng, D.; Ge, Q.; Wang, C.; Islam, M.M. Design method of structural retrofitting using viscous dampers based on elastic-plastic response reduction curve. *Eng. Struc.* **2020**, *208*, 109917. [CrossRef]

26. Akehashi, H.; Takewaki, I. Comparative investigation on optimal viscous damper placement for elastic-plastic MDOF structures: Transfer function amplitude for double impulse. *Soil Dyn. Earth. Eng.* **2020**, *130105987*. [CrossRef]

27. Naeem, A.; Kim, J. Seismic performance evaluation of spring viscous damper cable system. *Eng. Struc.* **2018**, *176*, 455–467. [CrossRef]

28. Wei, M.; Rui, X.; Zhu, W.; Yang, F.; Gu, L.; Zhu, H. Design, modelling and testing of a novel high-torque magnetorheological damper. *Smart Mat. Struc.* **2020**, *29*, 025024. [CrossRef]

29. Farahmand Azar, B.; Veladi, H.; Talatahari, S.; Raeesi, F. Optimal design of magnetorheological damper based on tuning Bouc-Wen model parameters using hybrid algorithm. *Struc. Eng.* **2020**, *24*, 867–878.

30. Petrini, F.; Giaralis, A.; Wang, Z. Optimal tuned mass-damper-inerter (TMDI) design in wind-excited tall buildings for occupants' comfort serviceability performance and energy harvesting. *Eng. Struc.* **2020**, *204*, 109904. [CrossRef]

31. Masnata, C.; Matteo, A.D.; Adam, C.; Pirrotta, A. Smart structures through nontraditional design of tuned mass damper inerter for higher control of base isolated systems. *Mech. Res. Comm.* **2020**, *105*, 103513. [CrossRef]

32. Love, J.S.; McNamara, K.P.; Tiat, M.J.; Haskett, T.C. Series-type pendulum tuned mass damper-tuned sloshing damper. *J. Vib. Ac.* **2020**, *142*, 011003. [CrossRef]

33. Gandomi, A.H.; Yang, X.S.; Talatahari, S.; Alavi, A.H. Metaheuristic algorithms in modeling and optimization. *Metaheuristic Appl. Struct. Infrastruct.* **2013**, 1–24.

34. Kaveh, A.; Pirgholizadeh, S.; Khadem, H.O. Semi-active tuned mass damper performance with optimized fuzzy controller using CSS algorithm. *Asian J. Civil Eng.* **2015**, *16*, 587–606.

35. Lee, C.L.; Chen, Y.T.; Chung, L.L.; Wang, Y.P. Optimal design theories and applications of tuned mass dampers. *Eng. Struct.* **2006**, *28*, 43–53. [CrossRef]

36. Marano, G.C.; Greco, R.; Trentadue, F.; Chiaia, B. Constrained reliability-based optimization of linear tuned mass dampers for seismic control. *Int. J. Solids Struct.* **2007**, *44*, 7370–7388. [CrossRef]

37. Salvi, J.; Rizzi, E. Optimum tuning of Tuned Mass Dampers for frame structures under earthquake excitation. *Struct. Control Health Monit.* **2015**, *22*, 707–725. [CrossRef]

38. Tigli, O.F. Optimum vibration absorber (tuned mass damper) design for linear damped systems subjected to random loads. *J. Sound Vib.* **2012**, *331*, 3035–3049. [CrossRef]

39. Wong, K.K. Seismic energy dissipation of inelastic structures with tuned mass dampers. *J. Eng. Mech.* **2018**, *134*, 163–172. [CrossRef]

40. Nigdeli, S.M.; Bekdaş, G.; Sayin, B. Optimum tuned mass damper design using harmony search with comparison of classical methods. *AIP Conf. Proc.* **2017**, *1863*, 540004.

41. Bertero, V.V. *Strength and Deformation Capacities of Buildings under Extreme Environments*; Prentice-Hall: Upper Saddle River, NJ, USA, 1980.

42. Vamvatsikos, D.; Cornell, C.A. Incremental dynamic analysis. *Earthq. Eng. Struct. Dyn.* **2002**, *31*, 491–514. [CrossRef]

43. Han, S.W.; Chopra, A.K. Approximate incremental dynamic analysis using the modal pushover analysis procedure. *Earthq. Eng. Struct. Dyn.* **2006**, *35*, 1853–1873. [CrossRef]

44. Sánchez-Olivares, G.; Tomás Espín, A. Design of planar semi-rigid steel frames using genetic algorithms and component method. *J. Cons. Steel Res.* **2013**, *88*, 267–278. [CrossRef]

45. Sánchez-Olivares, G.; Tomás Espín, A. Improvements in meta-heuristic algorithms for minimum cost design of reinforced concrete rectangular sections under compression and biaxial bending. *Eng. Struc.* **2017**, *130*, 162–179. [CrossRef]

46. Bilondi, M.R.S.; Yazdani, H.; Khatibinia, M. Seismic energy dissipation-based optimum design of tuned mass dampers. *Struct. Multidiscip. Optim.* **2018**, *58*, 2517–2531. [CrossRef]

47. Mazzoni, S.; McKenna, F.; Scott, M.H.; Fenves, G.L. *The Open System for Earthquake Engineering Simulation (OpenSEES) User Command-Language Manual*; Pennsylvania State University: University Park, PA, USA, 2006.

48. FEMA-356. *Prestandard and Commentary for the Seismic Rehabilitation of Buildings*; American Society of Civil Engineers: Reston, VA, USA, 2000.

49. Connor, J.J.; Laflamme, S. Applications of Active Control. In *Structural Motion Engineering*; Springer: New York, NY, USA, 2014; pp. 347–386.

50. Naderpour, H.; Naji, N.; Burkacki, D.; Jankowski, R. Seismic response of high-rise buildings equipped with base isolation and non-traditional tuned mass dampers. *Appl. Sci.* **2019**, *9*, 1201. [CrossRef]

51. Khatami, S.M.; Naderpour, H.; Razavi, S.M.N.; Barros, R.C.; Jakubczyk-Gałczyńska, A.; Jankowski, R. Study on methods to control interstory deflections. *Geosciences* **2020**, *10*, 75. [CrossRef]

52. Eskandar, H.; Sadollah, A.; Bahreininejad, A.; Hamdi, M. Water cycle algorithm–A novel metaheuristic optimization method for solving constrained engineering optimization problems. *Comput. Struct.* **2012**, *110*, 151–166. [CrossRef]

53. Siemaszko, A.; Jakubczyk-Gałczyńska, A.; Jankowski, R. The idea of using Bayesian networks in forecasting impact of traffic-induced vibrations transmitted through the ground on residential buildings. *Geosciences* **2019**, *9*, 339. [CrossRef]

54. Ibarra, L.F.; Krawinkler, H. *Global Collapse of Frame Structures under Seismic Excitations*; Pacific Earthquake Engineering Research Center: Berkeley, CA, USA, 2005; pp. 29–51.

55. Ibarra, L.F.; Medina, R.A.; Krawinkler, H. Hysteretic models that incorporate strength and stiffness deterioration. *Earthq. Eng. Struct. Dyn.* **2005**, *34*, 1489–1511. [CrossRef]

56. Lignos, D.G.; Krawinkler, H. Deterioration modeling of steel components in support of collapse prediction of steel moment frames under earthquake loading. *J. Struct. Eng.* **2011**, *137*, 1291–1302. [CrossRef]

57. Krawinkler, H.; Zareian, F.; Lignos, D.G.; Ibarra, L.F. Prediction of collapse of structures under earthquake excitations. In Proceedings of the 2nd International Conference on Computational Methods in Structural Dynamics and Earthquake Engineering COMPDYN 2009, Rhodes, Greece, 22–24 June 2009; paper no. CD449.

58. Kanai, K. An empirical formula for the spectrum of strong earthquake motions. *Bull. Earthq. Res. Inst.* **1961**, *39*, 85–95. (In Japanese)

59. Mohebbi, M.; Shakeri, K.; Ghanbarpour, Y.; Majzoub, H. Designing optimal multiple tuned mass dampers using genetic algorithms (GAs) for mitigating the seismic response of structures. *J. Vib. Control* **2013**, *19*, 605–625. [CrossRef]

60. Tajimi, H. A statistical method of determining the maximum response of a building structure during an earthquake. In Proceedings of the 2nd World Conference on Earthquake Engineering, Tokyo and Kyoto, Japan, 11–18 July 1960; pp. 781–797.

61. Wu, J.; Chen, G.; Lou, M. Seismic effectiveness of tuned mass dampers considering soil–structure interaction. *Eart. Eng. Struc. Dyn.* **1999**, *28*, 1219–1233. [CrossRef]

62. *UBC-97, Uniform Building Code*; International Conference of Building Officials: Whittier, CA, USA, 1997.

63. Kamgar, R.; Rahgozar, R. Determination of critical excitation in seismic analysis of structures. *Earth. Struc.* **2015**, *9*, 875–891. [CrossRef]
64. Wong, K.K.; Johnson, J. Seismic energy dissipation of inelastic structures with multiple tuned mass dampers. *J. Eng. Mech.* **2009**, *135*, 265–275. [CrossRef]

MDPI
St. Alban-Anlage 66
4052 Basel
Switzerland
Tel. +41 61 683 77 34
Fax +41 61 302 89 18
www.mdpi.com

Applied Sciences Editorial Office
E-mail: applsci@mdpi.com
www.mdpi.com/journal/applsci